MASS LOSS FROM STARS

ASTROPHYSICS AND
SPACE SCIENCE LIBRARY

A SERIES OF BOOKS ON THE RECENT DEVELOPMENTS

OF SPACE SCIENCE AND OF GENERAL GEOPHYSICS AND ASTROPHYSICS

PUBLISHED IN CONNECTION WITH THE JOURNAL

SPACE SCIENCE REVIEWS

VOLUME 13

MASS LOSS FROM STARS

PROCEEDINGS OF THE SECOND TRIESTE COLLOQUIUM ON ASTROPHYSICS, 12–17 SEPTEMBER, 1968

Edited by

MARGHERITA HACK

D. REIDEL PUBLISHING COMPANY

DORDRECHT-HOLLAND

ISBN-13:978-94-010-3407-4 e-ISBN-13:978-94-010-3405-0
DOI: 10.1007/978-94-010-3405-0

EDITOR'S NOTE

The present volume contains the papers presented during the Second Colloquium on Astrophysics organized by the Astronomical Observatory of Trieste.

I wish to thank:

Professor Abdus Salam and Professor Paolo Budini, Director and Deputy Director of the International Centre of Theoretical Physics of Trieste, for having given us hospitality;

the participants, who have animated the colloquium with their contributions and with lively discussions;

Professor Zdeněk Kopal, Editor-in-chief of *Astrophysics and Space Science*, and the publishing house D. Reidel, Dordrecht, for their interest in the publication of these Proceedings;

the International Astronomical Union, which has sponsored the colloquium and granted some travel-ships for young astronomers.

The Editor,

MARGHERITA HACK

Trieste, January 1969

TABLE OF CONTENTS

SESSION III

MASS LOSS FROM BINARY STARS. OBSERVATIONS

SESSION IV

MASS LOSS FROM CLOSE BINARIES. THEORIES

SESSION V

MASS LOSS FROM UNSTABLE STARS

LIST OF PARTICIPANTS

A. Abrami, Trieste
C. Aydin, Ankara
A. Baglin, Paris
M. J. Baines, Reading
J. Balazs-Detre, Budapest
G. Barbaro, Padova
G. T. Bath, Brighton
C. Blanco, Catania
A. Boury, Liège
H. Brancewicz, Krakow
P. Broglia, Merate
F. Caputo, Frascati
C. Casini, Merate
V. Castellani, Frascati
F. Catalano, Catania
S. Catalano, Catania
B. Cester, Trieste
C. Chevalier, Paris
A. W. Cremin, Reading
R. Cristaldi, Catania
N. Dallaporta, Padova
M. de Groot, Utrecht
L. Detre, Budapest
A. J. Deutsch, Pasadena
D. Ezer, Ankara
R. Faraggiana, Trieste
J. Field, London
M. G. Fracastoro, Torino
M. Friedjung, Paris
P. Giannone, Roma
G. Godoli, Catania
N. Gökkaya, Trieste
L. Gridelli, Trieste
S. Grzȩdzielski, Warsaw
A. Guarnieri, Bologna

E. A. Gussmann, Potsdam
M. Hack, Trieste
D. S. Hall, Nashville
B. E. Helt, Copenhagen
T. Herczeg, Hamburg
R. Herman, Paris
L. Houziaux, Mons
J. B. Hutchings, Victoria, B.C.
H. E. Jorgensen, Copenhagen
M. Kitamura, Tokyo
D. Koelbloed, Amsterdam
S. S. Kumar, Charlottesville
S. Kutter, Rochester
G. Larsson-Leander, Lund
D. Lauterborn, Göttingen
M.-C. Lortet, Paris
A. Mammano, Asiago
A. Martini, Asiago
A. Martini, Milano
A. Masani, Milano
M. McCarthy, Castelgandolfo
D. C. Morton, Princeton
G. S. Mumford, Medford, Mass.
K. Nariai, Greenbelt
L. Nobili, Padova
A. Noels, Liège
P. R. Owen, Brighton
L. Pasinetti, Merate
M. Plavec, Ondřejov
S. Refsdal, Oslo
A. Renzini, Bologna
M. Rodono, Catania
W. K. Rose, Cambridge, Mass.
L. Rosino, Asiago
C. Ryter, Gif-sur-Yvette

J. Sahade, La Plata

W. L. W. Sargent, Pasadena

Th. Schmidt-Kaler, Bonn

L. Secco, Padova

P. Stenner, Trieste

C. Summa, Padova

P. J. Treanor, Castelgandolfo

A. B. Underhill, Utrecht

C. J. van Houten, Leiden

F. Van 't Veer, Paris

R. Viotti, Frascati

K. Walter, Tübingen

W. Wenzel, Thüringen

I. P. Williams, Reading

F. B. Wood, Philadelphia

J.-P. Zahn, Paris

P. Zlobec, Trieste

MASS LOSS FROM STARS: A REVIEW

ARMIN J. DEUTSCH

Mount Wilson and Palomar Observatories
Carnegie Institution of Washington,
California Institute of Technology

For many decades astronomers have realized that some stars lose mass in the cata-strophic events that produce supernovae, ordinary novae, and planetary nebulae. Only recently have we learned that less conspicuous mass-loss processes occur throughout much of the lifetimes of most normal stars. My purpose in this Introduc-tion is to summarize briefly the kinds of evidence we have for this conclusion; also what can be said about the properties of the various kinds of flow, and about their significance relative to stellar evolution.

Perhaps it would be appropriate to begin by setting the theoretical context. This would seem to insure that if we had not observed mass loss from stars, we should have had to invent it. For what else we have observed about the stars, and what theories we have for them, all seem to demand that mass loss must occur.

The solar corona, for example, is known to have a temperature of the order of a million degrees. PARKER (1958) was the first to point out that solar gravity cannot retain the coronal gas at so high a temperature; for lack of a sufficient pressure on the corona by the surrounding interstellar gas, the corona cannot be in hydrostatic equilibrium. Expansion must occur; and, unless the gas is too hot, the flow will generally accelerate it from a small subsonic velocity near the star to a larger super-sonic velocity at great distances from the star. DAHLBERG (1964) and PARKER (1965) have recently reviewed the mathematical theory, and DESSLER (1967) has given a summary of the essential physical processes that govern such flows. WEBER and DAVIS (1967) have described some of the complications of the flow that result from the rotation of the sun and from its magnetic field.

The theory of the solar wind predicts how mass will be lost from a star with a corona for which the temperature, density, and other physical conditions are known at some level. The theory does not ordinarily suffice to predict how these boundary conditions are determined. However, the high temperature of the solar corona has been accounted for by attribution to a non-radiative energy flux, which penetrates the reversing layer as magneto-acoustic waves and/or internal gravity waves. The argument, originally due to BIERMANN (1946), is that these waves steepen as they dissipate their energy in the gas above the reversing layer. The chromosphere is the region (Figure 1) of steeply increasing temperature that lies between the photosphere, where the radiative temperature gradient prevails, and the very much hotter corona, where high thermal conductivity maintains a much flatter temperature gradient.

The hydrogen convection zone is held to be responsible for the non-radiative flux through the solar photosphere. Since all stars later than type F5 are thought to possess deep zones in which the radiative gradient is unstable and convection is well-developed, we should, therefore, expect an appreciable non-radiative flux at the photospheres of all such stars. They should also have chromospheres then, as the sun does, and coronas and stellar winds.

Unfortunately, in most cases, these high atmospheric levels are likely to have properties of a kind that makes observational detection difficult or impossible. But in the spectra of many stars emission lines are seen, of a kind that arise in the solar

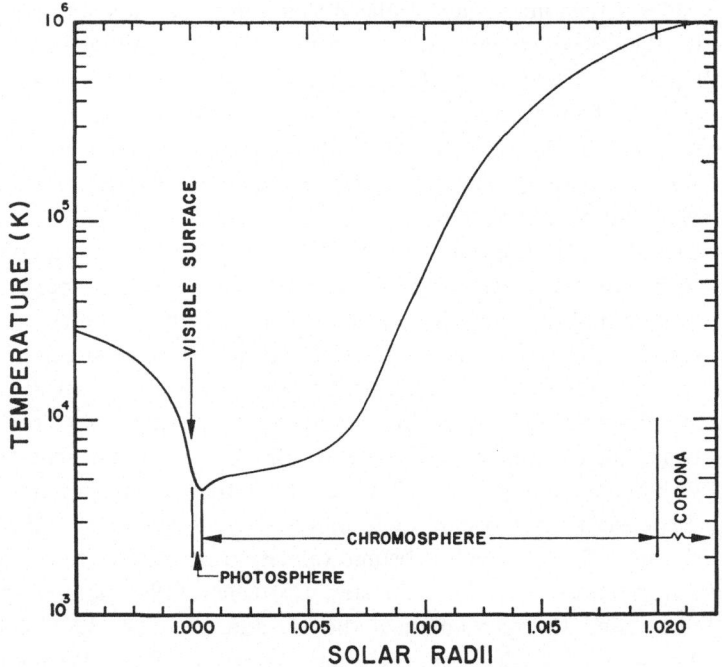

Fig. 1. The temperature profile of the sun in the vicinity of the visible surface. The visible surface is defined as occurring at unit optical depth for 5000 Å radiation. Note the temperature minimum above the visible surface. Heat is supplied to the chromosphere largely by conduction downward from the corona. After A. J. DESSLER, *Rev. Geophys.* **5**, 1967, 1.

chromosphere, and especially Ca II H and K. These emission lines originate chiefly in the borders of the chromospheric network on the sun (SIMON and LEIGHTON, 1964), and their intensity in the flux correlates well with the extent and intensity of solar activity.

The coolest stars among the red giants also show circumstellar absorption lines of a kind that could arise in the outer parts of a stellar wind. We shall examine these observations in more detail below; suffice to say here that they seem consistent with the *qualitative theoretical prediction that mass loss occurs from most stars later than F5.* The theory is inadequate for accurate quantitative predictions of the mass flows from various types of such stars; but some attempts have been made to calculate the structure of their chromospheres and coronas (DE JAGER and KUPERUS, 1961; WEYMANN, 1960).

Another kind of theoretical argument for mass loss relates to various aspects of stellar evolution. Thus, in any star that was initially more massive than about $1.5 M_\odot$ the original nuclear resources would be sufficient to maintain quasi-static equilibrium at most for a time that is shorter than the age of the galaxy. After exhausting these resources, the star can find equilibrium only in a degenerate configuration. But even this is impossible unless the residual mass is less than $\sim 1.2 M_\odot$. Until recently, it was therefore held that appreciable mass loss must occur at some stage in the evolution of all stars that originally exceed this mass limit; for a review of the theory, see MESTEL (1965). However, from recent work by HOYLE et al. (1964), by WHEELER (1966), and THORNE (1967), it appears that one cannot easily dismiss the theoretical possibility that some stars of large mass collapse into their Schwarzschild singularities. Observations of white dwarfs in clusters, and mass ratios in single-line spectroscopic binaries provide some evidence that this occurs rarely, if at all (EGGEN and GREENSTEIN, 1965; TRIMBLE and THORNE, 1969).

With exhaustion of their nuclear fuel, most stars evolve away from the main sequence and eventually take up degenerate configurations. From a knowledge of the numbers of stars with different masses, and their nuclear lifetimes, several authors have estimated the rate at which these 'star-deaths' occur (e.g., ABELL and GOLDREICH, 1966; SCHMIDT, 1966; WEIDEMANN, 1968). The result is $\sim 2 \times 10^{-10}$ stars pc^{-2} yr^{-1} for the column density perpendicular to the galactic plane near the sun, and ~ 0.3 stars yr^{-1} over the whole galaxy. The total rate of mass loss is $\sim 4 \times 10^{-10} M_\odot$ pc^{-2} yr^{-1}. About half of this comes from stars with masses $> 2 M_\odot$. Since the nuclear lifetimes of these stars are always much shorter than the age of the galaxy, they must condense from the interstellar medium at the same rate as they die. This entails the circulation through such stars of 1–2% of the interstellar medium in 10^9 year.

From calculations of models to represent RR Lyrae stars and other variables that are known to be in advanced stages of evolution, we may find additional theoretical arguments for extensive mass loss. Thus, CHRISTY (1966a, b) has been able to give satisfactory theoretical models for RR Lyrae stars only for masses $M \simeq 0.5 M_\odot$; but in the whole age of the galaxy, stars initially of such low mass could not yet have evolved into the present configuration. To judge from the incidence of these variable stars in globular clusters, they must be metamorphs of main-sequence stars with masses $M \simeq 1.2 M_\odot$. MASANI et al. (1968) have studied models for the pulsation of Mira and other variable stars like it: Their conclusion also is that $M \simeq 0.4 M_\odot$, despite the apparently late evolutionary stage of Mira.

For many stars in binary systems, mass loss will have to occur even at an earlier evolutionary stage or at a much faster rate than would be necessary for a single star of similar type. The reason is that, as the primary star evolves off the main sequence, it will expand and overflow its limiting Lagrangian surface. Mass need not be lost to the system; it may simply be transferred to the companion star. For recent literature, see PEREK (1968) and DOMMANGET (1967).

With these theoretical expectations, let us now turn to the observational evidence. Relative to the last argument noted above, concerning the *a priori* necessity of mass loss from the evolving primaries of close binary systems, we may find many eclipsing binaries in which this process must have occurred. They are the Algol systems, having secondaries that are subgiants, overluminous for their mass, and often filling their critical Lagrangian surfaces (PLAVEC, 1968). Recently CONTI (1967) has described a spectroscopic binary which is physically similar to the Algol variables, but which fails to produce eclipses because its inclination is too high. He estimates that such semi-detached binaries should outnumber the low-inclination Algol variables by a factor of 5. Other evolutionary stages which are more difficult of recognition include a very rapid one, when the original primary loses mass at a rate exceeding $10^{-5} M_\odot$ yr^{-1}; and much more prolonged stages preceding and following the semi-detached configuration. CONTI (1967) has argued that the metallic-line stars are common proto-morphs of semi-detached pairs. HOYLE (1964) and MCCREA (1964) have proposed that the 'blue stragglers', like those above the main-sequence turn-off in the old open clusters M67, are metamorphs of semi-detached pairs, and some recent results by CANNON (1968) and DEUTSCH (1968) support this identification. VAN DEN HEUVEL (1968) has given arguments for similar, but different, identifications of the metamorphs of mass exchange in close binaries. Such objects probably have a high incidence among A and B stars in the main-sequence band, and it would be an important advance to distinguish them observationally from stars that have had the very different evolutionary history of single objects. Another important problem concerns the disposition of the mass lost from the original primary; what proportion escapes from the system altogether, and what proportion falls onto the original secondary?

If the secondary stars of Algol systems may be taken as evidence that mass loss has occurred, even though the principal mass-loss process itself has not been unambiguously identified, then the existence of white dwarfs may also be taken as evidence. Like the Algol subgiants, the white dwarfs can only be understood as remnants of stars that were originally more massive. For, had these objects condensed with their present masses, they could not yet have become degenerate during the whole age of the galaxy. The degeneracy of these objects can only have occurred in the interior parts of stars that were once appreciably more massive than the white dwarfs. For a recent review and a bibliography, see WEIDEMANN (1968).

Another indirect argument for widespread mass loss can be formulated on the observed regularities in the abundances of various elements among different stars. Many of these regularities can be well understood on the theory that most nuclides have been produced from simpler species, and chiefly in one of a small number of distinctive processes of nucleosynthesis (BURBIDGE *et al.*, 1957). These processes generally occur in stellar interiors at advanced stages of evolution. However, to account for the abundances that are observed, the theory always requires the circulation of processed material through successive generations of stars by way of the interstellar medium. Among recent reviews of this subject are those by BASHKIN (1965), REEVES (1968), and ARNETT *et al.* (1968). See also WAGONER *et al.* (1967) and BODANSKY *et al.* (1968).

Although we have now given a variety of compelling arguments for believing that mass-loss processes are ubiquitous among the stars, it remains to admit that the process has so far been unambiguously observed only in a relatively small number of particularly favorable cases. These are objects in which we can detect an outward flow with a velocity exceeding the (local) velocity of escape. The solar wind represents such a flow from the sun.

As DESSLER (1967) has recounted in a recent review, a variety of astronomical and geophysical effects have long been attributed to a more or less intermittent flow of 'corpuscular radiation' from the sun. The actual experimental detection of the solar wind, however, dates only from the years 1960–63, when several spacecraft first carried instruments outside the magnetosphere of the earth. In the present view, "The quiet solar wind velocity is of the order of 300–350 km s^{-1} near the orbit of Earth with a direction almost away from the Sun. ... Its density is of the order of 5 protons cm^{-3} and the temperature probably closer to 10^4 °K than to 10^5 °K" (LÜST, 1968). Measurements show that between the orbits of Venus and Mars, the flow velocity is nearly constant, and the temperature gradient is rather less steep than $r^{-4/3}$, which would correspond to adiabatic flow. The flow velocity has a small tangential component, which is associated with the rotation and large-scale magnitude field of the sun. The flow is really not steady; at a given point, its velocity, density, and other properties are subject to large fluctuations, and these are clearly associated with various kinds of solar activity. NESS (1968) has recently reviewed this subject.

The mean rate of mass loss from the sun probably lies in the range 10^{-13}–10^{-14} $M_\odot$ yr^{-1}, one or two orders of magnitude less than the mass loss by thermonuclear reactions. The magnetic field pervading the corona causes angular momentum to be transported outwards with relatively high efficiency; on the assumption that the sun itself rotates rigidly, WEBER and DAVIS (1967) estimated that the relaxation time of its angular momentum is 7×10^9 year.

In no other main-sequence stars have we actually observed mass loss to occur. But for the reasons noted earlier in this review, we know those later than F5 to have chromospheres, and we may infer that they also have stellar winds. Moreover, the recent work of WILSON (1963, 1966) and KRAFT (1967a, b) shows that chromospheric activity and rotation diminish sharply as a star ages within the main-sequence band. The implication is clearly that, for some short time after reaching the zero-age main sequence, most late-type dwarfs support winds that are much more massive than the solar wind, and much more effective in decelerating the stellar rotation.

KUHI (1964) has found additional evidence for relatively high rates of mass loss in T Tauri stars. In these objects, which are still contracting toward the main sequence, the emission lines at H and K are usually very strong and wide, and are bordered by absorption lines at the shortward edge. The Balmer lines have similar profiles. Kuhi concludes that the emission lines arise in a spherical H II region having a diameter several times that of the photosphere and enveloped by a thin shell that produces the absorption lines. "The ejection velocities range from 225 to 325 km/sec and 0.7 to 1.0 times the escape velocity." He also writes that therefore "... the material cannot be expected to leave the star completely, and so it should eventually return unless some other forces act. In the T Tauri stars observed, no sign of this returning material, if it exists, is present in the spectra. Thus, we are forced to think of the matter as being constantly pushed outward by the pressure of the material rising below it..." The profiles usually indicate that the flow is not accelerated outwards as in the solar wind, however; although "... it may be that as mass-ejection subsides in a solar-type star it comes to resemble a solar wind-type phenomenon in which accelerating forces play a dominant role". The rates of mass loss are estimated to be of the order of 4×10^{-8} $M_\odot$ yr^{-1}. In the aggregate, then, the T Tauri stars in unit volume of space will lose mass at a rate $\sim 10\%$ that of the neighboring stars in advanced stages of evolution.

In high-dispersion spectra of M giants, one can see violet-displaced absorption cores in the strong metallic lines of low excitation. These lines indicate that above the reversing layer is a gas of low temperature which is in steady expansion at speeds in the range between 5 and 25 km/sec. These speeds are small compared with the escape velocity at the photosphere, which is nearly 100 km/sec. However, observations of the visual binary α Herculis, and of several other M giants with visual companions, have shown many of the circumstellar absorption lines in the spectrum of the secondary star. These lines reveal that the expanding gas extends several hundred radii above the primary star, and that at these heights it exceeds the local escape velocity and must escape from the system.

DEUTSCH (1960) has found that among normal M giants, with $M_V > -2.5$, the strengths of the circumstellar lines increase as the spectral type grows later. He has estimated that the column density of circumstellar gas contributing to the lines increases by a factor of $\sim 2 \times 10^3$ between types M1 and M5 and that the corresponding rate of mass loss increases by a slightly smaller factor. At M5 III he has found the rate of mass loss to be of the order of $10^{-8} M_\odot$ yr^{-1}. In more luminous M stars, the rate is appreciably higher; WEYMANN (1962) found $4.5 \times 10^{-6} M_\odot$ yr^{-1} for α Orionis, "... with an uncertainty factor of 10 either way". DEUTSCH (1966) finds a rather greater uncertainty in his results, which he attributes chiefly to a lack of knowledge regarding the degree of double ionization in the metals and regarding the height over the photosphere above which the observed lines are formed.

Observations of circumstellar lines in cool giants and supergiants, and their interpretation, have been reviewed by DEUTSCH (1960, 1961) and by WEYMANN (1963). The flows are not well understood; the possibility exists that they represent a phenomenon closely analogous to the solar wind, but much more massive and at much lower temperatures. For, most stars show strong emission lines at H and K if they also show the circumstellar lines that indicate mass loss, and this may signify a magneto-acoustic flux sufficient to produce the flows that are observed. Radiation pressure has also been invoked as an important factor in the flow dynamics (WILSON, 1960; WICK-RAMASINGHE et al., 1966). Important information can now be obtained regarding the outer parts of these flows, as they may be observed in the spectra of visual companions. In particular, application should be made of the recent work by BAHCALL and WOLF

(1968) on the collisional excitation of fine-structure levels in the ground terms of FeI and TiII. Thus, for example, in the circumstellar spectrum of the companion of α Scorpii, these energy levels produce lines of TiII but not of FeI.

Circumstellar lines can usually not be seen in the spectra of giants earlier than M0 and less bright than luminosity class Ib; but DEUTSCH (1960) and WEYMANN (1963) believe that many of these objects may support mass flows which are too highly ionized for ready detection. The supergiants of types G and K often do exhibit features that probably arise in a massive stellar wind. In contrast to what is observed in the M stars, these supergiants often show strong time variations in the displaced lines. Further data are needed to secure their connection with mass-loss processes.

A variety of spectroscopic peculiarities have been taken as evidence for mass loss in earlier supergiants. SARGENT (1961) found displaced lines of low excitation in the spectrum of ϱ Cassiopeiae (F8 Ia), and from them deduced a non-steady mass loss with a maximum rate of $3 \times 10^{-5} M_\odot$ yr^{-1}. Displaced lines with variable expansion velocities of <200 km/sec also occur in the spectrum of 89 Herculis (F2 Ia) (BÖHM-VITENSE, 1956). In HR 8752 (G0 Ia) the Balmer lines indicate an outflow of material, and [NII] emission lines reveal a low-density envelop in slow expansion (SARGENT, 1965). In the unique variable star FU Orionis, which grew about 100 times brighter during 1936, HERBIG (1966) found two sets of absorption lines: one of type F2 I-II, and the other corresponding to much lower temperature and indicating an expansion of approximately 80 km/sec. In this star, Herbig has proposed that the 1936 brightening marked a pre main-sequence collapse, with associated loss of mass and angular momentum.

Supergiants of type A and earlier also show spectroscopic evidence of non-steady outflows. Variable radial velocities generally occur in lines that are produced in the reversing layer (ABT, 1957). Hα is often found to exhibit a P-Cygni profile, the expansion velocity from the absorption component occasionally being comparable with the escape velocity (ABT and GOLSON, 1966). From spectrograms made in the rocket ultraviolet, MORTON (1967a, b) has discovered strong circumstellar absorption lines indicative of massive outflows from several O and B stars of luminosity class III and higher. The circumstellar lines do not occur in δ Scorpii (B0 V) or π Scorpii (B1 V) (STONE and MORTON, 1967). In the giants and supergiants circumstellar lines arise from the ground terms of CIV, NV, SiIII, and SiIV. They indicate expansion velocities of about 1500 km/sec, or approximately twice the escape velocities of these stars. An excited line (6.5 eV) of CIII also occurs in ε and ζ Orionis, with an expansion velocity several hundred km/sec smaller than those from the resonance lines (MORTON, 1967b; MORTON et al., 1968). The rate of mass loss is estimated to be of the order of $1 \times 10^{-6} M_\odot$ yr^{-1}. Radiation pressure in the strong resonance lines has been found by LUCY and SOLOMON (1967) to be capable of producing mass flows that would resemble those observed, although slightly less massive.

CARRUTHERS (1968) and MORTON et al. (1968) have found displaced circumstellar lines in the rocket ultraviolet of ζ Puppis (O5 f) and γ^2 Velorum (WC7+O7). Similar lines were discovered by SAHADE (1955) and by UNDERHILL (1959) in the terrestrially accessible spectra of some Wolf-Rayet stars. UNDERHILL (1966, 1967, 1968) has recently described these; she gives expansion velocities which are usually in the range 1000–2000 km/sec. She has estimated the mass loss rate to be "... in the neighborhood of 10^{-6} solar masses per year". Deutsch (unpublished) has found that the metastable line HeI λ 3888 is generally the strongest absorption feature in the spectra of WN-B stars (HILTNER and SCHILD, 1966). It yields expansion velocities comparable with those by other authors, but in at least some of the WN-B stars both the velocities and the intensities are time-variable. Among the WC stars, the displaced line near λ 3888 has been found only in those WC stars that are known spectroscopic binaries, or that have nebular lines in their spectra.

A number of Wolf-Rayet lie near the centers of nebulous rings or other kinds of diffuse nebulae. The rings may be interpreted as the inner boundaries of shells swept clean of interstellar gas by the gas and radiation streaming from the star (SMITH, 1967; JOHNSON and HOGG, 1965). PACZYŃSKI (1968) has discussed the possibility that the Wolf-Rayet stars that are spectroscopic binaries have exchanged mass with their companions, in accord with the theory of KIPPENHAHN and WEIGERT (1967).

Among the B stars, nearly 20% show emission at Hα, at least intermittently. A number of them also show other emission lines and/or a shell-type absorption-line spectrum. These spectroscopic peculiarities generally arise in an extended atmosphere, with a radius only a few times larger than that of the photosphere. In other respects, the Be stars comprise a heterogeneous group. The most numerous objects are rapid rotators that populate a narrow-band parallel to the main sequence and about one magnitude above it (SCHMIDT-KALER, 1964a). These stars are only marginally stable against rotational disruption. Intermittent mass loss occurs because "... according to theory, the stars undergo a short contraction phase coinciding with the dying out of the convective core" (SCHMIDT-KALER, 1964b). Although few of these stars have shown direct evidence for expansion of matter at velocities sufficient for escape, "... it seems probable that these gases eventually escape from the star" (UNDERHILL, 1960). The mean rate of mass loss is estimated to be of the order of $10^{-7} M_{\odot}$ yr^{-1}. HAZLE-HURST (1967) and LIMBER (1967) have discussed the theory of rotationally forced ejection; a recent contribution is by LIMBER and MARLBOROUGH (1968).

Another group of Be stars are high-luminosity objects like β Orionis (B8 Ia,) χ^2 Orionis (B2 Ia), and similar B and A supergiants studied by ABT and GOLSON (1966). Probably these stars support massive stellar winds like those found by MORTON (1967a, b) from ultraviolet spectra of other early stars of high luminosity, including ε Orionis (B0 Ia) and δ Orionis (O9.5 II). We have no evidence that rotational instability plays a role in the generation of these flows; perhaps the theory of LUCY and SOLOMON (1967) will suffice to account for all of them.

A third group of Be stars are found in binary pairs that are close enough for the companion to affect the Be envelope. Some of these Be stars are probably nearly like single Be stars near the main-sequence band. The chief effects of the companion are just to lower the threshold of rotational instability, and to complicate the flow pattern in the gas that is spun off the Be star. An example might be the B2 star ζ Tauri

(UNDERHILL, 1966). In other pairs, the companion of the Be star is itself losing mass, and the Be envelope is supplied partly or chiefly from this material. The phenomenon is most often encountered in those eclipsing binaries where at least one component lies above the main sequence; an example is U Cephei (B8 V + G8 III), where the secondary star is a cool subgiant that fills its Roche lobe. Gaseous rings and streams occur in this pair, and in other where the secondary loses mass as a result of tidal disturbances or unusual prominence activity, even though it does not actually overflow its Roche lobe. Many close binaries show emission lines and/or other evidence of extended atmospheres, whether or not the pair contains a B star. SAHADE (1960) has reviewed the spectroscopic peculiarities seen in these binaries, and he has summarized the evidence for "... the escape of matter through the external Lagrangian points and the existence of large expanding shells in close binaries with gaseous streams".

No doubt some of the binary stars with gas streams or common envelopes are exchanging mass in evolutionary processes of the kind described by Kippenhahn, his collaborators, and other authors. Recent reviews of these complex processes, and related considerations, have been given in *I.A.U. Trans.*, **12B**, 1966, 423; PEREK (1968); DOMMANGET (1967), and HUANG (1963). The review by SAHADE (1960) points up the evidence for the exchange or circulation of gas in a wide variety of binaries, including many where one or both stars lie far off the main sequence. Well-known examples of luminous pairs would include β Lyrae, B9 II + A2 II (WOOLF, 1965); VV Cephei, M2 I + B (PEERY, 1966); and RZ Ophiuchi, F3 Ib + cK5 (LOHMANN, 1950). Low-luminosity pairs include the W Ursae Majoris stars (near the main sequence), and a variety of systems incorporating a hot subdwarf or white dwarf that exhibits explosive phenomena. These systems have recently been reviewed by KRAFT (1967a, b) and MUMFORD (1967). They include such dissimilar pairs as the symbiotic variables (SAHADE, 1960), the U Geminorum variables, the ordinary novae (KRAFT, 1964), the Mira variables R Aquarii (ILOVAISKY and SPINRAD, 1966) and o Ceti (DEUTSCH, 1958), and the irregular variable CH Cygni (FARAGGIANA, 1968; FARAGGIANA and HACK, 1968). Presumably the explosive instability in some or all of these pairs arises from the deposition on the hot component of mass that is transferred from the primary by some non-explosive process (CRAWFORD and KRAFT, 1956). Even the well-known Be star P Cygni, generally regarded as a very slow nova, may be a component of a close binary pair (MAGALASHVILI and KHARADSE, 1967) in which mass transfer from a companion has triggered an explosive instability (SCHATZMANN, 1965; WEIDEMANN, 1968; ALEXANDER and WALLERSTEIN, 1967; FERNIE, 1968; LUND, 1968; DE GROOT, 1969).

An ordinary nova is estimated to lose about 10^{-4} or $10^{-3} M_\odot$ in an outburst (*Colloque sur les Novae, Novoïdes et Supernovae, Ann. Astrophys.*, **27**, 1964). To judge from the strengths and velocities of the lines observed during the recurrent eruptions of explosive variables, somewhat smaller amounts of gas are also lost in these events. Interacting binaries that do not suffer explosions also appear to lose some mass to the interstellar medium. For example, on spectroscopic evidence, STRUVE (1958) estimated that the system of β Lyrae loses 10^{-5} or $10^{-6} M_\odot$ yr^{-1}. All these interacting binaries together, including the ordinary novae, contribute only a small fraction of the approximately $\frac{1}{2} M_\odot$ yr^{-1}, which must be ejected if most of the dying stars in the galaxy are transformed into white dwarfs. In any case, most of this mass is probably lost from stars that have no companion close enough to promote mass loss at any evolutionary stage.

Among such objects, HOYLE and FOWLER (1960) have proposed that some, having main-sequence masses in the range from $\sim 1.2 M_\odot$ to $\sim 1.5 M_\odot$, explosively disintegrate as Type I supernovae. The catastrophe "... results from the ignition of degenerate nuclear fuel...", such as C^{12}, O^{16}, and Ne^{20}. In advanced evolutionary stages of stars in the relevant mass range, such degenerate material must always be present. If the central temperature exceeds $\sim 1.5 \times 10^9$ °, it detonates and liberates $\sim 5 \times 10^{17}$ erg gm^{-1}. About half this energy is dissipated in radiation and in dispersal of the stellar gas with a velocity of the order of 3000 km/sec. The rest of the energy disintegrates Fe^{58} into $He^4 + n$, and nucleosynthesis follows by the r-process of BURBIDGE et al. (1957).

The total frequency of supernova explosions is usually taken as $\sim 3 \times 10^{-3}$ yr^{-1} galaxy^{-1}; roughly half of these are of Type I and half of Type II (ZWICKY, 1965; MINKOWSKI, 1964). KATGERT and OORT (1967) have obtained a total frequency about ten times higher than these authors, after taking account of incompleteness factors. The larger frequency is still an order of magnitude smaller than the current estimates of the rate at which stars leave the main sequence in the relevant mass range, ~ 0.3 stars yr^{-1} in the galaxy (ABELL and GOLDREICH, 1966). Evidently most such stars lose mass by other processes fast enough to keep the central temperature below the explosion point, until nuclear energy-generation stops and the star becomes a white dwarf.

Type II supernovae result from the dynamical instability of a massive star that has exhausted its nuclear resources. The observations are consistent with the hypothesis that all single stars must become such supernovae if they have main-sequence masses greater than a certain limit M_{min}. Depending on the rates of mass loss in red supergiants and luminous giants, the effects of neutrino emission on the evolutionary timescales of these objects, and the uncertain value for supernova frequency, the limiting mass is likely to lie in the range from 4 to $18 M_\odot$ (STOTHERS, 1963). FOWLER and HOYLE (1964) suppose the limit to be $\sim 10 M_\odot$. They argue that, in an advanced evolutionary stage of a star more massive than this, "... hydrogen is fused into successively heavier nuclei until the iron-group elements are reached". This occurs in the e-process of B^2FH at $T = 3.8 \times 10^9$ °. Nuclear-energy generation then ceases, and photodisintegration sets in. This rapidly produces an endothermic iron-to-helium-neutron phase change in an inner core, and a resultant implosion of the material. A shell of e-processed matter envelopes the inner core, and a shell of oxygen envelopes the iron. The burning of this oxygen to silicon provides the energy ($\sim 5 \times 10^{17}$ erg gm^{-1}) necessary to disperse the whole star. However, COLGATE and WHITE (1966) "... find that the dynamical implosion is so violent that an energy many times greater than the available thermonuclear energy is released from the star's core and transferred to the star's mantle by the emission and deposition of neutrinos". They argue that, however much mass is ejected, it will always contain $\sim 1\,M_\odot$ of unstable nuclei, the decay of which then energizes the outburst of radiation. The process may occur in any highly evolved star more massive than $\sim 2\,M_\odot$, and it always leaves as a remnant a neutron star of $\sim 1\,M_\odot$.

The literature of the last two decades contains many other theoretical models for supernovae. For example, FINZI and WOLF (1967) have recently shown that catastrophic implosion can occur in certain old, massive white dwarfs, due to inverse β-reactions in material that still contains some nuclear fuel.

Most single stars leaving the main sequence in unit time have masses in the range $1\text{-}2\,M_\odot$, and very likely most of these do become white dwarfs. Some of the requisite mass loss is known to occur in the red-giant evolutionary stage. Then, at the end of the red-giant stage, many of these stars – and perhaps all of them – also produce planetary nebulae by the impulsive detachment of $\sim 0.3\,M_\odot$. Stars initially more massive than $1.5\text{-}2\,M_\odot$ probably do not produce planetary nebulae; but the phenomenon might occur in some of these objects after they have already lost most of their original mass in other processes.

For the stars that do produce planetaries, unless we are to assume that the ejection velocity V_{es} equalled the escape velocity to within a few percent, we find that the present slow nebular expansions require V_{es} to have been very low, as in the most distended red giants. This argument is due originally to SHKLOVSKY (1956), and it has been rediscussed by ABELL and GOLDREICH (1966). The numbers of planetary nebulae, their distribution in the galaxy, and their evolutionary time-scales are also compatible with the hypothesis that they are metamorphs of red giants in an advanced stage of evolution. The transition stage has not yet been clearly identified; possibilities that have been examined in the recent literature include some of the symbiotic stars, the Mira variables, and certain unusual supergiants. Several authors have discussed various instabilities that could lead to the impulsive detachment of $\sim 0.3\,M_\odot$ from a highly evolved red giant; the details of this process are still obscure. On the other hand, it now seems possible to understand rather well the evolution of the stellar remnant after the removal of its envelope. Even before the detachment occurs, the interior part of the star closely resembles a white dwarf. Theory and observation together show that after mass ejection the luminosity and radius of the star diminish toward their quasi-static equilibrium values on a time-scale of 10^4 or 10^5 years. Recent advances in this subject are well-reviewed in a volume edited by OSTERBROCK and O'DELL (1968).

Our review has now touched upon most of the various distinctive processes that are effective in transporting mass from stars back into the interstellar medium. The ubiquity of mass loss is now established; we understand something of the forces that can produce it, and something of its relevance to stellar evolution and nucleosynthesis. But in virtually every instance where we know mass loss to occur, there still remain many questions relative to the circumstances of the flow, its efficiency, its origin, and its termination in the interstellar gas These are the problems we must now examine in some detail, here at Trieste in our Colloquium, and in our Observatories in the years to come.

References

ABELL, G. O. and GOLDREICH, P.: 1966, *Publ. Astron. Soc. Pacific* **78**, 232.
ABT, H. A.: 1957, *Astrophys. J.* **126**, 138.
ABT, H. A. and GOLSON, J. C.: 1966, *Astrophys. J.* **143**, 306.
ALEXANDER, T. and WALLERSTEIN, G.: 1967, *Publ. Astron. Soc. Pacific* **79**, 500.
ARNETT, W. D., HANSEN, C. J., TRURAN, J. W. and CAMERON, A. G. W.: 1968, *Nucleosynthesis*, Gordon and Breach, New York.
BAHCALL, J. N. and WOLF, R. A.: 1968, *Astrophys. J.* **152**, 701.
BASHKIN, S.: 1965, in *Stars and Stellar Systems*, vol. VIII (ed. by L. H. Aller and D. B. McLaughlin), p. 1.
BIERMANN, L.: 1946, *Naturwiss.* **33**, 118.
BODANSKY, D., CLAYTON, D. D., and FOWLER, W. A.: 1968, *Astrophys. J. Suppl.* **16**, 299.
BÖHM-VITENSE, E.: 1956, *Publ. Astron. Soc. Pacific* **68**, 57.
BURBIDGE, E. M., BURBIDGE, G. R., FOWLER, W. A., and HOYLE, F.: 1957, *Rev. Mod. Phys.* **29**, 547.
CANNON, R. D.: 1968, *Obs.* **88**, 206.
CARRUTHERS, G. R.: 1968, *Astrophys. J.* **151**, 269.
CHRISTY, R. F.: 1966a, *Astrophys. J.* **144**, 108.
CHRISTY, R. F.: 1966b, *Ann. Rev. Astron. Astrophys.* **4**, 353.
COLGATE, S. A. and WHITE, R. H.: 1966, *Astrophys. J.* **143**, 626.
CONTI, P. S.: 1967, *Astrophys. J.* **149**, 629.
CRAWFORD, J. A. and KRAFT, R. P.: 1956, *Astrophys. J.* **123**, 44.
DAHLBERG, E.: 1964, *Astrophys. J.* **140**, 268.
DE GROOT, M.: 1969, I.A.U. Colloquium on Non-Periodic Phenomena in Variable Stars, and *Bull. Astron. Inst. Neth.* **20**, No. 4 (in press).
DE JAGER, C. and KUPERUS, M.: 1961, *Bull. Astron. Inst. Neth.* **16**, 71.
DESSLER, A. J.: 1967, *Rev. Geophys.* **5**, 1.
DEUTSCH, A. J.: 1958, *Astron. J.* **63**, 49.
DEUTSCH, A. J.: 1960, in *Stars and Stellar Systems*, vol. VI (ed. by J. L. Greenstein), p. 543.
DEUTSCH, A. J.: 1961, in *I.A.U. Symp.* **12**, 283.
DEUTSCH, A. J.: 1966, in *Stellar Evolution* (ed. by R. F. Stein and A. G. W. Cameron), p. 377.
DEUTSCH, A. J.: 1968, in H. W. Babcock, *Ann. Rep.*, *C.I.W. Yrbk.* **67**.
DOMMANGET, J.: 1967, *Observatoire Roy. Belgique, Communications*, Ser B , No. 14.
EGGEN, O. J. and GREENSTEIN, J. L.: 1965, *Astrophys. J.* **141**, 83.
FARAGGIANA, R.: 1968, *Mem. Soc. Astron. It.* **39**, 291.
FARAGGIANA, R. and HACK, M.: 1968, *Astrophys. Space Sci.* **3**, 205.
FERNIE, J. D.: 1968, *Observatory* **88**, 167.
FINZI, A. and WOLF, R. A.: 1967, *Astrophys. J.* **150**, 115.
FOWLER, W. A. and HOYLE, F.: 1964, *Astrophys. J. Suppl.* **9**, 201.
HAZLEHURST, J.: 1967, *Z. Astrophys.* **65**, 311.
HERBIG, G. H.: 1966, *Vistas in Astron.* **8**.
HILTNER, W. A. and SCHILD, R. E.: 1966, *Astrophys. J.* **143**, 770.
HOYLE, F.: 1964, *Royal Obs. Bull.* No. 82, 90.
HOYLE, F. and FOWLER, W. A.: 1960, *Astrophys. J.* **132**, 565.
HOYLE, F., FOWLER, W. A., BURBIDGE, E. M., and BURBIDGE, G. R.: 1964, *Astrophys. J.* **139**, 909.
HUANG, S. S.: 1963, *Astrophys. J.* **138**, 471.
ILOVAISKY, S. A. and SPINRAD, H.: 1966, *Publ. Astron. Soc. Pacific* **78**, 527.
JOHNSON, H. M. and HOGG, D. E.: 1965, *Astrophys. J.* **142**, 1033.
KATGERT, P. and OORT, J. H.: 1967, *Bull. Astron. Inst. Neth.* **19**, 239.
KIPPENHAHN, R. and WEIGERT, A.: 1967, *Z. Astrophys.* **65**, 251.
KRAFT, R. P.: 1964, *Astrophys. J.* **139**, 457.
KRAFT, R. P.: 1967a, *Astrophys. J.* **150**, 551.
KRAFT, R. P.: 1967b, *Publ. Astron. Soc. Pacific* **79**, 470.
KUHI, L. V.: 1964, *Astrophys. J.* **140**, 1409.
LIMBER, D. N.: 1967, *Astrophys. J.* **148**, 141.
LIMBER, D. N. and MARLBOROUGH, J. M.: 1968, *Astrophys. J.* **152**, 181.
LOHMANN, W.: 1950, *Z. Astrophys.* **27**, 161.
LUCY, L. B. and SOLOMON, P. M.: 1967, *Astron. J.* **72**, 31.

LUND, L. S.: 1968, I.A.U. Colloquium on Non-Periodic Phenomena in Variable Stars, Budapest (in press).
LÜST, R.: 1968, in *I.A.U. Trans.* **13A**, 1004.
MAGALASHVILI, N. L. and KHARADSE, E. K.: 1967a, *Comm. 27 I.A.U. Inf. Bull. Var. Stars*, No. 210.
MAGALASHVILI, N. L. and KHARADSE, E. K.: 1967b, *Observatory* **87**, 295.
MASANI, A., MARTINI, A., and ALBINO, E.: 1968, *Konkoly Information Bull.* No. 251.
McCREA, W. H.: 1964, *Monthly Notices Roy. Astron. Soc.* **128**, 147.
MESTEL, L.: 1965, in *Stars and Stellar Systems*, vol. VIII (ed. by L. H. Aller and D. B. McLaughlin), p. 297.
MINKOWSKI, R.: 1964, *Ann. Rev. Astron. Astrophys.* **2**, 247.
MORTON, D. C.: 1967a, *Astrophys. J.* **150**, 535.
MORTON, D. C.: 1967b, *Astrophys. J.* **147**, 1017.
MORTON, D. C., JENKINS, E. B., and BOHLIN, R. C.: 1968, *Astrophys. J.* **154**, 1661.
MORTON, D. C., JENKINS, E. B., and BROOKS, N. H.: 1968, preprint.
MUMFORD, G. S.: 1967, *Publ. Astron. Soc. Pacific* **79**, 283.
NESS, N. F.: 1968, *Ann. Rev. Astron. Astrophys.* **6**, 79.
OSTERBROCK, D. E. and O'DELL, C. R. (eds.): 1968, in *Planetary Nebulae*, I.A.U. Symp. 34, D. Reidel, Dordrecht, Holland pp. 361, 440.
PACZYŃSKI, B.: 1968, in *Highlights of Astronomy* (ed. by L. Perek), D. Reidel, Dordrecht, Holland, p. 409.
PARKER, E. N.: 1958, *Astrophys. J.* **128**, 664.
PARKER, E. N.: 1965, *Space Sci. Rev.* **4**, 666.
PEERY, B. F.: 1966, *Astrophys. J.* **144**, 672.
PEREK, L.: 1968, *Highlights of Astronomy*, D. Reidel, Dordrecht, Holland, Section F.
PLAVEC, M.: 1968, in *Highlights of Astronomy* (ed. by L. Perek), D. Reidel, Dordrecht, Holland, p. 394.
REEVES, H.: 1968, *Stellar Evolution and Nucleosynthesis*, Gordon and Breach, New York.
SAHADE, J.: 1955, *Publ. Astron. Soc. Pacific* **67**, 348.
SAHADE, J.: 1960, in *Stars and Stellar Systems*, vol. VI (ed. by J. L. Greenstein), p. 466.
SARGENT, W. L. W.: 1961, *Astrophys. J.* **134**, 142.
SARGENT, W. L. W.: 1965, *Obs.* **85**, 33.
SCHATZMANN, E.: 1965, in *Stars and Stellar Systems*, vol. VIII (ed. by L. H. Aller and D. B. McLaughlin), p. 327.
SCHMIDT, M.: 1966, *Trans. I.A.U.* **12B**, 423.
SCHMIDT-KALER, TH.: 1964a, *Z. Astrophys.* **58**, 217.
SCHMIDT-KALER, TH.: 1964b, *Veröff. Bonn*, No. 70.
SHKLOVSKY, I. S.: 1956, *Astron. Zh.* **33**, 315.
SIMON, G. W. and LEIGHTON, R. B.: 1964, *Astrophys. J.* **140**, 1120.
SMITH, L. F.: 1967, *Astron. J.* **72**, 829.
STONE, M. E. and MORTON, D. C.: 1967, *Astrophys. J.* **149**, 29.
STOTHERS, R.: 1963, *Astrophys. J.* **138**, 1085.
STRUVE, O.: 1958, *Publ. Astron. Soc. Pacific* **70**, 5.
THORNE, K. S.: 1967, *High Energy Astrophysics*, Vol. 3 (ed. by C. De Witt, E. Schatzmann, and P. Veron), Gordon and Breach, New York.
TRIMBLE, V. L. and THORNE, K. S.: 1969, in press.
UNDERHILL, A. B.: 1959, *Publ. Domin. Astrophys. Obs.* **11**, 209.
UNDERHILL, A. B.: 1960, in *Stars and Stellar Systems*, vol. VI (ed. by J. L. Greenstein), p. 411.
UNDERHILL, A. B.: 1966, *The Early Type Stars*, D. Reidel, Dordrecht, Holland.
UNDERHILL, A. B.: 1967, *Bull. Astron. Inst. Neth.* **19**, 173.
UNDERHILL, A. B.: 1968, *Ann. Rev. Astron. Astrophys.* **6**, 39.
VAN DEN HEUVEL, E. P. J.: 1968, *Bull. Astron. Inst. Neth.* **19**, 11, 326.
WAGONER, R. V., FOWLER, W. A., and HOYLE, F.: 1967, *Astrophys. J.* **148**, 3.
WEBER, E. J. and DAVIS, L.: 1967, *Astrophys. J.* **148**, 217.
WEIDEMANN, V.: 1968, *Ann. Rev. Astron. Astrophys.* **6**, 351.
WEYMANN, R.: 1960, *Astrophys. J.* **132**, 380.
WEYMANN, R.: 1962, *Astrophys. J.* **136**, 844.
WEYMANN, R.: 1963, *Ann. Rev. Astron. Astrophys.* **1**, 97.

WHEELER, J. A.: 1966, *Ann. Rev. Astron. Astrophys.* **4**, 393.
WICKRAMASINGHE, N. C., DONN, B. D., and STECHER, T. P.: 1966, *Astrophys. J.* **146**, 590.
WILSON, O. C.: 1960, *Astrophys. J.* **131**, 75.
WILSON, O. C.: 1963, *Astrophys. J.* **138**, 832.
WILSON, O. C.: 1966, *Astrophys. J.* **144**, 695.
WOOLF, N. J.: 1965, *Astrophys. J.* **141**, 155.
ZWICKY, F.: 1965, *Z. Astrophys.* **27**, 367.

Discussion

Morton: Do statistics of stellar evolution agree with number of white dwarfs?

Kutter: Weidemann estimates (Prague 1967) that about 5–50% of all white dwarfs have passed through planetary nebula phenomenon. The remaining white dwarfs must then have been formed along other evolutionary paths, such as by mass exchange in the binary systems.

Morton: Can the white-dwarf density in the solar neighbourhood be accounted for by all the observed mass loss?

Deutsch: I am not aware of any problem here. However, one must know the past star-density as well as the mass-loss rates in order to predict the present white dwarf density. I suspect there would be a difficulty if one had to suppose that the present *observed* rate of mass loss represents the *actual* rate over the age of the galaxy.

Rose: Is there any observational evidence to indicate whether or not discrete shells of cold matter (mass $\gtrsim 0.1\ M_\odot$) are ejected from red giants? It would be of considerable interest to determine if such relatively massive shells could be ejected from red giants with a frequency that is greater than or equal to the observed frequency for the formation of planetary nebulae.

Deutsch: I think the evidence is rather strong that few such shells occur, if any; for they could be easily observable as shortward displaced absorption-cores in the resonance lines of many abundant elements.

MASS LOSS FROM SINGLE STARS
OBSERVATIONS

THE WOLF-RAYET STARS AND MASS LOSS

ANNE B. UNDERHILL

Sonnenborgh Observatory, University of Utrecht, The Netherlands

Abstract. Wolf-Rayet stars are defined, a summary is given of the properties of Wolf-Rayet stars, and a qualitative model of a Wolf-Rayet star is sketched. It is incontrovertible that Wolf-Rayet stars are losing mass, a typical rate of mass loss being near 10^{-5} $M_{\odot}$ per year. The outward directed velocity of the expanding shell has been estimated for 10 stars. The largest value found is 2500 km/sec; most values lie between 1000 and 1500 km/sec. Two outstanding problems are to understand how the observed high velocities are generated and to demonstrate quantitatively the effect of these velocities on the observed spectrum. Five questions raised by the fact that mass loss is observed to take place from Wolf-Rayet stars are discussed briefly in Section 5.

1. Introduction

'Do all stars lose mass?' is a question that may well have an affirmative answer. The fact that mass loss is occurring can be ascertained only if the escaping material generates an observable phenomenon which we can recognise to be due to mass loss. The Wolf-Rayet stars have long been recognised as outstanding members of the class of stars which lose mass because their spectra contain a few flat-topped emission lines and a few shortward-displaced absorption lines which can most simply be interpreted as being due to a spherical atmosphere expanding at a velocity greater than the velocity of escape. With other stars the fact that matter is escaping from the star is not so easily detected. For instance, observations of the normal solar spectrum will not indicate that matter is streaming from the sun, yet the solar wind is a well-attested phenomenon.

The term 'Wolf-Rayet star' means that the spectrum of the star in the region 3000–7000 Å consists of a faint continuum upon which rather wide, strong emission lines of He I, He II and lines of the third, fourth and fifth spectra of C, N and O are superimposed. A few other emission lines are sometimes present, as well as a few particular absorption lines, but no spectroscopically forbidden lines are seen. The spectral subtypes are assigned according to the details of the intensity pattern of the emission lines. In stars of types WC emission lines of C III, C IV, O III and O IV dominate while in stars of types WN lines of N III, N IV and N V dominate. Weak lines of C IV, O IV and O V are present in WN spectra but there is no C III. The lines of He I and He II appear in both classes of Wolf-Rayet spectra, the lines of He II usually being stronger in WN spectra than in WC spectra.

The spectroscopic criteria by which Wolf-Rayet stars are distinguished indicate that the atmosphere of a Wolf-Rayet star is a plasma with an electron temperature of the order of 30000–50000 K and a particle density of the order of 10^{12}. The particle density is certainly not greater than 10^{14}, otherwise Stark broadening of the He II lines would be significant, nor it is less than 10^{10}, for then forbidden lines would be seen.

In the Galaxy there are two types of objects which show Wolf-Rayet spectra:

M. Hack (ed.), Mass Loss from Stars. All rights reserved

(i) a few stars (124 known in the Galaxy) which are closely associated with OB stars, and (ii) some central stars of planetary nebulae.

Only the Wolf-Rayet stars of group (i) will be considered in what follows. They may be considered to belong to extreme Population I. The central stars of planetary nebulae are usually considered to be rather old stars belonging to the Disk Population. Little information is available about the central stars of planetary nebulae which have Wolf-Rayet spectra because of the difficulty of making adequate observations. Significant work on this project is in progress by Aller and his associates (ALLER, 1968).

The central stars of planetary nebulae seem to have visual absolute magnitudes of the order of 0, while stellar structure studies suggest that their masses are near $1 M_\odot$. Although their spectra closely resemble the spectra of extreme Population I Wolf-Rayet stars, the central stars of planetary nebulae which have Wolf-Rayet spectra are not similar in mass and luminosity to the Wolf Rayet stars of extreme Population I. The spectroscopic criteria which are used to define the class 'Wolf-Rayet' do not serve to isolate stars of similar mass, radius, luminosity and stage of development. The physical phenomenon which causes a Wolf-Rayet spectrum to appear seems to be able to happen early in the development of a star (the Population I Wolf-Rayet stars) and at a rather late stage of development (some central stars of planetary nebulae). Full references for the facts quoted in this introduction and later in this paper may be found in UNDERHILL (1968).

2. The Properties of the Population I Wolf-Rayet Stars

The Wolf-Rayet stars are associated with O and B stars and are closely confined to the plane of the Galaxy. No Wolf-Rayet star is known in the quadrant centered on the anti-centre direction. At least one-half of the galactic Wolf-Rayet stars are in open clusters, more WN stars being found in clusters than are WC stars. At least one-third of the WR stars are known to be spectroscopic binaries from the fact that the spectra of two stars are observed.

Twenty-seven Wolf-Rayet stars are known definitely to be associated with H II regions and such association appears to be probable for 12 more. Small ring-like nebulae are seen around some Wolf-Rayet stars. Miss SMITH (1966) has demonstrated that these nebulae are formed as the surrounding interstellar matter is pushed away from the Wolf-Rayet star by the expanding atmosphere and by radiation which comes from the Wolf-Rayet star. At least two of these ring nebulae have been observed at radio frequencies (JOHNSON and HOGG, 1965). The existence of these ring nebulae is definite evidence for mass loss from Wolf-Rayet stars.

The masses of the Wolf-Rayet stars can be estimated from the orbital elements of seven double-lined spectroscopic binaries. The companion is an O or early B type star. The results are very consistent in showing that the mass of the Wolf-Rayet star is about $\frac{1}{3}$ that of the OB star. A fair summary is to say that the mass of a Wolf-Rayet star is $10 \pm 2 M_\odot$, where ± 2 is the probable error. There is no direct evidence that the masses of Wolf-Rayet stars are greater than or nearly equal to those of O or early B type stars.

The most reliable estimates of the visual absolute magnitudes of Wolf-Rayet stars are those made by Miss SMITH (1966) using apparently single stars in the Large Magellanic Cloud. One may conclude that

$$M_V(\text{WR}) = -4.5 \pm 0.5 \text{ (p.e.) mag.}$$

for all types except, perhaps, WN7 and WN8, which may be as bright as -6.5. Thus the Wolf-Rayet stars have visual absolute magnitudes like those of O stars.

KUHI (1966), using moderate resolution photoelectric scans, has demonstrated that the continuous spectrum of a Wolf-Rayet star has a shape very like that of an O star. There is slight evidence that both WC and WN stars have a small infrared excess; the WN stars have a definite UV excess with respect to O stars beginning at about 4000 Å. The intrinsic B-V colour of a WC star, when correction for the presence of emission lines has been made, is about the same as that of a B2 star, while that of a WN star is like that of an O5 star.

Nothing is known directly about the bolometric correction for Wolf-Rayet stars. If the bolometric correction is like that of O stars it is near -3.0 mag. and the bolometric absolute magnitude of most Wolf-Rayet stars is near -7.5, which means that the total luminosity of a Wolf-Rayet star is about 8.3×10^4 that of the sun. STECHER (1968) has obtained observations in the far ultraviolet of the double-lined spectroscopic binary γ_2 Velorum (WC8 + O7) using a scanning spectrometer on a rocket. The intensity distribution of the continuous spectrum is similar in appearance to that of ζ Puppis, O5f, thus there is no evidence that this Wolf-Rayet star is significantly brighter than an O star in the part of the ultraviolet region where most of the flux is transmitted.

The best available information about the radii of Wolf-Rayet stars comes from the study of the eclipsing binary V444 Cygni by KRON and GORDON (1950) and the interferometer measurements of γ_2 Velorum by HANBURY BROWN et al. (1968). From these studies one may conclude that the photosphere has a radius like that of an O star, say about $7\,R_\odot$, while the region within which most of the strong emission lines are formed has a radius of the order of 5 times the radius of the photosphere, about $35\,R_\odot$.

The spectrum of a Wolf-Rayet star contains three types of line:
(i) rounded emission lines which have a more or less Gaussian shape,
(ii) flat-topped emission lines typical of a uniformly expanding spherical atmosphere,
(iii) shortward displaced absorption components, the displacement of which corresponds to the velocity of ejection deduced from the width of the flat-topped lines.

In the case of binary stars, the spectrum of an O star or an early B type star is frequently seen to be superimposed on the Wolf-Rayet spectrum, and rather sharp irregularly changing emission or absorption components (depending on the line) are often seen. The latter type of feature indicates that the binary system is enveloped in streams of gas. Insufficient information is available to make a definite picture of the distribution of gas in any one binary system which contains a Wolf-Rayet star, but it is very definite that such gas is present in all systems that have been studied in any detail and that its distribution and velocity component in the line-of-sight change from time to time. Most of the spectral variations which have been attributed to Wolf-Rayet stars refer to changes in the spectrum of the gas in the system. The basic Wolf-Rayet spectrum appears to have a rather stable appearance.

The lines of type (i) compose most of the spectrum. In the case of He II the shape is constant when expressed in velocity units no matter what the quantum numbers are of the line. This fact indicates that the line shape is due chiefly to motion and partly to electron scattering. One may conclude that the electron density may lie in the range 10^{11}–10^{12} particles per cm^3. The emission lines of type (i) are not usually accompanied by a shortward-displaced absorption component.

Only a very few flat-topped emission lines, type (ii), are seen. The most conspicuous are C III 5696 in WC stars and N IV 4058 in WN stars. Both lines appear to be excited by particular processes in an extended, low-density atmosphere where the radiation field is dilute. The strongest He I lines sometimes also have flat-tops, in particular He I 5876 in WN stars.

The lines of type (iii) are of two sorts. Either they are the type known to be strengthened in absorption due to departures from thermodynamic equilibrium in a high-temperature, low-density gas or they come from relatively low-lying levels of an abundant terminal ion. The first group contains lines such as He I 10830, 3888, 3187 and He I 5876, 4471 as well as the C III multiplet at 4650 Å and the N IV multiplet at 3480 Å. The second group contains lines like He II 4686, 5411 and 4541 as well as C IV 5801 and 5812 and N V 4603 and 4620. The lines of types (ii) and (iii) indicate that each Wolf-Rayet star is surrounded by a low-density, rather high-temperature expanding envelope. The electron temperature in the expanding atmospheres around WC stars appears to be lower than that in the expanding atmospheres around WN stars, for no absorption lines from excited levels having excitation potentials greater than 40 V are seen in WC spectra, whereas absorption lines from excited levels having excitation potentials as high as 56 V are seen in WN spectra.

3. A Qualitative Model for a Wolf-Rayet Star

A typical single Wolf-Rayet star, not disturbed by a companion, appears to be an object which has a mass near $10\,M_\odot$, a photosphere which radiates very like the photosphere of an O star in the spectral range 3500–7000 Å and which has a radius of about $7\,R_\odot$. Surrounding this photosphere there is a spherically symmetric atmosphere extending to about $35\,R_\odot$. The density of heavy particles is probably of the order of 10^{11}–10^{12} in this atmosphere, the electron temperature is high, say 30000–50000 K, and the particles may have large chaotic motions. A typical average component of velocity in the line-of-sight may be as much as 500 km/sec to 800 km/sec. The average velocity varies significantly from star to star or else the amount of broadening by electron scattering varies, for the widths of the rounded emission lines vary over a range of the order of a factor 5 from star to star. This dense shell is transparent in continuum frequencies but opaque in line frequencies. However, no 'normal' stellar absorption lines are seen. Outside the part of the atmosphere where most of the emission lines are formed, an exosphere or expanding atmosphere exists. It may be observed only in lines which are strengthened either in emission or in absorption by the particular radiative and collision processes which can occur in a rather low-density plasma illuminated by a typical O type radiation field and a strong flux of He II 303. So far no estimates have been made of a typical density for this exosphere, nor of its extent.

It is not possible to make a satisfactory generalised model of a Wolf-Rayet atmosphere giving particular details about how the velocity field and the electron temperature increase or decrease outward. To do this requires a satisfactory theory of the formation of the line spectrum. Many attempts have been made to set up suitable theories, but none are truly satisfactory, in particular those based only on radiative processes fail. Qualitatively it seems certain that much of what is known as a Wolf-Rayet spectrum is due to collisional excitation with fast moving protons, but it has not yet been possible to work out a quantitative theory. Consideration of dynamic processes appears to be necessary in order to obtain a quantitative model of a Wolf-Rayet atmosphere. However, it must be stressed that classical Wolf-Rayet spectra are stable from a secular point of view. The basic emission-line spectrum remains constant in appearance for periods of 40–50 years at least. There is only one case known, HD 45166, of a spectral development such as shown by shell stars or by novae, and there is some doubt if it is correct to classify this star as a Wolf-Rayet star.

4. Mass Loss from Wolf-Rayet Stars

The velocity of expansion of the outer layers of a Wolf-Rayet star can be estimated from the width of the flat-topped C III 5696 line and from the displacement of the absorption lines which are formed in the expanding atmosphere. Some typical results are given in Table I. The result for HD 68273 (γ_2 Velorum) is due to CODE and BLESS (1964); the other estimates have been made by the author.

The velocity of escape from a star of M solar masses and a radius of R solar radii is

$$v_{esc} = 618 \, (M/R)^{\frac{1}{2}} \text{ km/sec.}$$

Since the typical mass for a Wolf-Rayet star is $10 \, M_\odot$, while a typical radius for the outer part of the atmosphere is $35 \, R_\odot$, it is clear that the atmospheres surrounding most of the stars in Table I must be escaping. If N heavy particles per cm^3 of mass $\bar{m}$ are crossing a radius R with a velocity v, then the mass loss per year is $4\pi N t \bar{m} v R^2 / M_\odot$ solar masses, where t is the number of seconds in a year. A typical value is 3.5×10^{-5} $M_\odot$/yr. This number corresponds to 10^{11} particles per cm^3 having an average mass of 4.02×10^{-24} g crossing a radius of $30 \, R_\odot$ with a speed of 1000 km/sec.

No emphasis should be placed on the numerical value which has been estimated for the mass loss from Wolf-Rayet stars. The most uncertain factor is N, the number density of particles crossing the effective boundary of the atmosphere. A mass loss of the order of 10^{-5}–$10^{-6} \, M_\odot$/yr seems to be typical. There is no direct evidence

TABLE I

Expansion velocities from Wolf-Rayet stars

Star (H.D. No.)	Type	Velocity (km/sec)	Star	Type	Velocity (km/sec)
184738	WC9	− 580	92740	WN7	− 0:
192103	WC7	− 1200	93131	WN7	− 120
192641	WC7	− 1300	151932	WN7	− 500
68273	WC8	− 1300	192163	WN6	− 1400
193793	WC6	− 2500	191765	WN6	− 1600

that the mass loss occurs at a rate as high as $10^{-4} M_\odot$/year. Miss SMITH (1966) has shown that mass-loss rates of the order of 10^{-5}–$10^{-6} M_\odot$/year are consistent with the inferred lifetimes of the ring nebulae seen around some Wolf-Rayet stars.

5. Questions raised by Mass Loss from Wolf-Rayet Stars

(i) How much mass do all of the known Wolf-Rayet stars contribute to interstellar space in 10^6 years? If the mass-loss rate of $3 \times 10^{-5} M_\odot$/year is taken as an upper limit, then 100 Wolf-Rayet stars will have contributed something like $3000 M_\odot$ in 10^6 years. This mass will be spread around the plane of the Galaxy in the neighbourhood of the Wolf-Rayet stars. Since it is a small amount in comparison to the total amount of mass existing as dust and gas in interstellar space, the presently observable Wolf-Rayet stars probably have done little to change the composition of interstellar space during their lifetime.

(ii) Is there any evidence concerning the chemical composition of the material ejected from Wolf-Rayet stars? No. The ejected material is observed by means of a few lines from He atoms, He^+, C^{++}, N^{+++} and N^{++++} ions and there is no way of estimating the relative abundances of the elements from the few specially excited lines which are observed. Observations in the far ultraviolet of the resonance lines of most of the ions of the light elements would give a clearer picture of the composition of the material being ejected, because probaly most of the ions present emit or absorb the resonance lines, and the present lack of a reliable theory of line formation in extended atmospheres by means of which to estimate the distribution of atoms and ions over unobserved energy states will not be too serious. Probably it is best at present to assume normal Population I composition. Such an assumption implies that the composition of the outer layers of the star has not been changed by the energy-generating nuclear processes in the interior.

(iii) What is the origin of the high outwardly directed velocities which are observed? The largest known velocity of expansion is about 2500 km/sec and it is observed in the WC6 star HD 193793. Values of the order of 1000–1600 km/sec occur for many WC and WN stars, while values $\gtrsim 500$ km/sec are found for the sharp-lined WN7 and WN8 stars and at classes WC8 and WC9. Radiation pressure in the ultraviolet resonance lines is a possible accelerating force. The opacity of the atmosphere of a Wolf-Rayet star in the ultraviolet resonance lines of the second, third and fourth spectra of most of the light elements must be large, and if the apparent similarity between the continuous spectra of Wolf-Rayet stars and O stars in the 4000–7000 Å region means anything, one may expect the peak of the flux from Wolf-Rayet stars to occur between 1000 and 1500 Å. LIMBER (1964) has suggested that forced rotational ejection is an important factor for explaining the characteristic details of Wolf-Rayet spectra. There is no direct evidence that the photospheres of Wolf-Rayet stars are rotating rapidly; nor is there evidence to the contrary. Forced rotational ejection may be significant in shell stars (many of which are known to rotate rapidly), but in no shell star are velocities of ejection of the order of 1000 km/sec known to be the usual thing. Consequently it seems unlikely that forced rotational ejection is a primary factor in causing the ejection of Wolf-Rayet atmospheres. It should be recalled that the meager evidence about the masses and sizes of shell stars and of Wolf-Rayet stars indicates that so far as these two properties are concerned Wolf-Rayet stars and the hotter shell stars γ Cassiopeiae and φ Persei are very similar objects. The presence of high velocities in the atmospheres of Wolf-Rayet stars are probably an important reason why the spectra of the shell stars differ from Wolf-Rayet spectra.

(iv) Does the observed expansion velocity vary according to the ion which is observed and therefore, presumably, according to the distance from the photosphere at which the line (absorption or emission) is formed? Presently available information on Wolf-Rayet stars gives no clear answer to this question. MORTON (1967) from rocket ultraviolet spectra of OB supergiants has found slight evidence that the expansion velocity of different ions is different in the atmospheres of the B0 supergiants, while it has long been known – cf. BEALS (1950) – that the expansion velocity varies with the ion observed in the atmospheres of stars like P Cygni. A range of velocities similar to that of Wolf-Rayet stars is found for the B0 and O9 supergiants.

(v) Are magnetic fields an important factor for controlling the velocity field in Wolf-Rayet atmospheres? Most of the particles giving the observed Wolf-Rayet spectrum are ions, helium being the only conspicuous atom observed. There is evidence that hydrogen is partly neutral in the atmospheres of WC7 and WC8 stars as well as in WN7 and WN8 stars. The presence of a magnetic field would constrain the ions, and thus the most abundant constituents of a Wolf-Rayet atmosphere to move along the lines of force. Since the shapes of lines formed in the expanding atmosphere can be explained most simply by a uniformly expanding sphere, any action of possible magnetic fields must maintain spherical symmetry. The spectra of Wolf-Rayet stars, as presently known, contain no features demanding for their interpretation a localised distribution of magnetic fields such as might occur in star spots, or in an equatorial sheet. So far no method of observing directly a magnetic field in a Wolf-Rayet star, should it exist, has been derived.

6. Conclusions

In order to understand the excitation of the line spectrum of a Wolf-Rayet star it seems to be necessary to consider collisional excitation processes as well as radiative processes (cf. CODE and BLESS, 1964). The fact that an observable amount of matter is streaming from Wolf-Rayet stars at velocities which usually lie in the range 1000–1500 km/sec is incontrovertible. The outstanding problems are to understand how the high velocities are generated and to demonstrate quantitatively the effect of these velocities on the observed spectrum.

The phenomenon known as a 'Wolf-Rayet spectrum' is found among stars of extreme Population I, that is the classical Wolf-Rayet stars, and among stars of the Disk Population, that is among the central stars of planetary nebulae. The latter stars are presumably at a late stage of development, while stars of the first group have not evolved far. In fact it can be argued that the classical Wolf-Rayet stars are making their first approach to the main sequence and also that they are departing from the main sequence. Neither hypothesis can be rejected with present information, nor is either entirely satisfactory. Possibly both are valid. In any case, it is clear that the process of generating high velocities in the outer layers of a star can occur at at least two stages in the lifetime of a star. The occurrence of the phenomenon known as a Wolf-Rayet spectrum does not appear to be uniquely related to the mass of the star, for it occurs among stars of $10 \pm 2 M_\odot$ and among stars having masses near $1 M_\odot$.

References

ALLER, L. H.: 1968, in *Planetary Nebulae* (ed. by D. E. Osterbrock and C. R. O'Dell), D. Reidel Pub. Co. Dordrecht, pp. 339–354.
BEALS, C. S.: 1950, *Pub. Dom. Astrophys. Obs.* **9**, 1–137.
CODE, A. D. and BLESS, R. C.: 1964, *Astrophys. J.* **139**, 787–792.
HANBURY BROWN, R. *et al.*: 1968, in *Proc. Symp. on Wolf-Rayet Stars* (ed. by K. B. Gebbie and R. N. Thomas), Nat. Bur. Stds. No. 307, p. 79.
JOHNSON, H. M. and HOGG, D. E.: 1965, *Astrophys. J.* **142**, 1033–1040.
KRON, G. E. and GORDON, K. C.: 1950, *Astrophys. J.* **111**, 454–483.
KUHI, L. V.: 1966, *Astrophys. J.* **143**, 753–769.
LIMBER, D. N.: 1964, *Astrophys. J.* **139**, 1251–1266.
MORTON, D. C.: 1967, *Astrophys. J.* **150**, 535–542.
SMITH, L. F.: 1966, Thesis, Australian Nat. Univ., pp. 1–204.
STECHER, T. P.: 1968, in *Proc. Symp. on Wolf-Rayet Stars* (ed. by K. B. Gebbie and R. N. Thomas), Nat. Bur. Stds. No. 307, p. 65.
UNDERHILL, A. B.: 1968, *Ann. Rev. Astron. Astrophys.* **6**, 39–78.

Discussion

Dallaporta: You mentioned the percentage of binaries among the Wolf-Rayet stars. I would like to ask: What is the evidence that the Wolf-Rayet stars considered as single are really single?

Underhill: If one has no evidence of a second spectrum or if photometry gives no suggestion of the presence of a second star, the Wolf-Rayet star is considered to be single. Adequate radial-velocity studies of Wolf-Rayet stars have not been done.

Nariai: Is there a possibility that WR stars are He-rich?

Underhill: WC 7 and 8 stars do show evidence of hydrogen Balmer decrements. In hotter stars the ionization and excitation conditions are unfavorable for detecting hydrogen. One cannot say definitely that hydrogen is absent from the atmosphere.

Boury: How many masses do we know?

Underhill: Seven double-lined spectroscopic binaries are known, from which some mass estimates can be made.

Morton: Is the estimated average absolute bolometric magnitude of -7.5 too large for the average mass of $10 M_\odot$?

Underhill: Yes, according to stellar evolution theories of hydrogen-burning stars near the main-sequence.

Rose: The high L/M ratios that are observed for Wolf-Rayet stars may be explained if it is assumed that they are hydrogen-deficient helium core-burning stars.

Sahade: The question of the absolute magnitude of a WR object may be related with the question of whether or not they are all binaries, because the contribution of the companion should come in.

Underhill: The result quoted, $M_v = -4.5 \pm 0.5$ mag. was derived by Miss L. F. Smith using only stars for which there was no evidence of a companion contributing to the spectrum or to the light.

Houziaux: How did you determine the electron density quoted in your paper?

Underhill: The values given are estimates based on: (a) Knowledge of the electron density in the outer layers of OB model atmospheres. (b) Aller's statement that if $N_e < 10^{10}$ one should see forbidden lines as in planetary nebulae and such lines are not observed. (c) If $N_e \approx 10^{14}$ one would expect the He II lines to show Stark broadening. Lines with lower quantum numbers 3, 4, 5 and upper quantum numbers 4–16 or so are observed and they all have the same exponential shape, so Stark effect cannot be important.

Sahade: I thought that there was no doubt about the atmosphere being stratified because of the correlation between widths and ionization potential of the ions involved.

Underhill: The interpretation of the available information is by no means clear. One must be careful to consider together only lines formed more or less under the same circumstances. For instance it is not appropriate to compare the widths of lines formed in the expanding atmosphere of low density with lines formed in the inner more compact atmosphere because the cause of the line shape is different in the two cases. The small amount of material permitting to demonstrate a correlation between widths and ionization potential may not indicate stratification but merely reflect the point that the expanding atmosphere is observable in only a few lines.

Dallaporta: Is it possible to give some figures concerning the relative abundances of C, N, and O for the two sequences of WC and WN stars?

Underhill: No. Information about abundances can be obtained from observed line strengths only by using a reliable theory of line formation. Such a theory does not exist for WR stars.

MASS LOSS FROM P CYGNI

MART DE GROOT

Sonnenborgh Observatory, University of Utrecht, The Netherlands

Abstract. From a careful study of 35 high-dispersion spectrograms of P Cygni it is concluded that the spectroscopic data do not confirm the conclusion of Magalashvili and Kharadze that P Cygni is a WUMa type system. It is found that many of the absorption lines are double, the hydrogen absorption lines even triple. This is attributed to line formation in different shells. In the outer shell variations with a period of 114 days lead to observed radial-velocity variations between − 180 and − 240 km/sec. A preliminary conclusion about the velocity field in the atmosphere of P Cygni is drawn and the mass loss is estimated.

P Cygni, in the Henry Draper Catalogue classified as B1p, has been known as a variable star from the year 1600 when it was discovered as a third magnitude nova by the Dutch chartmaker, geographer and mathematician Willem Janszoon Blaeu. The early history of the light variation of this star is nearly unequalled and rather puzzling. However, since 1880 P Cygni has been of nearly constant brightness. Some observers have reported irregular light variations with an amplitude up to 0.2 magnitude (e.g. NIKONOV, 1936, 1937). About 1 year ago Magalashvili and Kharadze reported some interesting two- and three-color observations of P Cygni. From their observations made during the period 1951–60 they concluded that P Cygni is a WUMa system with a period of 0.500656 days and with amplitudes of $0^m.10$ and $0^m.08$ for the primary and secondary minimum respectively (MAGALASHVILI and KHARADZE, 1967a, b).

When these results were first reported in the *Information Bulletin on Variable Stars* (no. 210) P Cygni was put on a constant observation program for 5 nights by ALEXANDER and WALLERSTEIN (1967), who reported that their observations did not reveal any variations of the brightness of P Cygni and thus did not confirm the observations made by Magalashvili and Kharadze.

In this paper some facts pertaining to the character of these light variations are presented from a different point of view. We have been working upon a collection of high-dispersion spectrograms of P Cygni, covering the period 1942–64. From the study of some 35 spectrograms the following facts have been established:

(1) On most of the spectrograms the lines of hydrogen, many lines of He I and the strongest lines of Fe III show besides the nearly undisplaced emission line two shortward displaced absorption components. In the case of the hydrogen lines with Balmer number $n \geqslant 9$ there often are even three components with velocities of about −95, −125 and −210 km/sec (cf. Figure 1).

(2) The radial velocity of the most shortward displaced component of the hydrogen lines is not constant but shows variations which after a closer inspection have a period of 114 days. Other lines do not show this periodicity.

(3) There are variations in the relative intensities of different absorption components. These variations seem to be rather irregular.

M. Hack (ed.), Mass Loss from Stars. All rights reserved

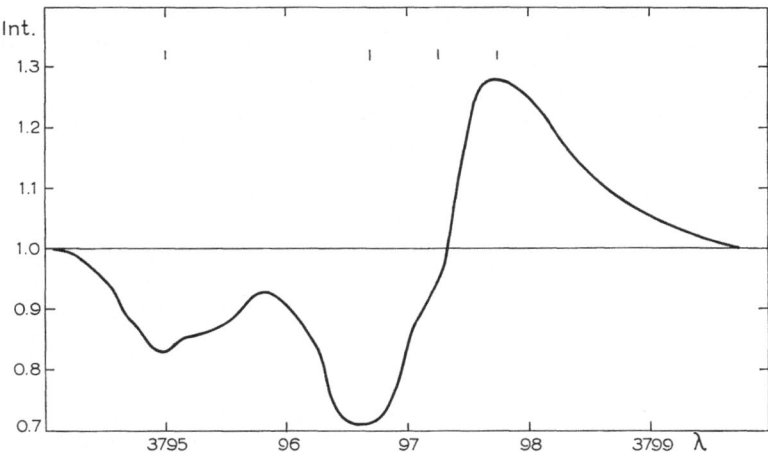

Fig. 1.   Profile of H10 λ 3797 showing three absorption components.

With this information let us consider again the conclusion of Magalashvili and Kharadze about the binary nature of P Cygni. Should the fact that the spectral lines often are double be regarded as a proof that P Cygni is a binary? The two absorption components which appear at the positions of the hydrogen and helium lines are of comparable strength. This means that a companion star should not be more than 1 magnitude fainter than the main star. If this statement were true, then also other spectral lines of the companion should be visible in the spectrum of P Cygni, this providing more double absorption lines. This is not the case. Only the hydrogen and some of the helium lines are double. One might think of a late B type companion with a few strong spectral lines except those of hydrogen. But then the Si ıı spectrum and the line of Mg ıı at λ 4481 should be more prominent than the lines actually observed in the spectrum of P Cygni.

Furthermore, the mean velocity of approach, as derived from the two absorption components of the hydrogen lines equals about − 170 km/sec. If the duplicity of the lines were a proof of the binary nature of P Cygni this figure would mean either that the system as a whole has a velocity of − 170 km/sec with respect to the sun, or that the WUMa binary is surrounded by a large expanding atmosphere. The first suggestion is not acceptable because it leaves unexplained the fact that all the emission lines lie at an average displacement of about − 15 km/sec. Also a velocity of − 170 km/sec is impossible to combine with the membership of P Cygni of the galactic cluster NGC 6871. The second suggestion is difficult to maintain because the two components always fall in the same limited radial-velocity intervals between − 180 and − 240 km/sec and between − 120 and − 160 km/sec respectively, but their relative intensities change. If these were two lines from the spectra of different stars their radial velocities should pass through all values between say − 120 and − 240 km/sec.

In order to find out if the intensity ratio between the two absorption components or their radial velocities show any correlation with the phase of the light variations given by Magalashvili and Kharadze (1967a), the phases of all the plates of this study were determined and in Figure 2 are shown plotted against the intensity ratio of the two absorption components at −210 and at −125 km/sec. The intensity ratios for the Balmer lines H9, H10, H11 and H12 were used, since these lines are essentially

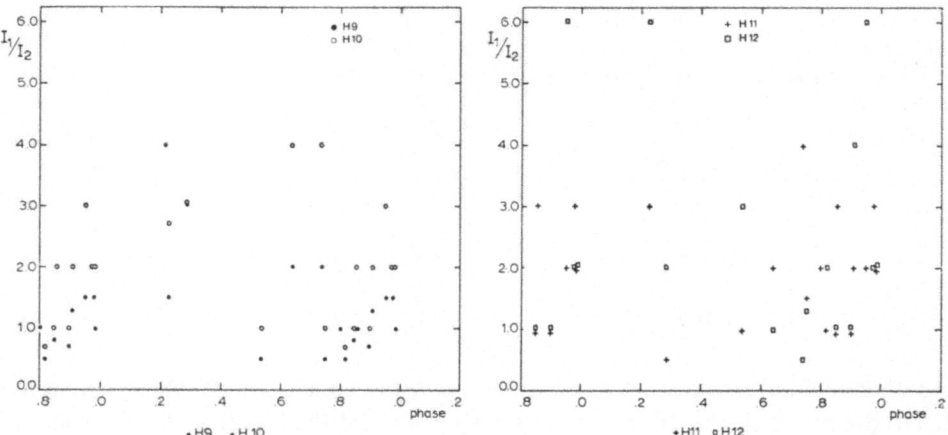

Fig. 2.   Intensity ratio of second and third components of hydrogen lines against phase of Magalashvili and Kharadze.

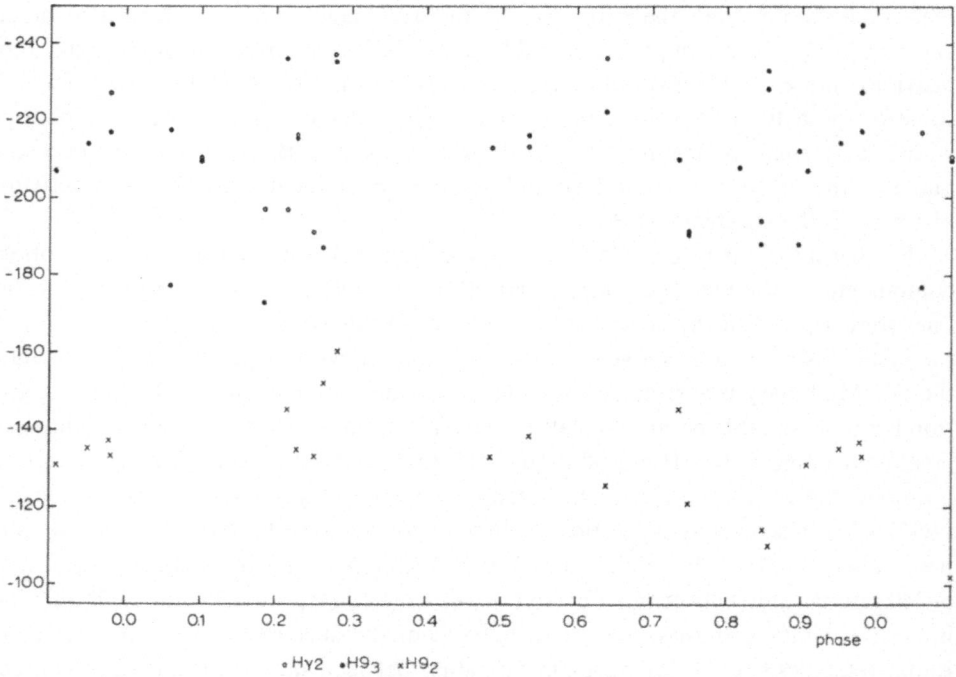

Fig. 3.   Radial velocity of some hydrogen absorption lines against phase of Magalashvili and Kharadze; open circles: second component of Hγ; dots: third component of H9; crosses: second component of H9.

free from blends and nearly always show the two components concerned. The same phases are also shown plotted against the radial velocities of the components of Hγ and H9 at about −210 km/sec and of H9 at about −125 km/sec (see Figure 3).

In both figures there is much scatter. In Figure 2 this is caused by the roughness of the visual intensity estimates that were made on the spectrograms while measuring them for their radial velocity. In Figure 3 much of the scatter is introduced by unresolved double or triple absorptions. No convincing evidence appears of a change either in the intensity ratio or the radial velocity in a period of 0.500656 day.

One must conclude that the conclusion of Magalashvili and Kharadze that P Cygni is a WUMa system, though very interesting from the points of view of stellar evolution and of explaining nova outbursts, is not supported by the spectroscopic information.

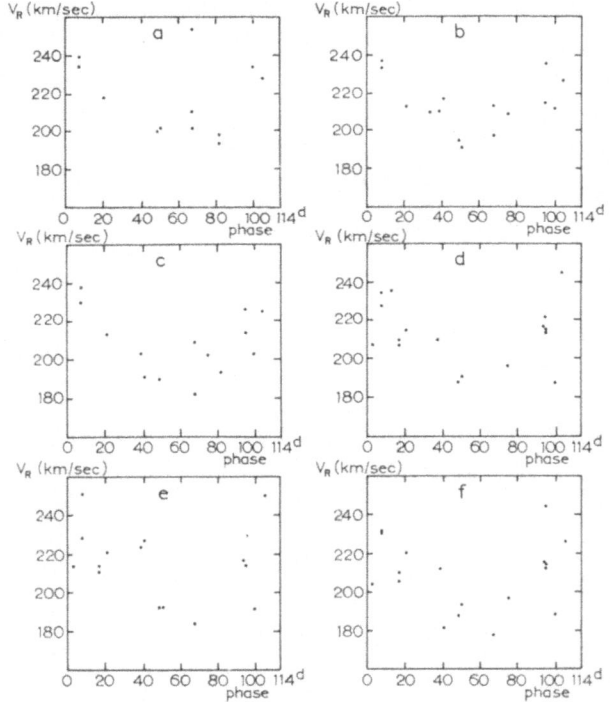

Fig. 4. Radial velocity of the most shortward displaced absorption components of various hydrogen lines against phase in the 114-day period; a: Hβ λ 4861; b: Hγ λ 4340; c: Hδ λ 4101; d: H9 λ 3835; e: H10 λ 3797; f: H11 λ 3770.

As is indicated above the radial velocity of the most shortward displaced component of the hydrogen lines shows variations with a 114-day period. This result is more fully illustrated in Figure 4, which shows the radial velocities of Hβ, Hγ, Hδ, H9, H10 and H11 against their phase in the 114-day period. It is found that all these lines show very much the same variations with corresponding phases and amplitudes. In evaluating Figure 4 one should keep in mind that many of the points in the lower part of the diagram at small and at large phases are from dates on which the H-lines did not show all three components. These points then are either the result of a blend between the third and second component, or they are only the second component the third being absent. In both cases these points give lower limits to the radial velocity of the third component.

Not only are the phases and amplitudes of these variations about the same, but also the mean value around which the radial velocity varies is strikingly similar for the various lines studied. If one assumes a unique relation between radial velocity and level in the stellar atmosphere, which in fact is a unique relation between radial velocity with respect to the star and the distance from the stellar surface, Figure 4 could be explained in either of two ways:

(1) At some high level in the atmospheres of P Cygni there is a layer which shows periodic velocity fluctuations. The velocity of that particular part of the atmosphere varies with a 114-day period between −180 and −240 km/sec.

(2) The velocity field in the stellar atmosphere is fixed. The variations are introduced by variations in the opacity of the atmosphere. Sometimes we can only see as deep as the layer with a velocity of −240 km/sec and half a period later we see a deeper layer with a velocity of −180 km/sec.

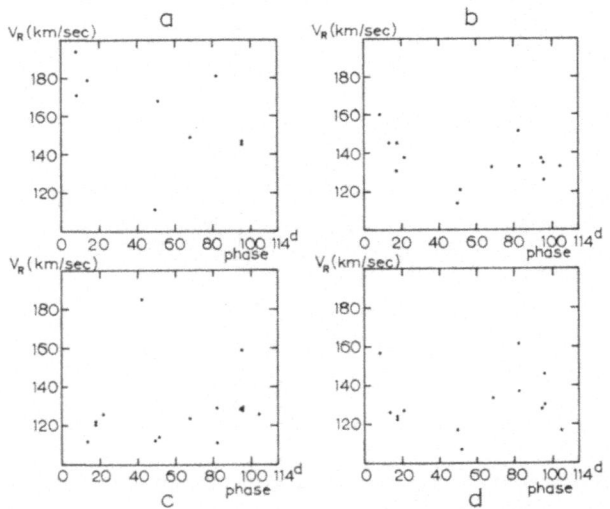

Fig. 5.   Radial velocity of the second absorption component of various hydrogen line against phase in the 114-day period; a: Hδ λ 4101; b: H9 λ 3835; c: H10 λ 3797; d: H11 λ 3770.

Before trying to decide which of these explanations should be chosen it is investigated whether similar variations are found in the behaviour of other spectral lines.

This has been done for the second absorption component of Hδ, H9, H10 and H11; the results are shown in Figure 5. It is clear that the general pattern of Figure 4 is not retained. The variations are more at random. This means that these second absorption components are formed in a layer where no radial-velocity fluctuations or opacity variations of the stellar atmosphere occur.

The same results are obtained for the radial velocities of the helium lines. From different series the best measured lines were selected and their radial velocities plotted against the phase in the 114-day period in Figure 6. The lines at $\lambda\lambda$ 3964, 4471, 4387

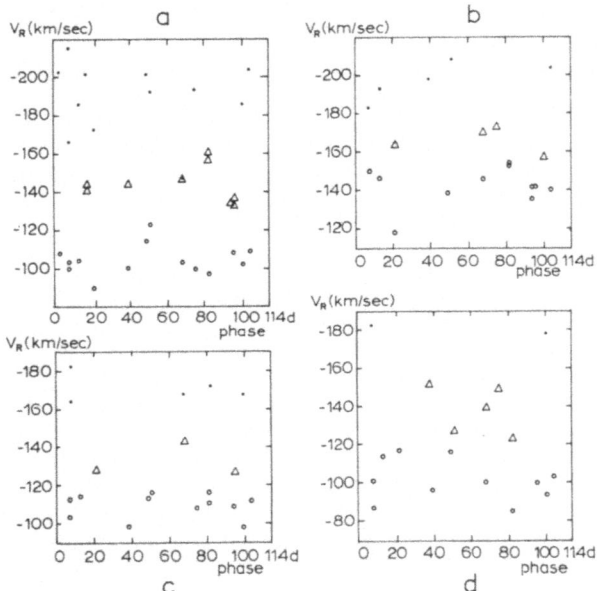

Fig. 6. Radial velocity of all components of some Heɪ lines; a: λ 3964; b: λ 4471; c: λ 4387; d: λ 4120. First components are indicated by open circles, second components by dots, and unresolved pairs by triangles.

and 4120 are used for this purpose. There are no indications of variations in that part of the atmosphere where these helium lines are formed. The second components of the lines at λ 4387 and at λ 4120 have radial velocities of about -180 km/sec and this value is well below the value found in the case of the varying velocity of the third components of the hydrogen lines. For the two other lines, $\lambda\lambda$ 3964 and 4471, the second components have radial velocities of nearly -200 km/sec. This value is about equal to the velocity minima of the third components of the hydrogen lines. That no variations are found in the case of λ 4471 may be due to the small number of measured second components. For λ 3964 the mean velocity of the second component is -193 km/sec, whereas the third hydrogen absorption component with smallest radial velocity, H11, still gives -208 km/sec. The conclusion is that even the radial velocity of λ 3964 is not subject to variations because this line is formed just below the layer of the atmosphere in which the variations occur.

The influence of the emission lines upon the measured radial velocities has been investigated also. The tendency is that a strong line fills in a larger part of the adjacent absorption and thus will cause a larger absorption velocity to be measured. By studying the radial velocities and the line profiles simultaneously it is possible to separate this 'emission-line effect' from the influence upon the radial velocity of the velocity gradient of the atmosphere. It appears that the corrections to be applied in correcting for the emission-line effect are always smaller than 15 km/sec. The effect of the stratification of the atmosphere which can be determined from a study of the radial velocities of lines from ions with different ionization potentials is much larger than the emission-line effect.

The results obtained by previous investigators (STRUVE, 1935; KHARADZE, 1936) about the dependence of the radial velocity upon the ionization potential are confirmed. This dependence is found best to show up if the ionization potential is plotted against the velocity difference absorption minus emission instead of plotting it against the absorption velocity only (cf. Figure 7).

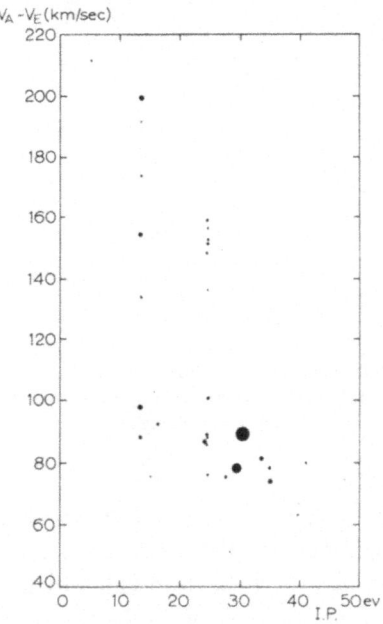

Fig. 7.   Absorption minus emission radial velocity against Ionization Potential. The size of the dots is a measure for their weight.

If we now combine all these results into one general picture of the atmosphere of P Cygni we find the following: Material from the stellar surface is driven away from the star. While moving outward it is accelerated unto a maximum velocity of about 240 km/sec. Beyond that point the velocity stays constant or may even decrease a little. The matter in the extended atmosphere is concentrated into three spherical shells of gas each giving rise to one of three absorption components. These shells are stationary; the particles move outward through the shells with high velocity. In the velocity range −80 to −140 km/sec the velocity increases not very much with the

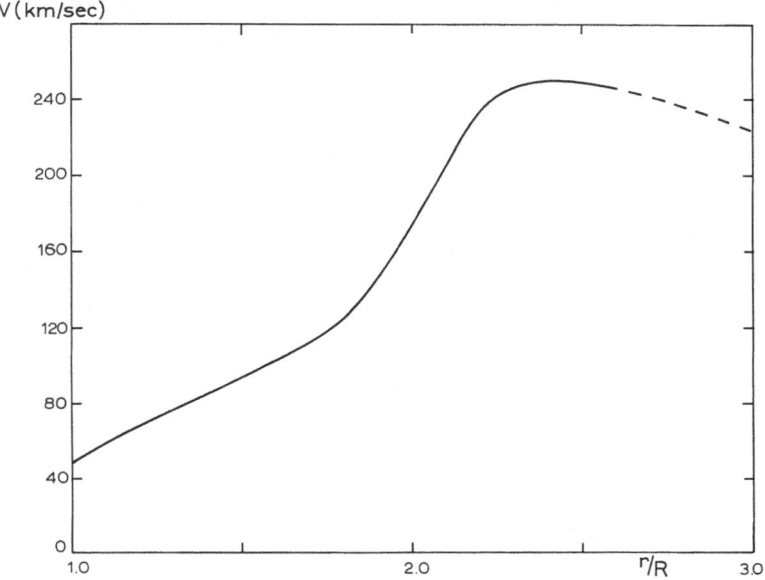

Fig. 8.   Tentative picture of the outward velocity in the extended atmosphere of P Cygni against the
distance to the stellar surface.

distance to the star but at higher levels (where the velocity is between -180 and
-240 km/sec) the velocity changes more rapidly (see Figure 8). If now a varying
opacity according to our previous second assumption permits one to see deeper
into the atmosphere the result is that at high velocities one really sees into a layer
with smaller velocity, while in the deeper layers one sees about the same velocity. This
explains why the velocity of the third component is varying while the first and second
components only scatter about their mean value.

The next step is to find out if and how it is possible to fulfil the equation of continuity in this case. Furthermore, it is possible from spectroscopic criteria about the
relation between emission intensity and dilution factor to give a more accurate height
scale to Figure 8. From the study of absorption equivalent widths it is then possible
to deduce values for the densities of the different shells which will complete the
present provisional picture. This work is hoped to be completed in the next few months.

Finally we estimate the mass loss from P Cygni with the aid of the normal formula

$$dm = 4\pi R^2 \varrho(R) \, v(R) \, dt \, .$$

By virtue of the equation of continuity we can use the values of R, $\varrho(R)$ and $v(R)$
for any one of the shells. It seems quite reasonable to assume for R the same value as
quoted by Underhill for the atmospheres of the Wolf-Rayet Stars in a previous paper
of this conference. So we put $R = 30 \, R_\odot$ and $v(R) = 125$ km/sec for the second shell.
From the last visible Balmer line a density $N_e = 10^{12}$ is estimated. These values give
a mass loss $dm = 2 \times 10^{-5} \, M_\odot$ year.

This value is of the same order as the one for the Wolf-Rayet stars and it may seem unexpectedly high. A final conclusion from this paper is that we will have to accept a rather high rate of mass loss for P Cygni and probably for the P Cygni type stars in general.

Acknowledgements

I am indebted to the Mount Wilson and Palomar Observatories, to the Dominion Astrophysical Observatory, to the Lick Observatory and to the Haute-Provence Observatory for the spectrograms which form the underlying material for this investigation. The stimulating remarks and comments of Prof. Anne B. Underhill I gratefully acknowledge.

References

ALEXANDER, TH. and WALLERSTEIN, G.: 1967, *Pub. Astron. Soc. Pacific* **79**, 500.
KHARADZE, E. K.: 1936, *Z. Astrophys.* **11**, 304.
MAGALASHVILI, N. L. and KHARADZE, E. K.: 1967a, *Comm.* **27** *I.A.U. Inf. Bull. Var. Stars*, No. 210.
MAGALASHVILI, N. L. and KHARADZE, E. K.: 1967b, *Observatory* **87**, 295.
NIKONOV, V. B.: 1936, *Abastumani Bull.* **1**, 35.
NIKONOV, V. B.: 1937, *Abastumani Bull.* **2**, 23.
STRUVE, O.: 1935, *Astrophys. J.* **81**, 66.

Discussion

Friedjung: Why are there three different absorption maxima? Are these different shells?

De Groot: There are three different shells, the maxima are due to the density variation with distance.

Friedjung: The maxima are similar to the absorption systems of novae.

De Groot: That is because also in P Cygni the gas is really moving out.

Friedjung: I believe ejection in novae is continuous, the process may be similar.

Underhill: A simple expanding atmosphere with a density which decreases as $1/r^2$ or in a uniform manner, produces a smooth profile sharp on the violet side and winged longwards. You observe bumps on the profile, indeed the wing is shortward. This must indicate that line-density fluctuations exist in the atmosphere of P Cygni. The velocity-radius curve you have sketched must be combined with a density distribution.

Morton: How constant have the hydrogen-line velocities been over the years this star has been observed?

De Groot: The 38 spectrograms cover the period 1942–64. In these years no deviations from the picture presented here are found.

Houziaux: Do you think the P Cygni shell is rather thin with uniform density located at $2 R_*$, or is it filling all the space between R_* and $2 R_*$ with one uniform density? To what shell do the 10^{12} particles cm^{-3} refer?

De Groot: The atmosphere is not very thin and fills up all the space between the star and say, $3 R_*$. However, the density does not seem to be uniform because then the absorption lines would just show an asymmetric violet wing. But in P Cygni three different components are found which indicate the presence of three different shells, i.e. regions where the density is higher.

Deutsch: If the velocity law is really like the curve you have drawn, the divergence theorem insures that there can be no density concentration of the kind you have described. Actually, we must allow v to *decrease* outwards faster than r^{-2} in order to produce density concentration. However, this sort of thing presumably might occur if as a result of cooling, the expanding gas flows through a succession of standing slabs.

Nariai: If the phenomenon is stationary, then you have to assume that the equation of continuity is fulfilled. Then the velocity curve you drew must be different?

De Groot: I may have forgotten to say that the present investigation has not been concluded yet, and that this Colloquium came as an interruption of my study. Therefore, although I have been thinking of the equation of continuity and am aware of some difficulties, the results have been presented as they are up till now.

ROCKET OBSERVATIONS OF MASS LOSS FROM
HOT STARS

DONALD C. MORTON

Princeton University Observatory, Princeton, N.J., U.S.A.

Abstract. Rocket observations have shown that the far-ultraviolet resonance lines have P-Cygni profiles in the spectra of many hot stars, including Of and Wolf-Rayet stars and OB supergiants. Velocity shifts as high as -3000 km sec^{-1} have been measured for the short-wavelength edges of some of the lines. Estimates of the rates of mass loss range from 10^{-8} to 10^{-6} $M_\odot$ year^{-1}.

Rocket observations of far-ultraviolet stellar spectra have provided evidence of mass loss from hot supergiants at remarkably high velocities. Absorption lines of the resonance transitions of Civ, Nv, Siiii, and Siiv have been found with shifts towards shorter wavelengths corresponding to Doppler velocities of 1000–2000 km sec^{-1} (MORTON, 1967; MORTON *et al.*, 1968). A tracing of the ultraviolet spectrum of ζ Orionis is reproduced in Figure 1 showing the P-Cygni profiles of the Civ and Siiv resonance lines and the displaced absorption lines of Ciii, Nv, and Siiii. In the expanding shell, where we presume these lines are formed, the dilution of the radiation and the low particle density leave few ions in excited states, so that we expect mainly absorptions from the ground states. Since the higher ion states present in these hot atmospheres all have their resonance lines in the ultraviolet shortward of the atmospheric limit, we can understand why large wavelength shifts have not been seen in ground-based spectra. Nevertheless WILSON (1958) did report evidence of weak,

Fig. 1. Densitometer tracing, on an intensity scale, of the far-ultraviolet spectrum of ζ Orionis, photographed by Princeton on September 10, 1966. The distribution of intensity with wavelength includes the unknown response of the spectrograph. Wavelengths increase towards the right from 1140 to 1630 Å. The Hi line is interstellar, but all the other identified absorption features are circumstellar with large Doppler shifts to shorter wavelengths.

very broad emission lines of Heii, Ciii, and Niii in the visual spectra of some OB supergiants.

More recent rocket observations by several investigators have confirmed and extended the data on the large wavelength shifts of ultraviolet absorption lines. CARRUTHERS (1968) of the U.S. Naval Research Laboratory has photographed spectra of stars from Orion to Vela with a windowless image intensifier; STECHER (1967) of the Goddard Space Flight Center has recorded spectra of several stars with a scanning spectrometer; and MORTON *et al.*, (1969) of Princeton have reported on the spectra of ζ Puppis and γ^2 Velorum obtained with an all-reflective objective

M. Hack (ed.), Mass Loss from Stars. All rights reserved

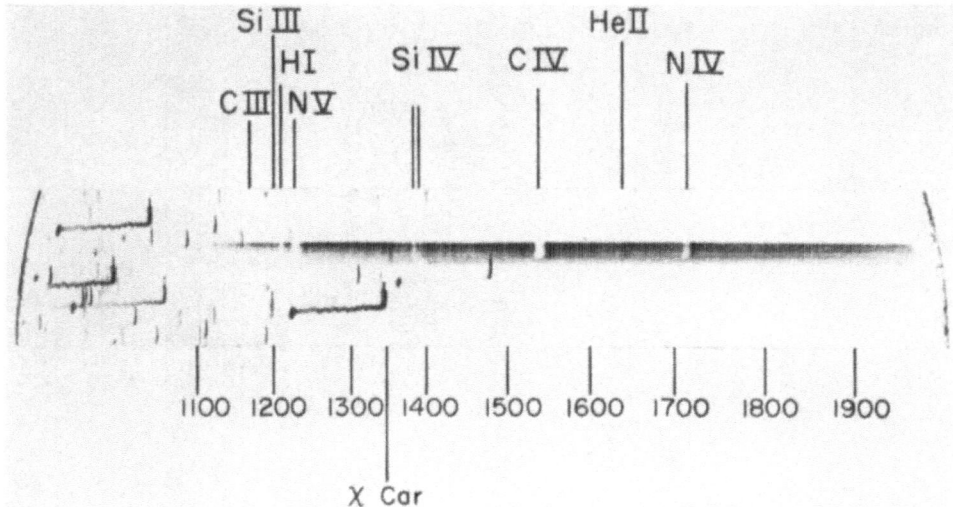

Fig. 2. Photograph of the far-ultraviolet spectrum of ζ Puppis from 1100 to 1965 Å obtained by Princeton on November 1, 1967. The labels identify the interstellar H I Ly-α line and the displaced circumstellar features. The tails on the zero-order images resulted from a failure of the fine stabilization 100 sec after the beginning of the exposure.

spectrograph. Figure 2 is a photograph of the Princeton spectrum of ζ Pup showing the shifted circumstellar lines. The He II, N IV, N V, and C IV features have both emission and absorption components, while the C III, Si III, and Si IV lines appear only in absorption.

Table I summarizes the data presently published on displaced absorption lines in ultraviolet stellar spectra, with the best estimates of the Doppler velocities of the line centers. Since the Princeton spectra have the highest resolution, their velocities are quoted in most cases when available; otherwise Carruthers' results are listed. An 'e' indicates an emission line also is present at approximately its laboratory wavelength. The surface escape velocity is on the order of 600 km sec^{-1} for the supergiants and perhaps twice this for γ^2 Vel and ζ Pup. Since all the stars have some lines shifted in excess of the escape velocity, mass loss must be occurring.

We see that in some of the stars the shifts also have been detected in absorption lines of He II, C III, and N IV from excited levels. Except for the C III line in the two hottest stars γ^2 Vel and ζ Pup, the excited lines tend to have smaller outward velocities than the resonance lines. The excited lines must be formed lower in the atmosphere, in the acceleration zone, where the density is higher. The velocities are more negative with decreasing excitation potential similar to the correlation found for the much smaller shifts in the visual spectra of P Cyg and some OB supergiants.

There now is evidence of high-velocity mass loss in enough OB stars for us to conclude that the phenomenon probably occurs in all hot supergiants and in at least some bright giants and giants B0.5 and earlier. However, the evidence for the shifted N V line in γ Ori (B2 III) is very uncertain since the line is poorly defined in the Princeton spectrum and Carruthers found no measurable shift. Stecher has a spectrum of

TABLE I

Velocity shifts of far-ultraviolet absorption lines in hot stars (in km sec⁻¹)

Ion Star	MK	M_V	C III 1175.7 6.46	Si III 1206.5 0	N V 1240.1 0	Si IV 1393.8 0	Si IV 1402.8 0	C IV 1549.5 0	He II 1640.4 40.64	N IV 1718.5 16.13	Observers[a] (Source of velocity in italics)
γ² Vel	WC8+07	−6.2, −5.2	−1275	−1370	eᵇ		eᵇ	eᵇ	eᵇ	eᵇ	*NRL*, *PUO*, GSFC
ζ Pup	O5f	−6.0	−1860	−1640	−1550eᵇ	−1810	−2140	−1840ᵇ	−350ᵇ	−780ᵇ	NRL, *PUO*, GSFC
ι Ori	O9 III	−5.2	−560ᶜ	−770ᶜ	−1620ᶜ						*PUO*, NRL
ζ Ori	O9.5 Ib	−6.6	−1050	−1600	−1770	−1420	−1600ᵇ	−1280ᵇ			*PUO*, NRL
δ Ori	O9.5 II	−6.3	−480	−1600	−1360	−1120	−1770ᵇ	−1320ᵇ	0	0	*PUO*
ε Ori	B0 Ia	−6.9	−970	−1320	−1360	−1250	−1240ᵇ	−1180ᵇ	0	0	*PUO*
κ Ori	B0.5 Ia	−6.4	−1020	−1620	−1450				0		*NRL*
γ Ori	B2 III	−3.3			−1400ᶜ						*PUO*, NRL

Note: superscript letters e, b, c on data values correspond to footnotes b and c below (e.g. eᵇ, −1550eᵇ, −560ᶜ).

ᵃ GSFC = Goddard Space Flight Center; NRL = Naval Research Laboratory; PUO = Princeton University Observatory.

ᵇ e = Emission line also present at approximately its laboratory wavelength.

ᶜ Velocities somewhat uncertain.

β Ori (B8 Ia) showing the Mg II resonance doublet at 2800 Å with some velocity shift. It would be worthwhile to obtain additional ultraviolet observations of late B supergiants and A types such as α Cyg (A2 Ia) to further define the regions of the HR diagram where high-velocity mass loss occurs. In α Cyg, Hα has a P-Cygni profile with an absorption velocity of about -40 km sec^{-1}. The data for ζ Pup show that large Doppler shifts also occur in stars relatively close to the main sequence if they are sufficiently hot. Farther down the main sequence, however, the spectra, of δ Sco (B0 V) and π Sco (B1 V) obtained by MORTON and SPITZER (1966) show no evidence of line shifts greater than 150 km sec^{-1}. The displacements in γ^2 Vel probably are due to the Wolf-Rayet component since evidence of high-velocity mass loss has been found even in the visual spectra of some of these stars, but we cannot be certain how much of the ultraviolet spectrum comes from the O7 companion.

MORTON (1967) has attempted to estimate the rate of mass loss from δ, ε, and ζ Ori using column densities obtained from the unsaturated lines of N V and Si IV. It was assumed that the lines were formed in a region of constant velocity beginning at a radius $R_* = 2 \times 10^{12}$ cm, the ionization was from the ultraviolet radiation of the star, and the elements were present in their cosmic abundance ratios. Application of the equations for continuity and ionization, and the assumption that the electron temperature was 10^4 K gave values for T_i the temperature of the ionizing radiation and $n_e(R_*)$ the electron density at the base of the constant-velocity flow. Typically $T_i = 26\,000$ K and $n_e(R_*) \sim 5 \times 10^9$ cm^{-2} giving dM/d$t \sim 10^{-6}\,M_\odot$ year^{-1} for all three stars. This rate is not enough to affect seriously the evolution of a $30\,M_\odot$ star, nor can it be a significant contribution to the general heating of the interstellar medium, whose energy must come from the ultraviolet photons of the much more numerous OB main-sequence stars. However, the gas flow could cause some heating in the vicinity of the stars, which might produce X-rays as suggested by BLESS et al. (1968).

The above model assuming radiative ionization does not explain adequately the observed strengths of the C IV and Si III lines. Consequently R. H. Sanders of Princeton has developed an alternative model assuming ionization by electron collisions low in the atmosphere before the acceleration begins. The state of ionization is believed to be frozen into the flow with only a few recombinations by the time the constant velocity region is reached. For an electron temperature of $60\,000$ K he found reasonable agreement for the strengths of C IV, N V, and Si IV and the rate of mass loss was dM/d$t \sim 10^{-8}\,M_\odot$ year^{-1}. Unfortunately, this model also fails to account for the relatively strong line of Si III, which apparently shares the velocity of the other features, unless the ionization temperature of the stellar radiation is less than $16\,000$ K, much lower than we should expect for O9.5 or B0 stars.

LUCY and SOLOMON (1967) have suggested that absorptions in the strong resonance lines of C III (977 Å), C IV and Si IV, provide a mechanism by which momentum can be transferred from the photons to the particles. The absorption of a photon at the top of the photosphere gives the ion a small acceleration. After de-excitation, the Doppler shift permits the ion to absorb a second time, at a shorter wavelength where the continuum flux is still strong, producing additional acceleration until a high velocity is obtained. This scheme requires that the absorption lines be negligible in the stationary part of the photosphere, for otherwise no photons are available to start the acceleration. Lucy and Solomon have attempted to calculate line profiles and a rate of mass loss from a consideration of the hydrodynamics of the mass flow. They predict that the absorption lines should have steep short-wavelength edges all at the same Doppler velocity, which is the maximum velocity of ejection. Table II lists the velocities of the short-wavelength edges well enough defined to be measured in the Princeton spectra. Although there is some scatter among the values for each star, the velocities probably are consistent with the prediction within the errors of measure-

TABLE II

Velocities of short-wavelength edges of shifted absorption lines

Star	C III 1175.7	Si III 1206.5	N v 1240.1	Si iv 1393.8	C iv 1549.5	N iv 1718.5
ζ Pup			−3000	−2400	−2900	−1300 km sec⁻¹
ζ Ori	−1810	−1860		−1970	−2280	
δ Ori	−1300			−1550	−2420	
ε Ori	−1660	−1810		−1760	−2090	

ment. The excited line N IV in ζ Pup is formed lower in the atmosphere and should not be expected to show the maximum shift like N v. The short-wavelength edges of most of the lines appear moderately steep, except for C III and Si III in ζ Pup and N v in the Orion stars. Higher resolution is necessary to check this prediction for the other lines. Lucy and Solomon have estimated $dM/dt \sim 10^{-7}$ to $10^{-8} M_\odot$ year^{-1}, which is not inconsistent with either of the Princeton results considering the uncertainties present in all the calculations.

Acknowledgement

I wish to thank Dr. Edward B. Jenkins for his contributions towards the preparations for the Princeton rocket flights, and his assistance in the analysis of the data.

References

BLESS, R. C., FISCHEL, D., and STECHER, T. P.: 1968, 'Of and Wolf-Rayet Stars as X-Ray Sources', *Astrophys. J.* **151**, L117.

CARRUTHERS, G. R.: 1968, 'Far Ultraviolet Spectroscopy and Photometry of Some Early-Type Stars', *Astrophys. J.* **151**, 269.

LUCY, L. B. and SOLOMON, P. M.: 1967, 'Mass Loss from O and B Supergiants', *Astronom. J.* **72**, 310.

MORTON, D. C.: 1967a, 'The Far-Ultraviolet Spectra of Six Stars in Orion', *Astrophys. J.* **147**, 1017.

MORTON, D. C.: 1967b, 'Mass Loss from Three OB Supergiants in Orion', *Astrophys. J.* **150**, 535.

MORTON, D. C. and SPITZER, L.: 1966, 'Line Spectra of Delta and Pi Scorpii in the Far Ultraviolet', *Astrophys. J.* **144**, 1.

MORTON, D. C., JENKINS, E. B., and BOHLIN, R. C.: 1968, 'Rocket Observations of Orion Stars with an All-Reflective Ultraviolet Spectrograph', *Astrophys. J.* **154**, 661.

MORTON, D. C., JENKINS, E. B., and BROOKS, N. H.: 1969, 'Far Ultraviolet Spectra of Zeta Puppis and Gamma² Velorum', *Astrophys. J.* **155** (in press).
STECHER, T. P.: 1967, 'Stellar Spectrophotometry from a Pointed Rocket', *Astronom. J.* **72**, 831.
WILSON, R.: 1958, 'Spectrophotometric Measurements of Early Type Stars', *Pub. Royal Obs. Edinburgh* **2**, 61.

Discussion

Nariai: By what do you explain the emission in these stars?

Morton: Radiation from excited atoms in a low-density shell.

Nariai: But is it possible that we assume the temperature reversal at the surface?

Morton: Yes.

Nariai: Well, let me make a comment along this line. In visual wavelength, the early-type supergiants show large turbulence. Therefore, we can expect the temperature reversal due to the dissipation of mechanical energy, which was generated in the turbulent layer. The problem now becomes how to create such a turbulence at such high effective temperature and at low gravity. The ordinary turbulence due to the thermal instability cannot be expected. The mechanism I wish to suggest here is the motion due to the radiation-pressure instability. In an atmosphere where g_{rad} is almost equal to GM/R^2, and if there is a density fluctuation in the horizontal direction, then the flux becomes non-uniform and the effective gravity is greatly disturbed from the static atmosphere. The motion of gas happens in the sense that the gas at the thin part goes up because of the larger flux, and vice versa.

Underhill: H. Lamers at Utrecht has been studying the spectrum of ε Orionis and his work, not yet completed, shows that the radial velocity does vary slightly with excitation potential. His results so far are:

Spectrum	E.P. (volts)	Rad. vel. (km/sec)	No. lines
H	10	−46	11
He I	21	−46	8
Si IV	21	−43	2
O II	25	−52	4
Ne II	30	−40	3
O III	33	−41	5

These results, combined with yours, show that there is a great change of velocity between the levels where the blue violet spectrum is formed, and those where the strong UV lines are formed.

Sahade: In γ_2 Velorum the He I 3888 line of the expanding outer shell yields also a velocity of −1300 km sec!

Morton: This is an interesting contrast with ζ Pup in which the He II 1640 Å line has a much lower velocity than the resonance line.

INTRINSIC POLARIZATION FOR OBJECTS WITH
EXTENDED ATMOSPHERES

A. KRUSZEWSKI, G. V. COYNE, and T. GEHRELS

Lunar and Planetary Laboratory, University of Arizona, Tucson, Ariz., U.S.A.

Abstract. Intrinsic polarization in B emission-line stars and in red variables is discussed. The peculiar wavelength-dependence of the polarization in Be stars, particularly the decrease in polarization in the ultraviolet, is explained by scattering from electrons in an asymmetrical envelope together with self-absorption in a hydrogen plasma.

Red variable stars show large polarizations in the ultraviolet. Unlikely configurations of the scattering envelope, with the opaque cloud in front of the star, are required to explain the observed polarizations by Rayleigh scattering on molecules in an asymmetric envelope. Difficulties with a model of elongated graphite grains oriented by a magnetic field are also discussed.

Stars with extended atmospheres commonly exhibit intrinsic polarization, polarization, that is, which originates in the atmosphere of a star or in a circumstellar shell. Such polarizations are frequently variable. The range of objects which exhibit such polarizations extends from early-type B emission-line stars to cool infrared objects. Though the first stars which exhibited variable polarization were discovered several years ago (γ Cas, BEHR, 1956; μ Cep, GRIGORIAN, 1958), it is only in recent years that concentrated efforts have been made to investigate these variable polarizations. Such investigations should help considerably in understanding the physical processes in extended atmospheres and in understanding mass-loss phenomena.

The aim of this paper is to summarize the observational results on the intrinsic polarizations of two classes of objects; Be stars and red variables, and to offer some tentative interpretation of these observations.

We shall start with a discussion of the polarization in emission line B stars. Figure 1 shows the ratio of polarizations in yellow and ultraviolet light, P^V/P^U, vs. the ratio of polarization in yellow and blue, P^V/P^B. Objects possessing supposedly pure interstellar polarization are shown with filled circles and Be stars are shown with open circles. This diagram was first introduced by SERKOWSKI (1968), but we have enlarged the sample of Be stars. It can be seen that the cases of supposedly pure interstellar polarization lie along a narrow sequence, while Be stars are scattered with a general tendency to lie above the interstellar sequence in the upper left-hand section of this figure. We shall see in a later figure that the main reason for this is a deficiency in the ultraviolet polarization in Be stars. This diagram, together with the observed variability of polarization ranging from 0.2–0.4% in an appreciable fraction of Be stars, indicates that a significant fraction of the polarization in these stars is intrinsic.

Figure 2 shows the wavelength-dependence of polarization for a representative sample of Be stars (COYNE and KRUSZEWSKI, 1969). The observed values were normalized by setting the average polarization through our green and blue filters equal

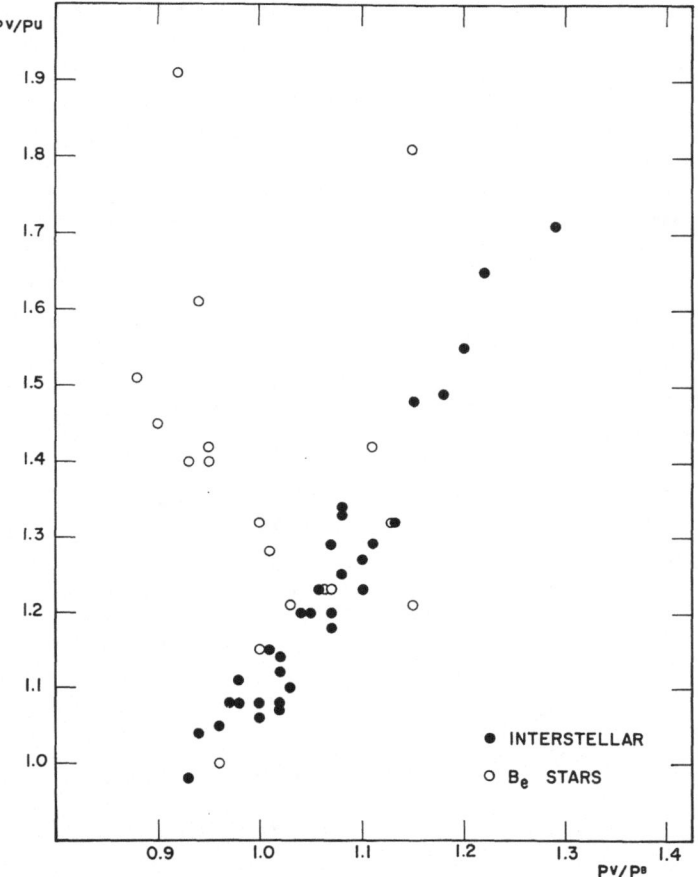

Fig. 1.   The ratio of polarizations ultraviolet to yellow plotted vs. blue to yellow.
Interstellar polarization with filled circles; Be stars with open circles.

to 100%. The actual value of this average is marked to the right side of each curve. For stars with variable polarization this average is an unweighted mean of the individual observations. A mean interstellar polarization curve (COYNE and GEHRELS, 1967) is drawn for comparison. In general the Be stars show a greater wavelength-dependence than the mean interstellar polarization. In most cases the polarization for the Be stars is the largest in the blue with a sharp drop into the ultraviolet. This drop in the ultraviolet polarizations explains why the Be stars are concentrated in the upper left-hand section of Figure 1. A secondary minimum of polarization occurs for the Be stars around 0.8 μ and there is a sharp increase around 1.0 μ. There are three stars, P Cygni, v Sagittarii and HD 11 606, which do not share this behaviour. The polarizations of the bright Be stars usually are small and a composite of intrinsic and interstellar effects. For P Cygni and HD 11 606 the interstellar polarization is probably dominant. v Sagittarii shows a wavelength-dependence different from any other object.

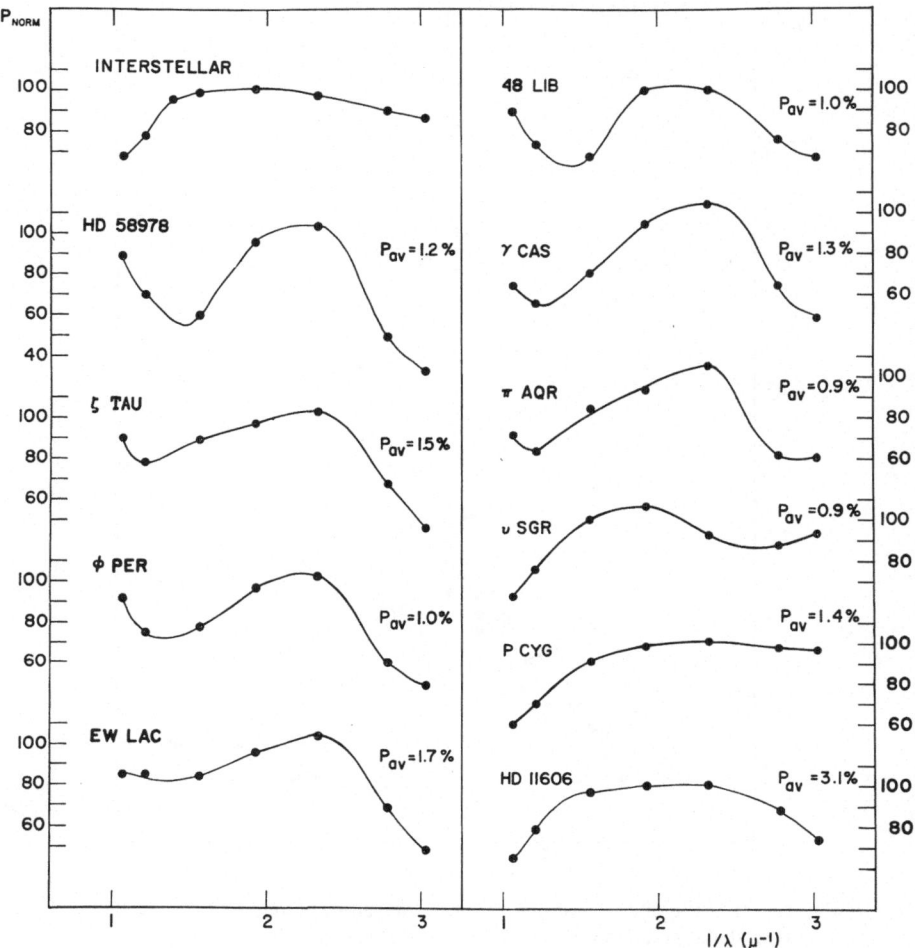

Fig. 2. Wavelength-dependence of polarization for a representative sample of Be stars. Interstellar curve is also plotted for comparison.

Figure 3 shows in the lower panel the average wavelength-dependence for 7 Be stars. The horizontal bars mark the half intensity widths of the filters. The upper panel shows the absorption coefficient of hydrogen plasma in arbitrary units for an electron temperature of 10 000 K. There is an indication of a correlation between the polarization and hydrogen absorption which might be explained by the following model. Suppose we have a star surrounded by a flat disk or ring of partly ionized hydrogen. Light scattered from the free electrons would be polarized independent of wavelength. However, the self-absorption of starlight before and, to a lesser degree after scattering, will modify the wavelength-dependence by suppressing the polarization in those spectral regions where the absorption is high. The observed intrinsic polarizations are small, usually around 1%, and therefore only a small part of stellar light need be scattered. Figure 3 shows that the qualitative agreement between the

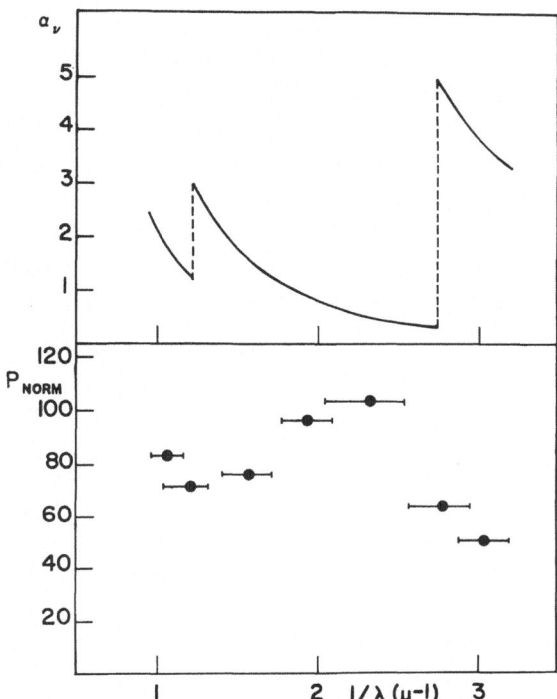

Fig. 3. Average wavelength-dependence of polarization for Be stars compared with the absorption coefficient of hydrogen plasma at $T_e = 10000$ K.

observations and the model is good. We should remark that v Sagittarii, the one Be star observed which clearly does not fit this model, is a hydrogen-poor star.

A few Ia and Ia+ early-type supergiants like χ^2 Orionis also show variable polarizations and peculiar wavelength-dependence. A number of eclipsing variables have been shown to vary in polarization. The polarization in β Lyrae (SHAKHOVSKOI, 1964), for instance, is evidently caused by scattering from free electrons in a flat ring, thus resembling Be stars. Little is known of the polarization of objects of intermediate spectral types. No variable polarization has been detected in Cepheids or RR Lyrae stars. Some RV Tauri stars like R Scuti were reported to vary (SHAKHOVSKOI, 1963). T Tauri objects usually show very small polarization but there are some evidences of variability (EFIMOV, 1967, unpublished LPL data). An R Coronae Borealis type star, RY Sagittarii, has variable polarization (Serkowski, private communication). R Coronae Borealis itself shows little or no variable polarization when observed at maximum light.

A great deal of attention has been given lately to highly polarized red variable stars (SERKOWSKI, 1966a, b; ZAPPALA, 1967; KRUSZEWSKI et al., 1968).

Figure 4 shows the wavelength-dependence of polarization for a representative sample of red variables (COYNE and KRUSZEWSKI, 1968; KRUSZEWSKI et al., 1968; KRUSZEWSKI, 1968). The polarizations are normalized in the same manner as in Figure 2. The most striking feature is the increase of polarization towards the ultraviolet. The typical wavelength-dependence of polarization for red variables is shown by three M-type Mira variables on the right. Examples of different shapes are shown by V Canum Venaticorum and μ Cephei. Fragmentary data are also shown for the infrared objects NML Tau and CIT-6 and also for a carbon star V Coronae Borealis. Notice the flattening in the yellow-blue region and a very sharp increase in red in V Coronae Borealis. The very steep wavelength-dependence, sometimes approaching a λ^{-4} law, could be explained by means of Rayleigh scattering on molecules in asymmetric envelopes. The simple model of such envelope gives a maximum polarization around 5.5% (KRUSZEWSKI et al., 1968). However, there are objects such as

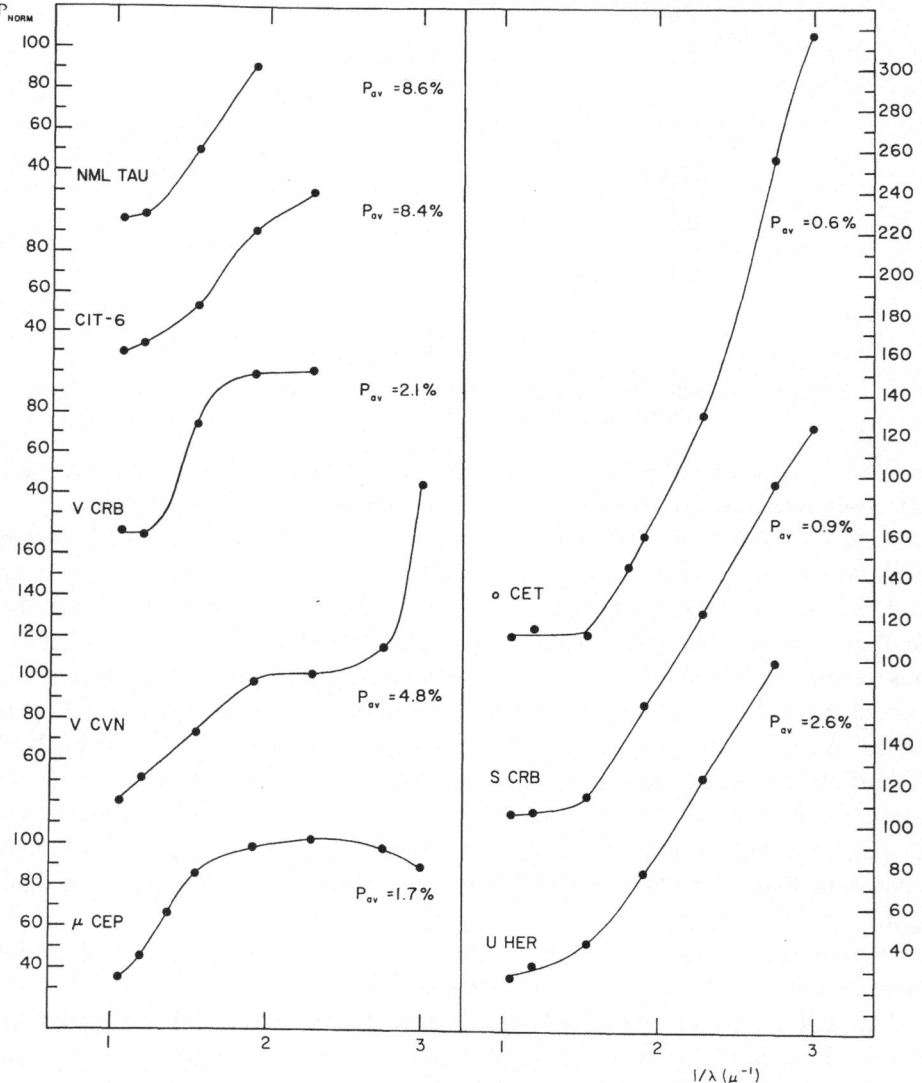

Fig. 4.   Wavelength-dependence of polarization for a sample of red variables. The scales are the same as in Figure 2.

V Canum Venaticorum, L_2 Puppis, NML Tauri and CIT-6 which show much larger polarizations than this.

Figure 5 shows as an example the degree of polarization as a function of time in different filters for two infrared objects. Such large polarizations as those for NML Tau and CIT-6 can be explained by a scattering envelope only after assuming a very special configuration of the envelope with the opaque cloud in front of the star. Such special configurations are unlikely. Another explanation of the large polarization observed is provided by elongated grains aligned by magnetic fields in an expanding circumstellar envelope. The existence of graphite flakes in the atmosphere of R Coronae

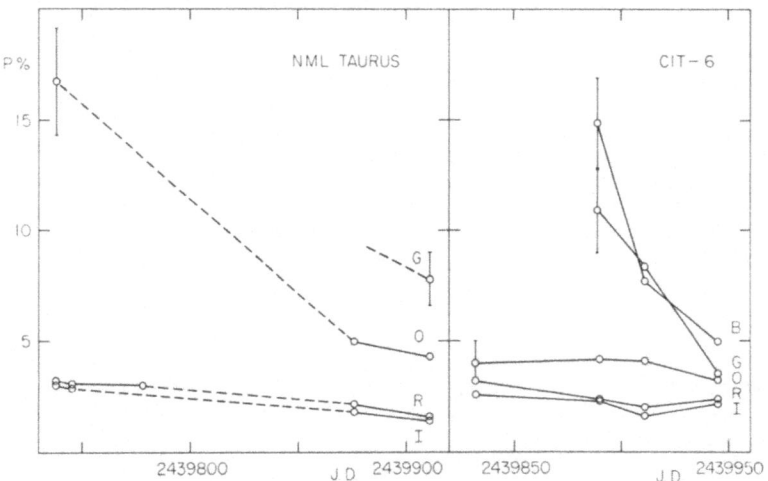

Fig. 5. Time-dependence of polarization as measured with different filters for NML Taurus and CIT-6.

Borealis and carbon stars was suggested long ago (O'KEEFE, 1939; CAYREL and SCHATZMAN, 1954: HOYLE and WICKRAMASINGHE, 1962). The discovery of expanding circumstellar envelopes required a mechanism for pushing out the atmospheric material. Solid particles pushed out by the radiation pressure and dragging the gas along could cause such expanding circumstellar shells. However, there are some difficulties with respect to this model of elongated graphite grains in an expanding circumstellar shell. While graphite particles can be easily produced in carbon stars, this is not the case when we consider M-type stars, which have oxygen-rich, carbon-poor atmospheres. Another difficulty is that for a particle to be aligned the collisions between gas and grains cannot be too frequent; and yet frequent collisions are needed for the coupling of the macro-motions of the gas and grains needed to get the expanding shell.

Until now little attention has been paid to polarization-time dependence and wavelength-dependence of red variables. Hopefully, coordinated polarimetric, photometric and spectroscopic observations of some crucial objects can help to decide if there are solid particles in red variables atmospheres which produce polarization and cause outward motion of the envelope. The presence of infrared excesses around 5–10 μ was reported by JOHNSON (1967a, b) for a wide variety of objects. It is interesting that many of the objects with infrared excesses also show intrinsic polarization. Among such objects are the Be stars γ Cas, ζ Tau, φ Per, υ Sgr and the red variables μ Cep, o Cet, R Leo. However, it is not always true and there are examples with an infrared excess and with apparently constant polarization (P Cyg, κ Dra).

Acknowledgements

In conclusion, we thank the National Science Foundation for the support of this work.

References

BEHR, A.: 1956, *Veröff. Göttingen*, No. 126.
CAYREL, R. and SCHATZMAN, E.: 1954, *Ann. Astrophys.* **17**, 555.
COYNE, G. V. and GEHRELS, T.: 1967, *Astron. J.* **72**, 887.
COYNE, G. V. and KRUSZEWSKI, A.: 1968, *Astron. J.* **73**, 20.
COYNE, G. V. and KRUSZEWSKI, A.: 1969, *Astron. J.* (to be published).
EFIMOV, Y. S.: 1967, *Izv. Krymsk. Astrofiz. Observ.* **37**, 251.
GRIGORIAN, K. A.: 1958, *Burakan Obs. Contr.* **25**, 45.
HOYLE, F. and WICKRAMASINGHE, N. C.: 1962, *Monthly Notices Roy. Astron. Soc.* **124**, 417.
JOHNSON, H. L.: 1967a, *Astrophys. J.* **149**, 345.
JOHNSON, H. L.: 1967b, *Astrophys. J.* **150**, L39.
KRUSZEWSKI, A.: 1968, *Publ. Astron. Soc. Pacific* **80**, 560.
KRUSZEWSKI, A., GEHRELS, T., and SERKOWSKI, K.: 1968, *Astron. J.* **73**, 677.
O'KEEFE, J.: 1939, *Astrophys. J.* **90**, 294.
SERKOWSKI, K.: 1966a, *Astrophys. J.* **144**, 857.
SERKOWSKI, K.: 1966b, *I.A.U. Information Bull. on Variable Stars*, No. 141.
SERKOWSKI, K.: 1968, *Astrophys. J.* **154**, 115.
SHAKHOVSKOI, N. M.: 1963, *Astron. Zh.* **40**, 1055.
SHAKHOVSKOI, N. M.: 1964, *Astron. Zh.* **41**, 1042.
ZAPPALA, R. R.: 1967, *Astrophys. J.* **148**, L81.

MASS LOSS FROM OB SUPERGIANTS

J. B. HUTCHINGS

Dominion Astrophysical Observatory, Victoria, Canada

Abstract. Three main types of evidence in the spectra of OB supergiants are discussed, which lead to conclusions on the extent and mass motions of the outer envelopes of the stars. Illustrations are given from two southern stars. A method of computing strong line profiles in extended moving atmospheres is outlined, and it is shown how these profiles support the proposed atmospheric structure of the stars. Preliminary general conclusions are drawn on the phenomenon of mass loss from OB supergiants.

1. Introduction

The envelopes of hot supergiant stars have for many years been considered to be extensive and probably in some form of mass motion. Theoretical studies of early-type stellar atmospheres have shown that even main-sequence stars have envelopes which depart considerably from the assumed thin atmosphere in LTE. In the supergiants the gravitational fields, pressures and densities are much lower, while photospheric radii and radiation pressure are much higher. Thus conditions deviate even further from the hydrostatic, local thermodynamic and radiative equilibrium cases which can be dealt with theoretically. It is therefore of the greatest importance to develop theoretical approaches which will bridge the gap between existing LTE model atmosphere and the high-dilution, low-density planetary nebulae. It is also important to form some ideas on what sort of structure and conditions prevail in some of the stars with known expanding atmospheres in order that the theoretical problem be better defined. It is in this connection that recent work on OB supergiants is particularly relevant. Most notable is perhaps the discovery (MORTON, 1967) from rocket UV spectra that resonance and strong lines of Si, C, and N show velocities of expansion of some thousands of km/sec in the Orion supergiants. I wish now to describe the evidence for such expansion and consequent mass loss which exists in the visual and near UV region in some stars, and an approach by which a picture can be formed of an accelerating outer envelope of these stars.

2. The Observational Evidence

There are three types of evidence in the photographic region of the spectrum relating to motions in the outer atmospheric layers. These may be entitled: (a) emission lines, (b) strong line asymmetry and displacement, and (c) systematic differences between lines of different ions.

Let us discuss each of these briefly.

(a) The very existence of an emission line is an indication of an extended envelope. Emission lines occur when electron de-excitation is predominantly radiative, rather than collisional. In a thin atmosphere, firstly, pressure and density are high, so that

de-excitation is predominantly collisional; secondly, only the facing half of the atmosphere is seen, so that even completely radiative de-excitation would give rise to
a net absorption line. In a low-density extended atmosphere there are two main
causes for the formation of emission lines. Firstly, mass motion (rotation or expansion) can separate the absorption and emission components. Secondly, there are
non-equilibrium processes such as overpopulation of metastable levels and fluorescence, leading to emission cascades.

Given a line with an emission component, it is not a simple matter to interpret it
in terms of atmospheric structure. Consider the extreme example of an expanding
shell of negligible thickness. At a distance of (say) two stellar radii this gives rise to a
profile such as that in Figure 1.

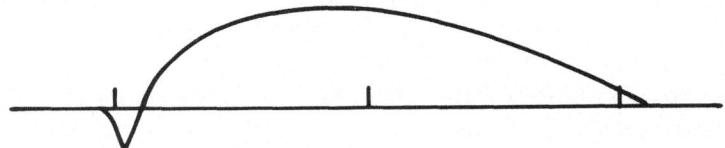

Fig. 1.   Profile originated by an expanding shell of negligible thickness.

In a continuous atmosphere this profile is modified in the following ways:
(1) attenuation of longward emission by photospheric occultation.
(2) density variation with height.
(3) electron-scattering of longward emission by lower atmospheric layers.
(4) absorption over a radial velocity range increasing with depth.

It can be shown that in the case of a uniform non-accelerating atmosphere moving
outwards, it is possible to derive the true velocity from the profile. As in practice both
velocity and density change with height, we can get only a first approximation to
the expansion velocity, which can be improved by iteration as I shall describe below.

In order even to form this first approximation it is necessary to know how different
lines behave in extended atmosphere conditions, and at present there is no really
satisfactory work on this. However, it is possible to get a reasonable idea by using
the results of work, originally by MENZEL and BAKER (1937, 1938), relating to the
Balmer lines, and by WELLMANN (1955) relating to strong HeI lines. From these
results it is possible to form an idea of the radiation dilution in a low-density atmosphere at which various lines are formed, given the emission component strength
of the line. Thus, a first approximation picture of an outer atmosphere can be built up,
knowing the apparent absorption velocities and emission component strengths of the
strong HeI and Balmer lines in a star showing P Cygni type profiles.

(b) The second line of evidence is the actual profile of a strong absorption line,
which may or may not have an emission component. A strong line core is formed in
the outer layers of the atmosphere, within limits governed by the effective excitation
temperature, while the weaker wings, if any, are formed lower down. Thus a velocity
and density field may show up as a line asymmetry. For example, in an accelerating

supergiant atmosphere, where line wings are unimportant, this appears as a long shortward wing. The second way in which a velocity field may be apparent is in a series of lines whose strength varies systematically. The obvious example is the Balmer series, whose strength (and hence atmospheric depth) decreases from Hα down. Here an accelerating atmosphere will show a systematic change in line core velocity, as shown in the case of HD 152236 in Figure 2.

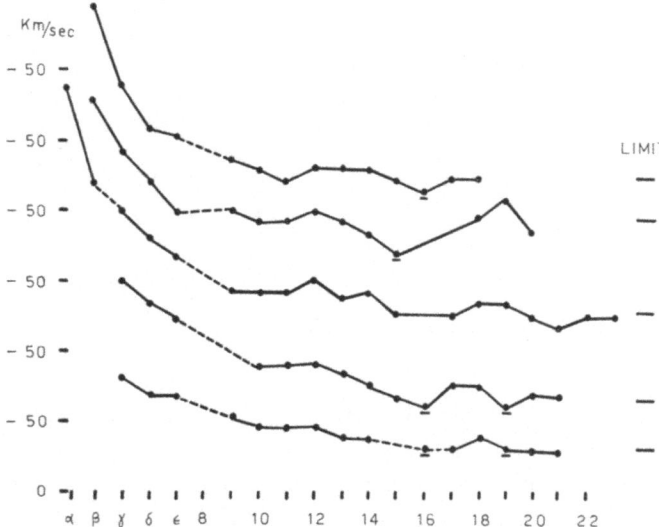

Fig. 2.   Balmer-line velocities from 5 plates of HD 152236.

The effect is also seen in HeI, where strong lines such as those at 4471 Å, 4026 Å have velocities different from the weaker HeI lines.

(c) Finally there is evidence which deals with the innermost part of an expanding atmosphere. The medium-strength lines are formed in well-defined volumes where the excitation temperature gives maximum level populations for the lines. If the velocities of all medium-strength lines are measured, an accelerating atmosphere in which T_{ex} varies with height will show a relation between excitation potential and line velocity. This at present can only give us a qualitative idea of whether and how rapidly the inner atmospheric layers are moving, as we do not have a theoretical physical model relating height and excitation temperature. However, the effect is seen quite markedly in some supergiants and with high-dispersion plates this is a powerful tool in making some sense of the atmospheric structures. It also explains why many of the radial-velocity determinations of supergiant stars are apparently unreliable – the velocities do in fact vary with time and depend on the particular lines measured. An example of this sort of evidence is shown in Figure 3, relating to the Of star HD 152408. This approach can also lead to the recognition of overpopulated states, as in the cases of NIII and SiIV in this star. The ionic velocities fit at a much

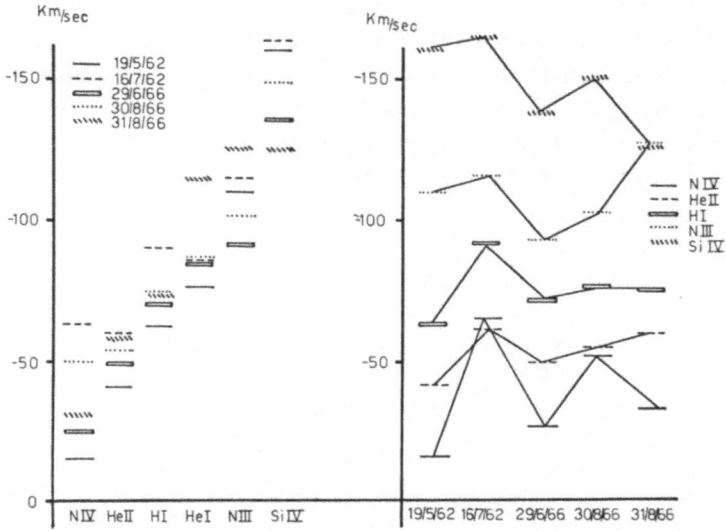

Fig. 3.   Absorption-line velocities for different ions in HD 152408.

lower T_{ex} than expected, or high in the extended atmosphere. Subsequent profile computations tended to confirm this conclusion.

Finally I should mention two other important indicators of a moving extended atmosphere. The first of these is the He I line at 3888 Å, which arises from the highly metastable level 2^3S (also the line at 3187 Å, which is more difficult to observe). Normally blended with H8, the line is formed in absorption at great heights in the atmosphere, where expansion velocities are great enough to separate it from the hydrogen line.

Secondly, there are the broad faint emission bands associated with the He II 4686 Å and N III 4640 Å emission lines, first shown by WILSON (1958) to exist in many O type stars. In the two O stars studied these extend over a wide range and occasionally have faint shortward absorption dips, corresponding to velocities of the order of 2000 km/sec.

3. Line Profile Computation

I shall now outline a system whereby the line profiles can be computed for strong transitions in extended atmospheres with mass motions. There are three aspects which make these computations different from other line profile computations:

(1) line formation extends to several radii from photosphere,

(2) the presence of mass motions,

(3) transparency of the outer layers, so that more than the facing hemisphere must be taken into account.

In the case of stars expanding or rotating at several hundred km/sec it can be shown (HUTCHINGS, 1968) that the profiles of strong lines are almost entirely dominated by the velocity field and the intrinsic line-absorption strength (i.e. the *gf* value). They are sensitive to a lesser extent to the density-height relation. It is in any case unnecessary (as well as a major theoretical undertaking) to solve the transfer equation to form an atmospheric model at this stage of the work. Thus I have concentrated more on geometrical rigour than model atmosphere considerations in these calculations.

Four line formation processes are taken into account:

(1) absorption between the photosphere and the observer,

(2) emission, over all atmosphere except region occulted by the photosphere,

(3) wavelength-independent electron-scattering over all atmosphere except occulted region,

(4) line-of-sight absorption subsequent to (3).

So far profiles have been computed for the elements Si, He and H, with level populations given by the Saha-Boltzmann equations, modified in the latter two cases to the extent indicated by the work mentioned of Menzel and Wellmann. The only broadening mechanism used is Doppler, corresponding to the excitation temperature. In fact, the final profiles are also very insensitive to the shape of the Doppler core.

Geometrically, the stellar atmosphere is divided into a number of shells and each shell into a number of equal sectors. Typically, there may be 350 sectors per shell and 20 shells per stellar radius, although these are freely variable parameters. The atmos-

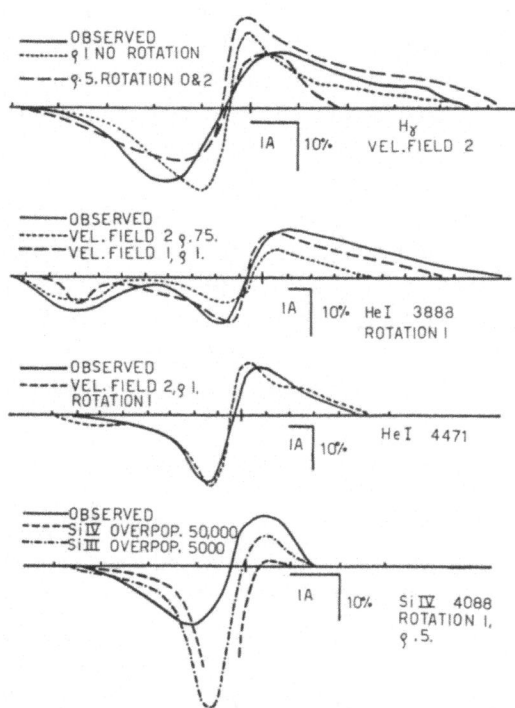

Fig. 4. Observed and computed line profiles in HD 152 408.

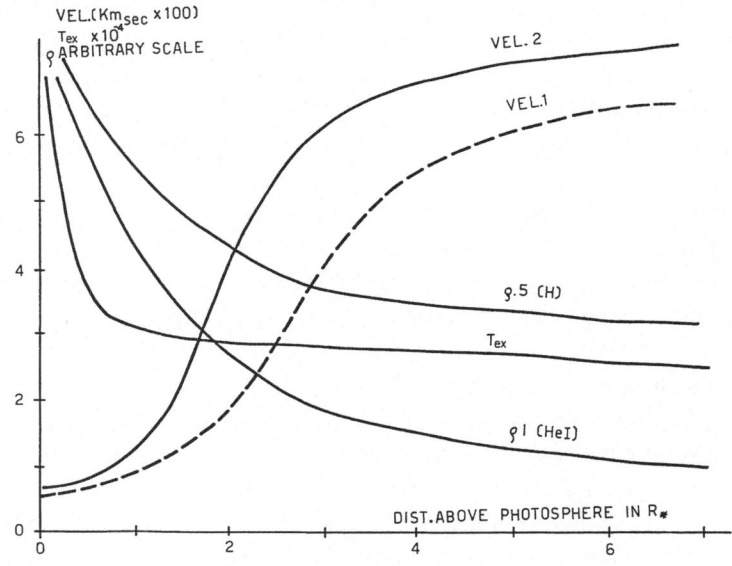

Fig. 5.   Outer atmospheric model of HD 152408.

pheric parameters are taken to be constant throughout each sector, and radial velocities are computed for any given velocity field, rotation and axial inclination.

The atmospheric structure itself is the final touch, which may be altered empirically to match profiles. In practice the supergiant outer atmospheres derived correspond closely to gas expanding adiabatically away from the star. The temperature is initially deduced from the observations, but like all parameters is variable within limits, so that iteration to the final profile allows all necessary degrees of freedom. Although this procedure cannot lead to a uniquely defined structure, a high degree of probability is ensured when the structure is capable of matching several different profiles.

Figures 4 and 5 show some typical profiles for the star HD 152408, and the finally adopted outer atmosphere. Models such as these have been produced for the stars HD 151804, 152236, 152408 (HUTCHINGS, 1968) and P Cygni (paper in preparation), and all show basically similar structure, although they vary widely in extent and density.

4. General Conclusions

Since studying the three bright Southern stars I have looked in detail at some 15 Northern supergiants, of which P Cygni is the only one showing as marked a mass-loss effect in the photographic region of the spectrum. The density of the P Cygni envelope appears to be considerably greater than that of the B star HD 152236, and it may be some 0.3m brighter, making it one of the most luminous stars known in the galaxy. The mass-loss rates obtained for these stars are about 10^{-5} $M_\odot$ per year for the Southern stars, and some 20 times greater for P Cygni. However, it should be noted that uncertainties in the absolute values of the densities and radii are about one order of magnitude.

Several other stars have shown some of the three observational effects, less strongly than the three Southern stars, and some of them show effects which are slightly different. The Northern stars are nearly all intrinsically fainter than their Southern counterparts, and the plates measured are mostly at 15 Å/mm, compared with the 7 Å/mm of the Southern ones. It is evident that higher dispersion data are required, and I am now taking a series at 2.5 Å/mm, together with high resolution rapid photo-electric scans of individual lines. A picture is beginning to emerge, in which it appears that both absolute luminosity and temperature determine the rate of mass loss of early-type stars. Bearing in mind the UV observations, the H-R diagram is tentatively divided into the regions shown in Figure 6 in which mass loss from OB supergiants may be expected.

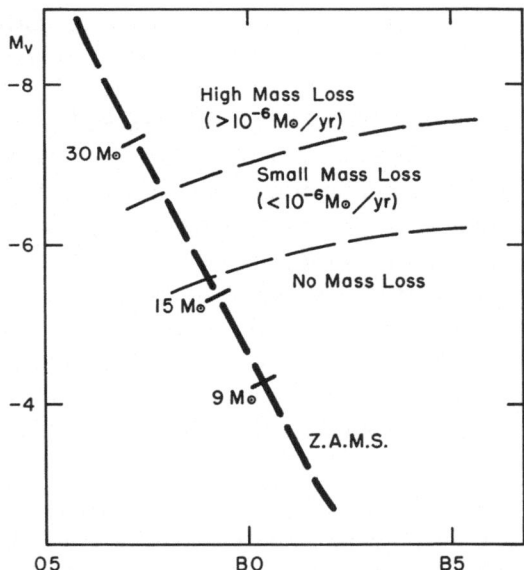

Fig. 6.   Tentative division of H-R diagram into areas of comparable mass loss by atmospheric expansion.

A full account of the investigation described in this section will be published in due course.

References

HUTCHINGS, J. B.: 1968, *Monthly Notices Roy. Astron. Soc.* **141**, 219 and 329.
MENZEL, D. H. and BAKER, J. G.: 1937, *Astrophys. J.* **86**, 70.
MENZEL, D. H. and BAKER, J. G.: 1938, *Astrophys. J.* **88**, 52.
MORTON, D. C.: 1967, *Astrophys. J.* **147**, 1017.
WELLMANN, P.: 1955, in *Vistas in Astronomy*, vol. 1, Pergamon Press, London–New York. p. 303.
WILSON, R.: 1958, *Edinburgh Publ.* **2**, 61.

Discussion

Nariai: Could you please explain the double absorption profile with your calculations?

Hutchings: In the case illustrated part of the lower velocity absorption is due to blended H8. Part of it is the absorption of He I 3888 produced at the maximum of the contribution curve for the line. The high-velocity dip is produced by less strong absorption which is enhanced by being formed along the horizontal branch of the velocity curve through great geometrical depths.

Underhill: The interpretation of line profiles is difficult, for one must not only obtain a representation that fits the observed line profile, but one should also demonstrate that the model used is meaningful not only as regards the geometrical aspects of the situation (the extent and shape of the atmosphere, density and state of motion) but also the physics of the possible interaction between radiation and matter. You have been able to use numerical techniques to handle a fairly complicated geometrical situation, and this is an advance of note, but your representation of the physical situation is not satisfactory. At the expected densities, the purely radiative dilution theories (Menzel, Aller and Hebb for H, as well as Struve and Wurms, Wellmann for He I) are of doubtful value. Collisions must be taken into account. Deutsch asked how the excitation temperature decreased outwards, but it seems to me that your representation is too simple to permit a meaningful answer to this question. Furthermore I believe you treated electron scattering as coherent – the direction of the light was changed but not its frequency –, whereas it is the noncoherence of electron scattering which is important for line shapes, cf. attempts to explain the shapes of Wolf-Rayet emission lines in this way. Thus, although you have been able to show how various velocity fields and rotations of the star might affect line profiles, you should hesitate to conclude that you have a thorough understanding of the line shapes in early-type supergiants.

Hutchings: As far as the line profile computations are concerned, the main object has been to demonstrate the importance of geometrical considerations in a moving atmosphere. While purely radiative line formation processes may not be completely applicable at the densities encountered, they seem to be the best approximation at the present time. As far as such considerations are valid, it is significant that the empirically derived atmospheric structure resembles an adiabatically expanding gas. My treatment of electron scattering was not made exact, as its contribution to strong line profiles of these stars is small. While we are still far from understanding the physical processes in extended atmospheres, these results indicate the type of gross atmospheric structure present in extreme cases of expansion.

EVIDENCE FOR MASS LOSS FROM THE F-TYPE
SUPERGIANT, 89 HERCULIS

WALLACE L. W. SARGENT* and PATRICK S. OSMER

*Mount Wilson and Palomar Observatories, Carnegie Institution of Washington,
California Institute of Technology*

Abstract. We describe changes in the spectrum of 89 Herculis (F2 Ia) in the interval 1961–67, which are ascribed to variable loss of mass from the surface. The star is remarkable in that some of the blue-displaced circumstellar absorption features have velocities of 150 km/sec. This is very close to the estimated escape velocity at the stellar surface of 200 km/sec. It is probable that the rate of mass loss is too small to affect the subsequent evolution of F supergiants.

1. Introduction

The star 89 Herculis (F2Ia) is one of the few supergiants known to occur at high galactic latitudes. A previous study of its spectrum by SEARLE *et al.* (1963) indicated that it has a normal composition and an absolute visual magnitude of −7.1. They also showed that it could have originated in the galactic plane as a 'runaway' B star.

Irregular variations in radial velocity were found by BÖHM-VITENSE (1956) and they prompted WORLEY's (1956) discovery of light fluctuation. Böhm-Vitense also noted that the hydrogen and sodium D lines have variable blue displaced absorption components. An expanding circumstellar envelope is probably the cause of these components.

In this paper we show that circumstellar envelope of 89 Her is remarkable because:

(1) The shell is expanding at a rate close to the gravitational escape velocity.

(2) The D line-absorption components are constant in shape for such a long time that it is hard to understand the place of and the mechanism for their formation. (The Balmer lines change in a shorter time that is easier to understand.)

(3) There are a number of narrow and unusual emission lines in the spectrum which are nearly stationary with respect to the photosphere.

2. Observations

The observational material consisted of 35 spectrograms, of dispersion ranging from 4.5 Å/mm to 15 Å/mm, obtained with the coudé spectrographs at Mount Wilson, Mount Palomar, Herstmonceux and Mount Hamilton between 1960 and 1967. All the plates were photometrically calibrated. The emulsion used were IIaO, IIaD, IIaF and IN to cover the appropriate region of the spectrum.

3. The Circumstellar Absorption Features

The Balmer lines exhibit the most striking variations in the spectrum of 89 Her. The

* Alfred P. Sloan Research Fellow.

M. Hack (ed.), Mass Loss from Stars. All rights reserved

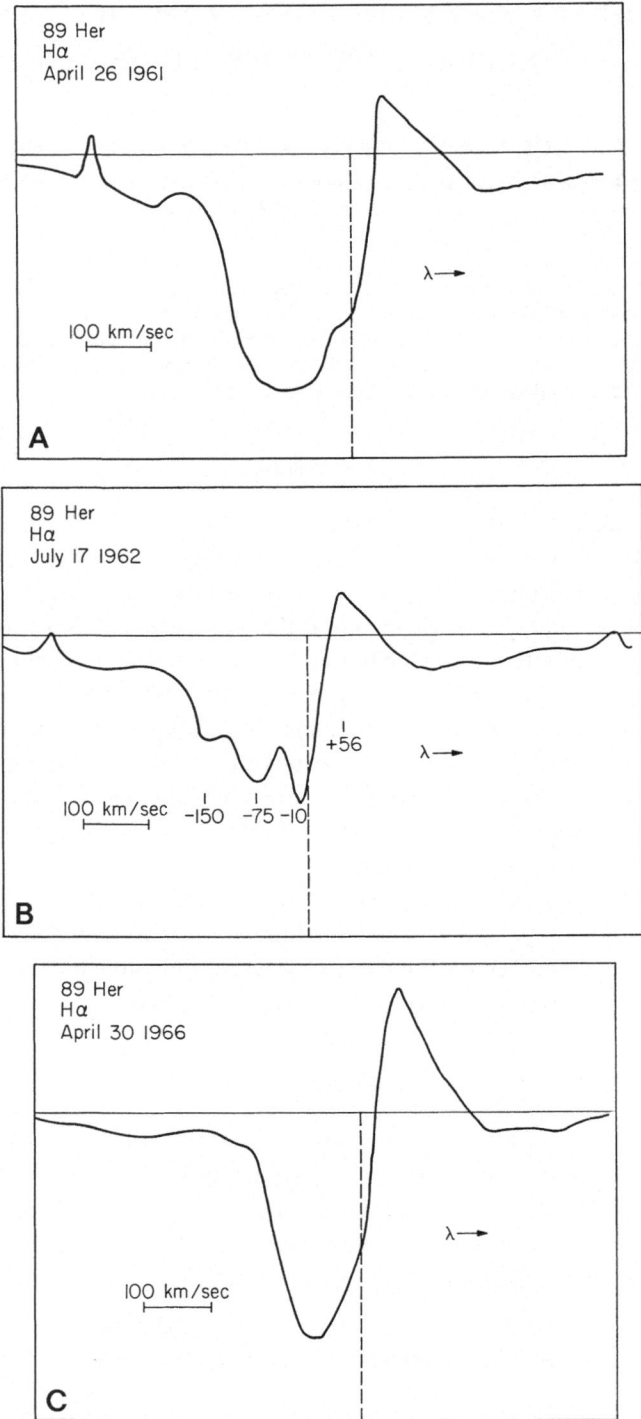

Fig. 1.   Profiles of Hα obtained at different times. The vertical dashed line is the zero of velocity defined by the photospheric absorption lines. This zero fluctuates by about 10 km/sec. Note the remarkable changes in the strength and structure of the blue-displaced components. The red displaced emission shows less evidence for changes.

Hα profiles of Figure 1 represent the range in appearance of the line. It was found that the blue-shifted components can change definitely in a month's time. The higher Balmer lines show the same absorption features but no emission. The D lines, which are shown in Figure 2, also have blue-shifted absorption components, but they vary

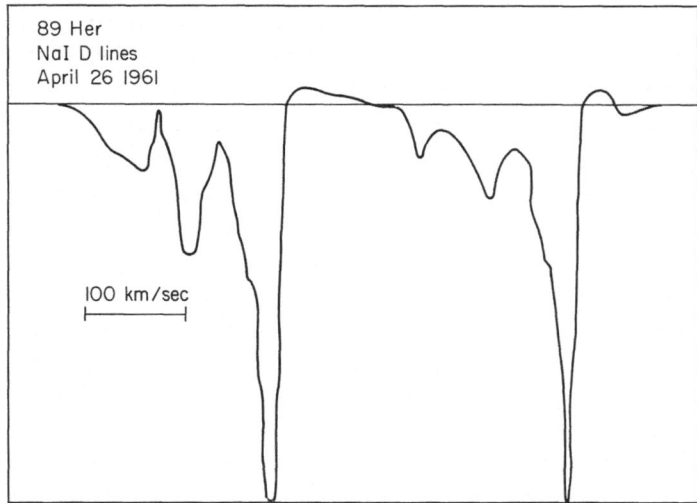

Fig. 2. A typical profile of the Na I D lines. The blue-displaced absorption components coincide in velocity with those seen at Hα in Figure 1b. However, the D lines change their appearance at a much slower rate than Hα, and there is some doubt as to whether they can be formed in the same regions.

in appearance at a much slower rate. Our material indicates that they change in a period of roughly 5 years. While the number is uncertain, the fact that the ratio of the two variation times is at least 10 is strong evidence that the Balmer- and D-line components are not formed in the same place. The stronger of the shifted D lines has a velocity with respect to the photosphere of about 80 km/sec. The Balmer components, which are not as well defined, have a comparable velocity. (Weaker components of both sets of lines have been observed with speeds up to 150 km/sec.) The H and K lines of Ca II do not show circumstellar components but the blue wing of each line has an unusual, square appearance.

We can derive the gravitational escape velocity at the surface of 89 Her from the atmospheric parameters obtained by Searle et al. and from the evolutionary tracks calculated by HASELGROVE and HOYLE (1959). Taking $M_{bol} = -7.3$ and $T_e = 7240$ K then the stellar radius is 155 $R_\odot$ and the mass turns out to be 16 $M_\odot$. The resulting escape velocity is 200 km/sec, greater than any velocity observed in the outflowing material.

We can also use the velocity of the material and the time of variation to estimate how far it travels. We are supposing that a shell of gas is ejected from the star and that it produces lines which remain visible until it has dispersed to a certain point. If the lifetime of the shell is the 1 month implied by the fluctuations of the Balmer lines, then it travels about three stellar radii at 80 km/sec. This is consistent with the idea that the lines are formed fairly close to the surface. However, the D-line components can move 120 stellar radii in their 5-year lifetime and it is hard to imagine that they are formed near the star. This approach does not give conclusive information about where the shell is formed but it does show that the D-line and Balmer-line absorption components are probably not formed in the same place and that the D lines may arise at large distances from the star. An alternative explanation is that some sort of continuous flow exists and that the ionization is such that the lines are produced only in a certain region of space.

Simplified calculations of the Hα profile produced by an expanding shell give qualitative agreement with the observations. If the shell is optically thick in the line and scatters the stellar flux, very little light is scattered forward by the material between us and the star. The blue wing will therefore appear in absorption. The material behind the star, which is moving away from us, will reflect some of the light in the forward direction, and emission will be present on the red wing. The maximum emission attainable in this model is about 15% of the continuum. Other processes must be responsible for the strong emission observed in some stars. In 89 Her the emission ranges from 20 to 30% of the continuum and a combination of processes must be involved.

We can get an estimate of the amount of circumstellar material in the line of sight by applying curve of growth techniques to the absorption components. In the process we also obtain Doppler parameters. Using a Schuster-Schwarzschild curve of growth for the Balmer lines gives the results: N (Balmer) $= 10^{13.44}$ atoms/cm^2 and $V_{\text{Doppler}} = = 20$ km/sec. Unfortunately there seems to be no way of establishing the radius of the shell and the degrees of excitation and ionization within it. This prevents us from determining the rate of mass loss.

If we assume that the circumstellar hydrogen is neutral, that it has an excitation determined by the stellar radiation field and that the material is close to the surface, the mass-loss rate is roughly 10^{-8} $M_\odot$/year. The assumptions are so idealized that the result only shows that the mass-loss rate could be quite small.

Application of the STROMGREN (1948) curve of growth to the circumstellar D lines gives the value $N(\text{Na\,I}) = 10^{12.6}$ atoms/cm^2 and $V_{\text{Doppler}} = 4$ km/sec. Again the difficulties with the degree of ionization make the column density hard to convert to the total density of sodium, but it should be noted that the D lines have a much smaller Doppler parameter than do the Balmer lines. Any theory about the mass-loss process will have to explain the differences in behavior and width of the Balmer and sodium lines.

4. The Emission Lines

Spectra of 89 Her taken on red sensitive emulsions show a number of narrow emission lines in addition to the emission at Hα. At first all the emission lines seemed to be longward of 5900 Å, but a search of the blue plates turned up two weak ones near 4730. Table I lists the wavelengths, identifications, equivalent widths, and information on the type of transition involved. All the lines result from neutral metals and are very

TABLE I

Emission-line data

λ	Element	Multiplet	E.P. upper level	Type of trans.*	$-\log W/\lambda$		
					Ce 15607	Ec 4860	Ec 4959
4727.936	Co I	15	3.04	I			
4732.051	Co I	15	3.12	I			
5956.702	Fe I	14	2.93		5.10	5.17	5.26
6007.313	Ni I	42	3.72	I			
6108.121	Ni I	45	3.69	I	4.91	5.00	5.15
6128.990	Ni I	42	3.68	I			
6191.186	Ni I, Fe I	45, 169	3.66	I, N	4.90	4.71	5.13
6252.561	Fe I	169	4.37	N			
6280.625	Fe I	13	2.82	I	5.43	5.29	5.51
6314.666	Ni I	67	3.88	I	5.44	4.93	5.38
6327.603	Ni I	44	3.62	I			
6335.335	Fe I	62	4.14	N			
6358.692	Fe I	13	2.80	I	5.21	5.01	5.41
6400.335	Fe I	13	2.84	I	5.18	5.13	4.75
6430.851	Fe I	62	4.09	N	5.32	5.45	5.35
6450.23	Co I	37	3.62	N	5.34	5.26	5.08
6498.950	Fe I	13	2.85	I	4.89	5.09	4.95
6546.276	Ti II	102	3.31	N	5.48	5.56	5.33
6554.226	Ti II	102	3.32	N	5.05	–	5.61
6572.781	Ca I	1	1.88	I	5.34	–	5.49
6574.238	Fe I	13	2.86	I	5.09	5.10	5.14
6586.33	Mn I:	51	6.28	N			
6599.112	Ti II	49	2.77	N			
6743.124	Ti II	48	2.73	N			

* I = Intercombination line. N = Normal permitted transition.

rarely seen in emission in stellar spectra. Rho Cas, an F supergiant with a circumstellar envelope, is the only other star known to us that shows some of the same emission lines.

Radial-velocity measurements of the emission lines and normal photospheric lines indicate that both sets fluctuate with an amplitude of 4–5 km/sec, and that the means of the two differ by less than 1 km/sec. On the average the region producing the emission is stationary with respect to the photosphere. The amplitude of the velocity variations is average for a supergiant.

Although large variations in equivalent width are noted for some lines, the average difference in width of all the lines for any two plates is 10%, which is not significant. The lines should probably be monitored with higher resolution equipment before any conclusions about variability are made.

As the lines are approximately twice as wide at half maximum as the instrumental profile, it is possible to estimate their intrinsic Doppler widths. If the stellar and instrumental profiles are Gaussian in shape, then

$$(\text{intrinsic half width})^2 = (\text{obs. H.W.})^2 - (\text{instr. H.W.})^2$$

The atoms producing the line are presumed to have the radial velocity distribution $N(V) \, dV \propto \exp(-V^2/V_D^2) \, du$, where $N(V)$ is the number of atoms with radial velocity between V and $V+dV$, and V_D is the Doppler-velocity parameter. Using these two relations we find that

$$V_D = \frac{1}{\sqrt{\log 2}} \left(\tfrac{1}{2} \text{ intrins. full width at half maximum in km/sec}\right).$$

Since the line width is affected by thermal motions, rotation, microturbulence and macroturbulence, V_D is a measure of the sum of all these effects. The mean V_D is 9 km/sec; the range is 7–14 km/sec. This value is close to the microturbulent velocity of 7 km/sec that Searle *et al.* derived for the photospheric lines.

The only property that the emission lines seem to have in common is a low transition probability. A visual comparison of spectra of 89 Her and *v* Aql (F2Ib) shows that with one exception no lines, absorption or emission, appear in *v* Aql at the position of the emission lines. The exception is an absorption line in *v* Aql near 6191 Å.

5. Discussion

The data described above indicate that the mass-loss process is very complex in 89 Her. The great difference in variation times for the Balmer and D lines and the stationary nature of the emission are evidence that three different regions of space around the star produce the different features, but we can determine very little about the physical conditions within the regions.

A comparison of 89 Her and ϱ Cas, the only other F supergiant with circumstellar features that has been analyzed (SARGENT, 1961) shows that the two have little in common. ϱ Cas does have some of the same metal emission lines, but they are not stationary with respect to the star. It does show circumstellar components of metallic lines in absorption, but none are visible in the D lines or Balmer lines. Hα is apparently filled in by emission.

A search for other F supergiants with definite circumstellar features has been unsuccessful and it seems that 10% of them or less are losing mass. We conclude that massive stars probably do not lose a significant fraction of their mass during the time they spend as F supergiants. However, the fact that such high velocities of expansion are observed in the spectrum of 89 Her implies that very energetic phenomena are occurring in its atmosphere.

References

BÖHM-VITENSE, E.: 1956, *Publ. Astron. Soc. Pacific* **68**, 57.
HASELGROVE, C. B. and HOYLE, F.: 1959, *Monthly Notices Roy. Astron. Soc.* **119**, 112.
SARGENT, W. L. W.: 1961, *Astrophys. J.* **134**, 142.
SEARLE, L., SARGENT, W. L. W., and JUGAKU, J.: 1963, *Astrophys. J.* **137**, 268.
STROMGREN, B.: 1948, *Astrophys. J.* **108**, 242.
WORLEY, C. E.: 1956, *Publ. Astron. Soc. Pacific* **68**, 62.

Discussion

Morton: What was the result of your earlier investigation whether 89 Her could live long enough to reach such a high galactic latitude?

Sargent: We found that 89 Her could just have reached its inferred height above the galactic plane in its evolutionary life-time if it had been expelled from the galactic plane with a z-component of velocity characteristic of a 'run-away' OB star.

Underhill: Is there any possibility that 89 Her is a shell star? I ask this because many of the lines used as spectroscopic criteria for M_v are strengthened in shells, including the O I lines you used, and because the variations which you find in the profiles of the Na D lines and of Hα are rather similar to the striking variations Geuverinck and I find in the spectrum of the shell star 48 Librae in 1968.

Sargent: The spectrum of 89 Her appears to be clearly separated into two components, a 'photospheric' component and a 'circumstellar' component with displaced lines. The analysis of the 'photospheric' component by Searle, Sargent and Jugaku did not reveal any effects which could be ascribed to dilution.

Houziaux: In connection with this effect that the emission lines seem to appear longward of λ 5900, I should like to mention that in shell stars there is a lot more emissions in the red and infrared part of the spectrum than in the blue region. This is particularly striking when one considers homologous lines of Balmer and Paschen series.

Viotti: As for your figure on the dependence of Log W-Log *gf* on the excitation potential, I found in XX Ophiuchi and Eta Carinae a similar linear dependence for the permitted Fe II emission lines with excitation temperature around 8000 K.

De Groot: I would like to make a remark about the coincidence of the Hα absorption components with the interstellar components in the spectrum of 89 Her. It seems to me that many of the interstellar components have a reasonable explanation. The component at − 10 km/sec could be due to another interstellar cloud. This has been found in many other spectra as well. It would seem that − 150 km/sec is kind of a maximum velocity in the atmosphere of 89 Her, independent of the other velocity variations. At the outer edge of the atmosphere this velocity could be transferred to the circumstellar medium. Then only the component at − 75 km/sec is left unexplained and coincides with one of the Hα absorption components.

Morton: How can the 10 km sec^{-1} components of the Na I lines be interstellar, as you suggest, when there are Balmer lines with the same velocity which cannot possibly be interstellar?

De Groot: I did not mean to say that the Hα absorption at − 10 km/sec is interstellar. I only tried to offer an explanation for the Na I line.

A NEW OUTBURST OF THE SHELL STAR 48 LIBRAE

ANNE B. UNDERHILL and H. G. GEUVERINK

Sonnenborgh Observatory, University of Utrecht, The Netherlands

Abstract. Spectrograms taken in 1967 and in 1968 of 48 Librae show that the shell spectrum is at present very strong with strongly asymmetrical lines. The cores of the Balmer lines are sharp on the longward side but winged shortward and emission is visible for the first few series members. The Fe II lines show similar asymmetry and the strongest lines are accompanied by emission. The radial velocity shown by different lines varies from about -130 km/sec to -3 km/sec. The observed velocities do not fit in with the cyclical pattern of changes observed between 1938 and 1962. A new outburst is indicated.

The significance of the changes in the hydrogen lines between 1967 and 1968 is discussed briefly. The Na D lines, Ca II K and some of the strong Fe II lines are double consisting of a strong broad diffuse component which is strongly displaced shortward, and a sharp weak component which is less displaced. The observations suggest that there is a strong radial-velocity gradient in the shell. The sharp components due to Na I and Ca II are probably formed in a circumstellar envelope which is escaping from the star. The Balmer-line profiles and radial velocities suggest a velocity field which is decreasing outward.

1. Introduction

Studies of the star 48 Librae (cf. the review by UNDERHILL, 1966) have shown that it is rotating rapidly with $v_e \sin i$ of the order of 350 km/sec to 400 km/sec, that it is a B3 main-sequence star which belongs to the Scorpio-Centaurus cluster, thus is at a distance of about 175–200 pc, and that it should have a radial velocity of about -6 km/sec. Indeed the radial velocity was constant at about -6 km/sec until at least 1931. In 1935 the shell was noted to be active and since then the shell velocity has varied from values of the order of -50 to -80 km/sec to values of the order of $+30$ km/sec. There is a considerable spread in the velocities found for different lines when the velocity has large negative values; the spread is not large when the velocity lies in the range ± 20 km/sec. The apparent period of the radial-velocity variations has been close to 10 years; the range of the variations was not closely the same in each cycle which has been observed in the years since 1935.

The line shapes are observed to change during the radial-velocity cycle. In the past the lines have been asymmetrical when the velocity of the shell with respect to the star was close to 0 km/sec and symmetrical at either extreme. The asymmetry is most distinct in the hydrogen lines and the Fe II lines, one side of the line being very sharp, the other winged. The wing has appeared on the side to which the velocity of the shell was changing. This behaviour has been quite consistent and it was observed at every crossing of the zero-velocity line. The equivalent widths of the shell lines have varied throughout the 10-year cycle.

The radial-velocity variations between 1938 and 1962 compiled from the observations of Merrill and of Underhill are shown in Figure 1. The radial velocity of Hδ is given by the solid points, that of the Fe II lines by open circles and that of the Ti II lines by crosses. The high members of the Balmer series, $n \geqslant 15$, have velocities like the

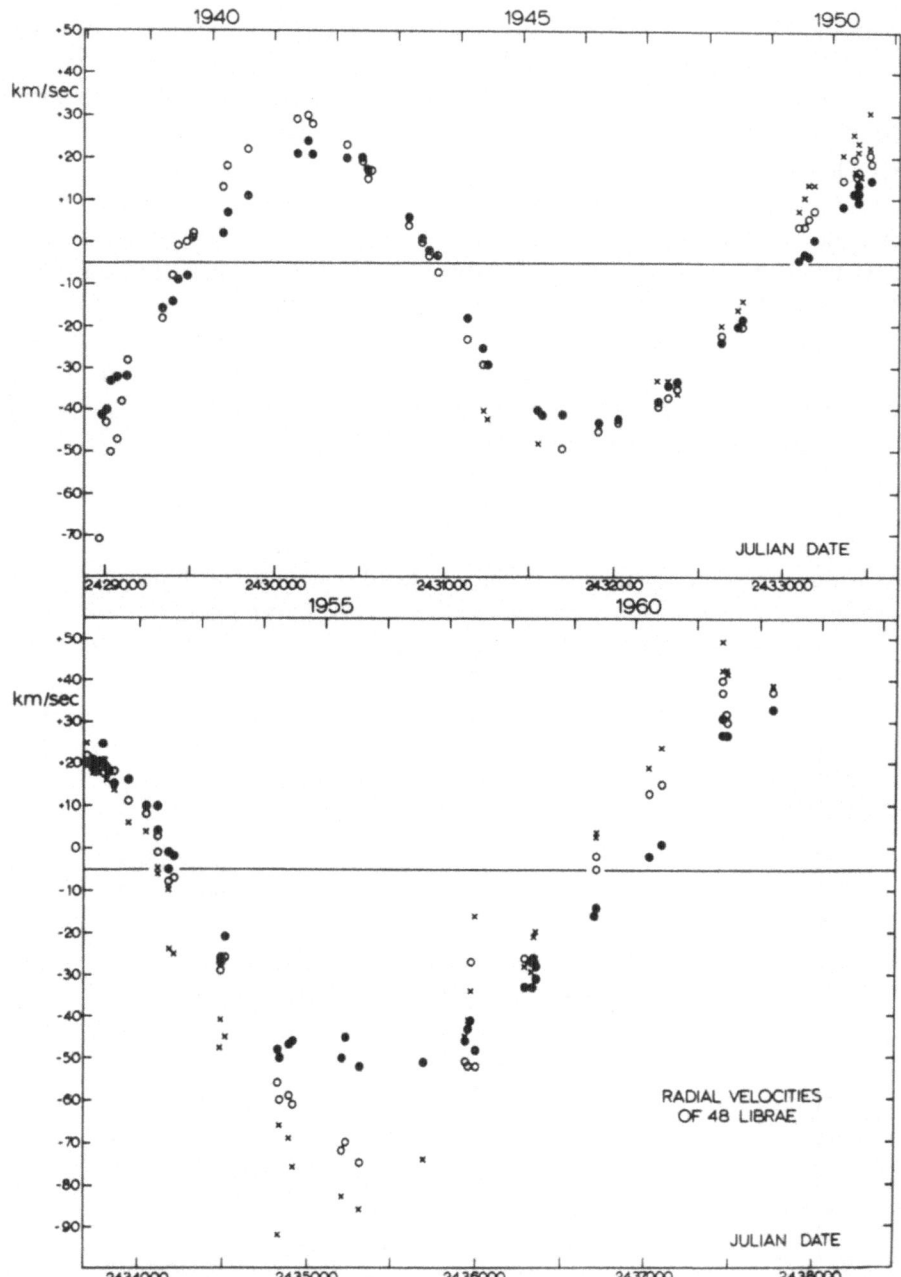

Fig. 1. The radial-velocity variations of 48 Librae between 1938 and 1962. The velocities from Hδ are indicated by filled points, those from the Feɪɪ lines by open circles, and those from Tiɪɪ by crosses.

Feɪɪ velocities; the Tiɪɪ velocities are typical of the weaker lines in the shell. The different lines in the shell rarely show the same velocity, the spread being greatest at velocity minimum.

If the radial-velocity oscillation shown in Figure 1 had continued unchanged velocities of the order of −45 to −50 km/sec would have been expected in 1967 and about −30 km/sec in 1968. In fact velocities of the order of −110 km/sec have occurred in 1967 and about −95 km/sec in 1968 indicating that after 30 years of more or less regular behaviour 48 Librae is again erupting violently.

In 1967 and 1968 there is a large spread in the velocities and many of the lines are extremely asymmetric. Some of the highlights of the observations will be presented here. Two blue-violet high-dispersion spectrograms were obtained June 20, 1967 at the Dominion Astrophysical Observatory and in 1968 blue-violet and red high-dispersion spectrograms were obtained at the Kitt Peak National Observatory on June 4 and 5 and at the Dominion Astrophysical Observatory on July 5, 6 and 17. Some radial-velocity measurements have been made by both authors and a preliminary study has been made of a few line profiles.

2. The Hydrogen Lines

The hydrogen lines are asymmetrical, the longward side of the absorption core being very steep, the shortward side distinctly winged. The central absorption of the lines Hβ to at least H22 in 1968 is 100%, so far as can be determined. In 1967 it is about

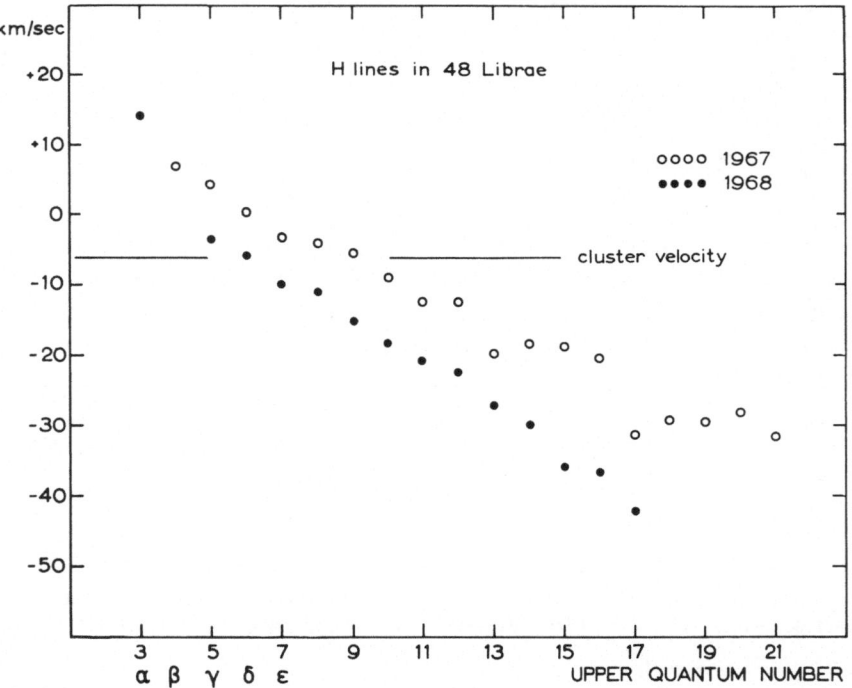

Fig. 2.   The radial-velocity progression of the sharp edge of the Balmer-line cores. The 1967 results are for June 20, the 1968 results for June 4. The Hα position is from a spectrogram taken on June 5, 1968.

the same, but the spectrograms, owing to guiding streaks, are unfortunately not suited to precise photometry.

There is a definite progression in the radial velocity *at which the sharp edge of the profile occurs*. This progression is steeper in 1968 than in 1967; it is shown in Figure 2. The implication is that a large range of velocity exists in the atmosphere, the layers in which the absorption cores of Hα, Hβ and Hγ are formed having chiefly positive velocities with respect to the star while the layers in which the cores of the higher series members are formed have negative velocities. The position of the very sharp longward edge of the hydrogen profiles can be measured very exactly on the spectrograms. This position is determined by the strength of the longward emission component and the velocity distribution in the material which is projected against the photosphere of the star.

The Balmer progression as defined by Merrill gives how the *midpoints* of the absorption cores vary in velocity as the upper quantum number varies. In the present case there is little systematic variation of the midpoint of the core with quantum number when $n \geqslant 5$. The first few Balmer lines tend to have more positive velocities than do the higher series members. This is probably due to blending with the emission components which are clearly visible for the first few members of the Balmer series. The radial velocities of the individual Balmer lines are listed in Table I. It is seen that

TABLE I

Radial velocities of the hydrogen lines in 48 Librae

	centre of the absorption core		longward edge of the core	
	(UT) 1967, June 20.246	(UT) 1968, June 4.268	(UT) 1967, June 20.246	(UT) 1968, June 4.268
α	–	− 2.8 km/sec	–	+ 13.9 km/sec[a]
β	− 19.3 km/sec	–	+ 6.7 km/sec	–
γ	− 32.5	− 54.9	+ 4.2	− 3.5
δ	− 54.8	− 54.4	+ 0.4	− 5.8
ε	− 56.3	− 50.3	− 3.2	− 10.1
8	− 44.2	− 54.1	− 4.0	− 10.9
9	− 57.2	− 51.9	− 5.3	− 15.2
10	− 56.3	− 58.1	− 8.9	− 18.2
11	− 53.0	− 58.7	− 12.4	− 20.8
12	− 49.2	− 61.3	− 12.4	− 22.5
13	− 63.2	− 58.1	− 19.7	− 29.2
14	− 63.4	− 61.6	− 18.2	− 29.8
15	− 56.6	− 64.2	− 19.0	− 36.0
16	− 61.4	− 62.6	− 20.4	− 36.8
17	− 54.5	− 61.2	− 31.6	− 42.1
18	− 61.9		− 29.3	
19	− 66.5		− 29.6	
20	− 57.0		− 28.2	
21	− 77.5		− 31.7	

[a] The Hα observation was obtained 1968, June 5.279 (UT).

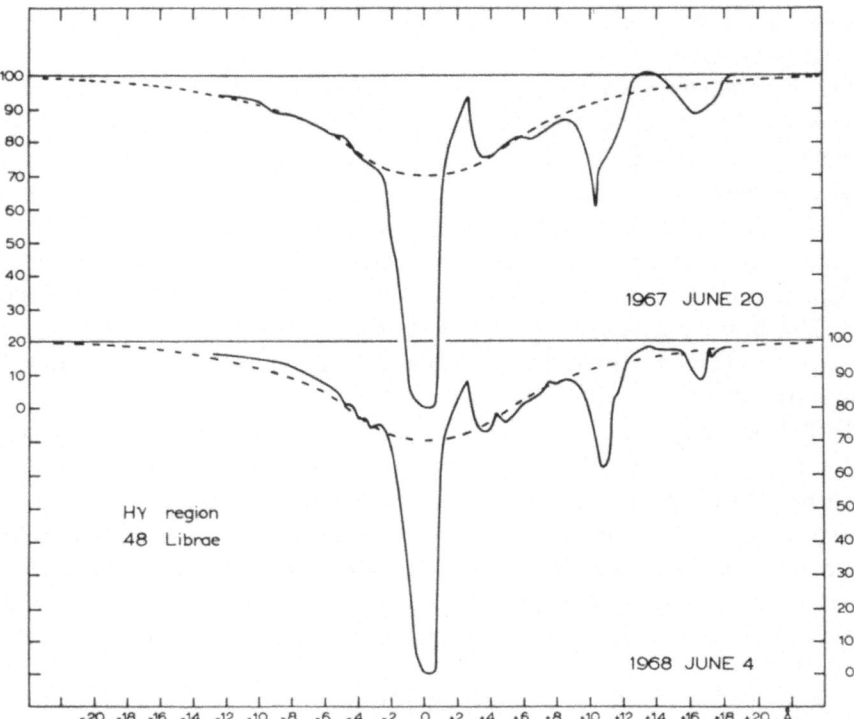

Fig. 3. The region near Hγ in 1967 and in 1968. The estimated stellar Hγ line is shown by a broken line. The zero of wavelength is at the centre of the stellar line.

the centres of the Balmer absorption cores correspond to velocities near −60 km/sec both in 1967 and in 1968, but the position of the sharp edge varies from line to line.

Profiles of the region near Hγ in 1967 and in 1968 are shown in Figure 3. The relative positions of the profiles take account of the fact that the sharp edge of Hγ in 1967 lies +4.16 km/sec from the rest velocity of Hγ with respect to the sun, while in 1968 it lies at −3.54 km/sec. The zero of wavelength is marked at the apparent centre of the *stellar*, rotationally broadened profile. The stellar profile is shown by a broken line. This profile was estimated by comparing both observed intensity profiles (which were obtained completely independently from the relevant spectrograms) and drawing a typical, symmetrical rotationally broadened profile. In addition to the stellar Hγ line one sees the deep shell absorption core, an emission component on the longward side of Hγ, a strong shell absorption line accompanied by emission due to Fe II 4351.76 and a weak shell absorption line due to Fe II 4357.57.

The equivalent width of the stellar Hγ line is 4.81 Å which compares well with the value 4.72 Å found previously by UNDERHILL (1953). According to the absolute magnitude calibration by PETRIE (1964), this $W(H\gamma)$ corresponds to $M_V = -2.8$ for a B3 star. According to WEAVER and EBERT (1964), the equivalent MK spectral type of 48 Librae should then be B3 IV.

The Hγ absorption core due to the shell in 1968 is narrower than what it was in 1967 and the sharp edge is displaced 0.11 Å shortward. Clearly the velocity field in the shell has changed, but more line profiles must be studied before even a schematic model can be made. The emission at Hα in 1968 is strong and broad, the longward component being slightly stronger than the shortward component. The total width of the Hα emission is about 35 Å, which is similar to the width found in 1953 (UNDER-HILL, 1953).

3. The FeII Lines

The FeII lines are strong in absorption in the shell spectrum of 48 Librae, the strongest lines being accompanied by longward emission components. The profiles are asymmetrical, much like the H lines. The profiles of FeII 4351.76 and FeII 4357.57 shown in Figure 3 are typical. The cores are definitely sharper and more asymmetric on 1967, June 20 than on 1968, June 4. The average radial velocity indicated by 18 FeII lines was −106.5 km/sec on 1967, June 20, while on 1968, June 4 the average value from 17 FeII lines was −91.8 km/sec. A few lines of TiII, CrII and NiII have been measured also, and they give somewhat different velocities. In general the radial velocities on 1968, June 4 are about 20 km/sec smaller than those found on 1967, June 20. This trend is in accord with a continuing radial-velocity oscillation of the sort shown in Figure 1.

In 1968 many of the stronger, broad FeII lines appear to be accompanied by a weak rather sharp component. The average radial velocity from seven such sharp components is −37.0 km/sec. It is possible that these weak lines represent the old shell. The sharp component of FeII 4351.76 can be seen in the lower section of Figure 3 as a small stillstand at about +11.7 Å from the centre of stellar Hγ. The sharp components of intrinsically stronger FeII lines such as λ 4233 are much clearer on the intensity tracings than is this component.

4. The Na D Lines

The Na D lines are strong in absorption in the shell spectrum of 48 Librae. On the 1968 spectrograms they are clearly seen to be double, the sharp component lying at −14.2 km/sec on June 5.279 (UT), while a broad strong line lies at −83.3 km/sec. Profiles of the Na D lines from a spectrogram obtained on July 6, 1968 are shown in Figure 4. The rather great strength of the Na D lines is evident. The region of the Na D lines was not observed in 1967.

According to MERRILL and SANFORD (1944), the Na D lines were double in 1943 and 1944. Merrill and Sanford were inclined to interpret the sharp component at −14 km/sec as an interstellar line, but because this sharp component is quite strong and wide, this interpretation hardly seems possible for 48 Librae is only 200 pc distant. Probably both lines are circumstellar in origin. It is quite remarkable that the shell spectrum produces at the same time strong lines from neutral sodium and strong lines from excited levels of SiII, MgII and HeI. The latter lines are typical of B type spectra.

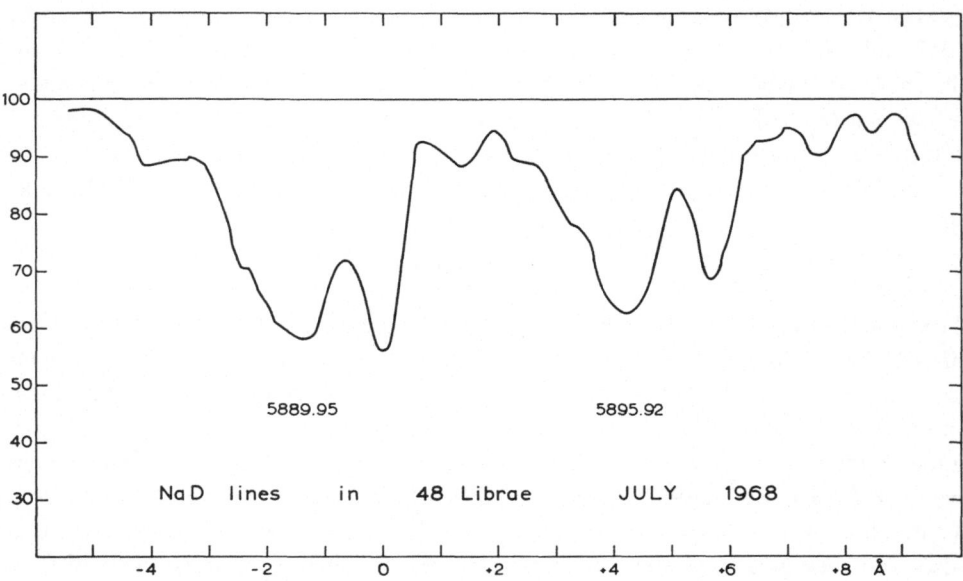

Fig. 4.   The profile of the Na D lines on 1968, July 6.

Spectrograms covering the region of the Na D lines were obtained by the senior author at the Dominion Astrophysical Observatory every year between 1950 and 1962, except in 1952. The dispersion is 30 Å/mm and the projected width is about 1 Å. The Na D lines do not appear double on any of these spectrograms, but they do seem to change in shape from time to time becoming occasionally rather diffuse. The resolution of these spectrograms is not adequate to show double lines unless the separation of the components is more than 50 km/sec.

Double absorption lines are also observed at the Ca II H and K lines on high-dispersion, high-resolution spectrograms. MERRILL and SANFORD (1944) have remarked upon these lines. On 1968, June 4.268 (UT) the radial velocity of the diffuse component of Ca II K was -78.5 km/sec while that of the sharp component was -9.5 km/sec. The components of Ca II H are blended with Hε and difficult to measure.

5. An Unidentified Emission Line

There is a distinct fairly wide, rather weak emission line visible at 6320 Å on both red spectrograms taken in 1968 (June 5 and July 7). This line cannot be identified. A similar line at the same wavelength was found in the shell spectrum of ζ Tauri by UNDERHILL and VAN DER WEL (1967). Weak unidentified emission lines at 6386.0 Å and 6387.6 Å also appear in the spectrum of ζ Tauri. A first survey of the 48 Librae spectrograms has not revealed these emission lines in the spectrum of 48 Librae. These unidentified emission lines are not accompanied by absorption lines. They have also been observed in ζ Tauri by Mme Herman and her group at the Meudon Observatory.

6. Conclusion

The radial-velocity variations shown by the shell spectrum of 48 Librae and the changes in profile indicate that there is a velocity field in this shell. The Balmer-line profiles in 1967 and 1968 suggest that a deceleration is working in the atmosphere. The presence of the sharp, shortward displaced lines due to Na I and to Ca II appear to give the clearest evidence that matter is leaving 48 Librae. A firm conclusion on this point must await development of a quantitative model for the shell.

Acknowledgements

We are grateful to the Kitt Peak National Observatory for the privilege of observing with the coudé spectrograph of the 84-inch telescope, and to the Dominion Astrophysical Observatory for the opportunity in 1967 and in 1968 to observe with the 72-inch telescope and spectrograph as well as with the 48-inch telescope and coudé spectrograph.

References

MERRILL, P. W. and SANFORD, R. F.: 1944, *Astrophys. J.* **100**, 14.
PETRIE, R. M.: 1964, *Publ. Dom. Astrophys. Obs.* **12**, 317.
UNDERHILL, A. B.: 1953, *Publ. Dom. Astrophys. Obs.* **9**, 363.
UNDERHILL, A. B.: 1966, *The Early Type Stars*, D. Reidel Publ. Co., Dordrecht, p. 238.
UNDERHILL, A. B. and VAN DER WEL, T.: 1967, in *Determination of Radial Velocities and their Applications* (ed. by A. H. Batten and J. F. Heard), Academic Press, London, p. 251.
WEAVER, H. F. and EBERT, A.: 1964, *Publ. Astron. Soc. Pacific.* **76**, 6.

MICROTURBULENCE IN STELLAR ATMOSPHERES
METHODIC REMARK

E. A. GUSSMANN

Astrophysical Observatory, Potsdam, G.D.R.

This is a short remark on a method which enables us to overlook the action of micro-turbulence on line profiles and equivalent widths in a simple way. Obviously, this action must be clarified, before the influence of large-scale turbulence or streams on profiles can be treated. Moreover, we will see that a generalization will allow us to account for emission features, whether they originate from a shell or a circumstellar medium.

We will use the weighting-function method in its modified form. For the moment, we will confine ourselves to the case of LTE. The depth of the line can then be written in the form

$$d(\mu) = \int_0^\infty g(\tau, \mu) \left[1 - \exp\{-\tau_\lambda/\mu\}\right] d\tau.$$

$\mu = \cos\theta$, $\tau =$ continuous optical depth, $\tau_\lambda =$ line optical depth. In the LTE-case the weighting function $g(\tau, \mu)$ is proportional to the gradient of the source function, while the expression in square brackets means a line profile originating in an absorption tube of thickness τ_λ/μ. The advantage of the above formula is that the qualities of the line enter it only once through the quantity τ_λ. This means that only the expression in square brackets needs to be regarded.

Microturbulence ξ enters the Doppler width in the usual form

$$\Delta\lambda_D = (2RT/\mu + \xi^2)^{1/2};$$

an increasing or decreasing ξ is reflected by an increase or decrease of $\Delta\lambda_D$. The optical line depth can then be written

$$\tau_\lambda \sim \int_0^\tau \frac{Nf}{\kappa} \frac{H(\alpha, v)}{\Delta\lambda_D} d\tau,$$

if the broadening of the line can be described by a Voigt-function $H(\alpha, v)$. ($\alpha = \gamma/2\Delta\lambda_D$, $\gamma =$ damping constant, $v =$ distance from line centre measured in Doppler-widths, κ continuous absorption coefficient, $N =$ number of absorbing atoms, $f =$ oscillator strength.)

Obviously, the behaviour of $H(\alpha, v)/\Delta\lambda_D$, i.e. the Voigt-function divided by the Doppler-width, has to be studied. In order to give the result in a lucid form, we change

M. Hack (ed.), Mass Loss from Stars. All rights reserved

over from the line depth to the equivalent width

$$w = \int_0^\infty d\tau \, g(\tau, \mu) \, 2\Delta\lambda_D \int_0^\infty dv \left[1 - \exp\{-\tau_\lambda/\mu\} \right]$$

and drop mathematical exactness by replacing τ_λ/μ by a convenient mean optical thickness $x/\Delta\lambda_D$.

In the case of

(a) pure Doppler broadening $H(\alpha, v)/\Delta\lambda_D = e^{-v^2}/\Delta\lambda_D$ we obtain for weak lines $(x \ll 1)$

$$w \sim x \left[1 - \frac{x}{2!\sqrt{2}} \frac{1}{\Delta\lambda_D} + \cdots \right],$$

and for strong lines $(x \gg 1)$

$$w \sim \Delta\lambda_D \left(\ln\{x/2\Delta\lambda_D\} \right)^{1/2}.$$

In the case of

(b) pure damping broadening in $H(\alpha, v)/\Delta\lambda_D$ the Doppler-width cancels out and w is independent of $\Delta\lambda_D$.

In the other regions of the Voigt-function numerical computation must be carried out.

The result is as follows: Microturbulence does not influence very weak lines and very strong lines with damping wings. The influence of microturbulence is strongly coupled with saturation in the lines: the stronger the line, the stronger is the influence of microturbulence. Moreover, a fixed x, that is a fixed number of absorbing atoms, will naturally give a greater equivalent width, if microturbulence increases. This result is not surprising, but it has been obtained up to now only for certain lines and certain atmospheric models.

The method can be generalized to the NLTE-case. The weighting function must be altered. Instead of one source-function in the LTE-case, two different source-functions for the continuum S_c and the line S_l must be considered. The weighting function then becomes

$$g(\tau, \mu) \sim \left[\frac{dS_l}{d\tau} + \frac{S_c - S_l}{\mu} \right] \exp\{-\tau_\lambda/\mu\}.$$

This expression means that according to the mutual behaviour of the source-functions S_c and S_l the minus-sign can dominate and emission occurs, so that emission features in line profiles as discussed in this Colloquium may be described by this method. A trouble is, of course, how to determine the source-function in the case of a special line, but it is to be hoped that in simple cases this trouble will be overcome by the work which is still going on. A short remark on this will appear in *The Observatory*.

Discussion

Underhill: You have given a formal extension of the theory of the curve of growth using weighting functions to the case when the lines are formed by NON-LTE. This requires defining the weighting function in terms of a line source function S_l and $(dS_l)/(d\tau)$, as well as a continuum source function S_c. One must consider the particular details of line formation for each line separately to find S_l, thus I see little advantage to retaining the weighting function formalism. If you have S_l and its dependence on depth, it is a relatively simple matter to compute the line profile directly and compare with observed profiles. Secondly *in practice* a curve of growth is constructed for a star by using *many* lines of varying spectroscopic properties to form one curve. With simple theory of line formation this procedure may be justified as a first approximation. In the case of NON-LTE line formation, the hypothesis that all lines can be brought to one curve of growth becomes invalid. Thus there is no sense in proceeding as you suggest. You will obscure the physics of the situation.

Gussmann: The extension is not confined to the curve of growth method but is an exact method for computing line profiles and equivalent widths. The advantage is that the method will give some physical insight, because the weighting function in the LTE and NLTE cases has a clear physical significance: it means that fraction of the continuous intensity available for line absorption. The curve of growth method has only served as a simple example, it cannot be proposed for the general case, of course.

SESSION II

MASS LOSS FROM SINGLE STARS.
THEORIES

RECENT STUDIES OF MODELS FOR UNSTABLE STARS

WILLIAM K. ROSE

Physics Department and Center for Space Research,
M.I.T., Cambridge, Mass., U.S.A.

Abstract. Recent studies of stellar models have uncovered instabilities associated with helium and hydrogen shell burning. The present paper describe some calculations of stellar models that include these unstable stages and discusses some of their possible consequences.

The influence of neutrino emission becomes very important during the helium shell-burning stage for stars more massive than $1 M_\odot$. Calculations are described that illustrate the influence of neutrino emission on stellar evolution up to the onset of carbon burning.

1. Introduction

Although it is apparent that some mass loss from stars can be caused by physical phenomena such as the solar wind, which can be described without reference to the overall structure of the star, it has been suspected for many years that some forms of mass loss may arise as a direct consequence of instabilities that involve the overall structure of the star. For this reason, there has been a great deal of effort in recent years to identify and understand these instabilities by means of stellar interior calculations. It has been hoped that the theory of stellar interiors would provide a basis for understanding such observed forms of mass loss as planetary nebulae and novae as well as explain the widely held belief that most stars initially more massive than the Chandrasekhar mass limit are able to evolve into the white-dwarf state. The methods of stellar interior calculations will not lead to a detailed understanding of all forms of mass loss (e.g., enhanced stellar winds). However, they may explain why particular stars evolve into configurations where physical phenomena such as enhanced stellar winds are likely to occur.

Today, I wish to describe some calculations of unstable stellar models. The purpose of these calculations is to construct theoretical models of stars whose behavior imitates some of the observed behavior of stars that are believed to be undergoing the final phases of evolution.

2. Unstable Helium-Burning Shells and the Origin of Planetary Nebulae

Recent investigations of stellar models have indicated that helium shell-burning stars are generally thermally unstable (SCHWARZSCHILD and HÄRM, 1965; WEIGERT, 1966; ROSE, 1966). It has been shown that the presence of thermal instability leads to a series of relaxation oscillations. During each relaxation oscillation, the rate of nuclear-energy generation goes through a relatively short interval of rapid increase that is followed by a relatively long interval of relaxation. These relaxation oscillations, which begin while the helium-burning shell is still non-degenerate, are described in Figure 1.

M. Hack (ed.), Mass Loss from Stars. All rights reserved

RELAXATION   OSCILLATIONS

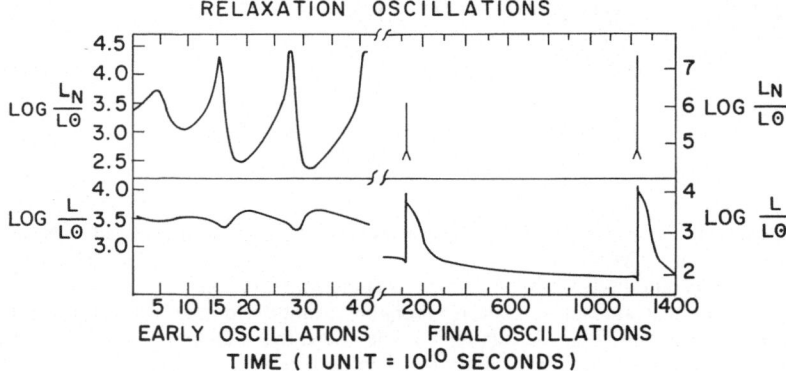

Fig. 1.   The nuclear-energy generation, L_N, and the luminosity, L, are shown as a function of time for a thermally unstable helium shell-burning star. The left-hand side of the figure shows a few of the early relaxation oscillations of the 0.75 $M_\odot$ star. The right-hand side shows the final two oscillations. The star contracts towards the white-dwarf state after the final oscillation.

The physical conditions necessary for the occurrence of thermal instability under non-degenerate conditions were first discovered and explained by Schwarzschild and Härm. Although detailed calculations of the equations that govern stellar evolution are necessary in order to prove the existence of thermal instability, the necessary conditions for thermal instability can be understood by means of reasonably straight-forward physical arguments. The first requirement for instability is that the source of nuclear energy be a shell source whose rate of energy generation is highly tem-perature-sensitive. In addition, it appears necessary that the initial stages of the instability take place under approximately isobaric conditions. Therefore, the unstable thermal mode that characterizes the instability cannot be homologous.

The calculated properties of thermal instability suggest that there is a physical similarity between thermal instability in stars and the theory of thermal explosions

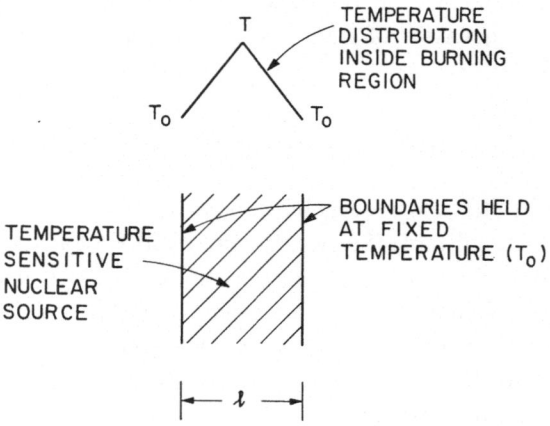

Fig. 2.   A simplified model for thermal instability is shown.

that is familiar to physical chemists. In order to pursue this point further, let us consider a highly temperature sensitive nuclear (or chemical) fuel that is contained inside a plane parallel region of space such as is shown in Figure 2. The temperature at the boundaries of this region is maintained at a constant value T_0, and it is assumed that changes in pressure and temperature occur under approximately isobaric conditions. If ε is the rate of nuclear energy generation in ergs/gm-sec inside the region, the equation of thermal conductivity becomes

$$\varrho c_p \frac{\partial T}{\partial t} = \sigma \frac{\partial^2 T}{\partial x^2} + \varrho \varepsilon, \tag{1}$$

where $\varepsilon = \varepsilon_0 \, e^{-\alpha/T}$. For the 3α process, $\varepsilon_0 = 3 \times 10^{11} \, \rho^2 x_4^3 \, f(10^8/T)^3$ (f, the correction for screening, is usually $\approx$ unity) and $\alpha = 44.1 \times 10^8$. For small variations in temperature, it is a good approximation to assume

$$\varepsilon \approx \varepsilon_0 \, e^{-\alpha/T_0} \, e^{\alpha(T-T_0)/T_0^2}. \tag{2}$$

The theory of thermal explosions asserts that the absence of stationary solutions to Equation (1) determines the necessary condition for a thermal explosion (i.e. instability implies that there are no solutions to Equation (1) if the left-hand side of the equation is put equal to 0). It can be shown (FRANK-KAMENETSKII, 1955) that this requirement leads to the condition

$$\lambda = \varrho \varepsilon_0 \, e^{-\alpha/T_0} \left(\alpha/T_0^2\right) \left(l/2\right)^2 / \sigma \geqslant \lambda_{cr} \approx 1 \tag{3}$$

for instability to take place. The above condition implies that if the mass of the fuel, size of the region and thermal conductivity are fixed, then instability (in this case explosion) is predicted if the temperature sensitivity of the fuel is sufficiently high. In the case of instability (i.e., $\lambda \geqslant \lambda_{cr}$) there exists no stationary temperature distribution (see Figure 2) that allows thermal energy to escape from the surface as rapidly as it is generated inside the region that contains the fuel.

It is not immediately obvious how conditions such as have been described above can be attained inside a star. The usual equation of radiative transfer in stellar interiors can be written

$$\frac{L}{4\pi r^2} = \frac{-4acT^3}{3\kappa\varrho} \frac{dT}{dr} = -\sigma \frac{dT}{dr}, \tag{4}$$

where σ = the effective thermal conductivity. The equation of energy conservation can be written

$$\tfrac{3}{2}\varrho^{2/3} \frac{d\left[P/\varrho^{5/3}\right]}{dt} = \varepsilon - \frac{dL}{dM_r} \tag{5}$$

where $\gamma = \tfrac{5}{3}$ for a fully ionized, non-relativistic gas, and $dM_r = 4\pi r^2 \varrho \, dr$. If the left-hand side of Equation (5) can be neglected and if r on the left-hand side of Equation (4)

can be assumed constant (a good approximation for a shell source), then Equations
(4) and (5) can be combined to give

$$\varepsilon = \frac{dL}{dM_r} = \frac{1}{4\pi r^2 \varrho}\frac{dL}{dr} = -\frac{1}{\varrho}\frac{d}{dr}\left(\sigma\frac{dT}{dr}\right) \qquad (6)$$

which reduces to Equation (1) if σ is approximately constant. The above arguments
suggest that Equation (3) represents an approximate necessary condition for thermal
instability. If the computed physical parameters for a helium shell-burning star that
is entering a stage of incipient instability are substituted into Equation (3), the condition for instability is found to be satisfied by a factor that is approximately unity.

The above discussion is, of course, only an approximate one. In testing an actual
stellar model for thermal instability, it is necessary to follow a more complicated
procedure. Having obtained a solution that satisfies the usual equations governing
stellar evolution (this solution can be described by the values for the dependent
variables P_i, T_i, L_i, R_i defined at a specific set of mass points labelled by the index i),
a new solution to the equations is sought by adding terms of the form $\partial P_i \, e^{t/\tau}$, $\partial T_i \, e^{t/\tau}$,
$\partial L_i \, e^{t/\tau}$, $\partial R_i \, e^{t/\tau}$ to the existing initial solution. If such a new solution can be found
with a positive τ, then the initial solution is said to be thermally unstable and τ, the
eigenvalue, characterizes the growth rate of the thermal instability. In carrying out
the calculations, the eigenvalue, τ, is determined first and then the corresponding
eigensolutions are calculated. If the eigensolutions were homologous (i.e.
$\partial P/P = -4\partial r/r$, $\partial \rho/\rho = -3\partial r/r$, $\partial T/T = -\partial r/r$), relatively small changes in temperature
would mean relatively large changes in pressure. Moreover, changes in temperature
and density would have similar sign. In his pioneering study of the secular (i.e.,
thermal) stability of main-sequence stars, Jeans assumed that the eigensolutions
were homologous. However, the eigensolutions for a thermally unstable helium
shell-burning star are not homologous. The calculations (ROSE, 1966) indicate
that $\partial P/P$ is small as compared with the corresponding changes in temperature.
Therefore, $\partial T/T \approx -\partial \rho/\rho$, if the instability takes place under non-degenerate conditions.
Of course, thermal instability can also take place when the helium-burning shell is
semi-degenerate. Moreover, the well-known helium core flash is also an example of a
thermal instability. If a thermal instability takes place under degenerate conditions,
then changes in temperature can take place without significant changes in either
density or pressure. This is true because the equation of state for degenerate matter
is nearly independent of the temperature.

SCHWARZSCHILD and HÄRM (1965) have shown by means of an evolutionary
sequence of $1\,M_\odot$ models that shortly after the onset of helium shell burning (for a
population II star this stage of stellar evolution is expected to occur after the star
has evolved across the horizontal branch), the helium shell becomes thermally unstable
and the star, which is a red giant, undergoes a series of relaxation oscillations. At this
point, the helium-burning shell is still non-degenerate. The appearance of thermal
instability during the helium shell-burning stage of evolution appears to be a very
general property of helium shell-burning stars (WEIGERT, 1966; ROSE, 1966, 1967).

HAYASHI *et al.* (1965) have obtained models for a star that contains both a helium- and hydrogen-burning shell. These solutions indicate that a star may evolve into a state such that helium shell burning is re-ignited under degenerate conditions. Calculations of models for stars with helium-burning shells but without hydrogen-burning shells (ROSE, 1966, 1967) indicate that for these stars, helium shell burning will not take place under degenerate conditions. However, if a star has both a helium-burning shell and a hydrogen-burning shell, it would appear possible for helium shell burning to occur under degenerate conditions. The calculations indicate that this type of solution may be reached if the helium-burning shell evolves more rapidly toward the surface than the hydrogen-burning shell and thereby reaches sufficiently low temperatures for helium burning to stop. The helium-burning shell can evolve more rapidly than the hydrogen-burning shell primarily because of the much lower energy content of helium (6×10^{17} erg/gm) as compared with hydrogen (6×10^{18} erg/gm). After helium burning has stopped, the mass interior to the hydrogen-burning shell will contract and the hydrogen-burning shell must supply most of the luminosity that is required to support the structure of the star. As the hydrogen-burning shell advances closer to the surface and the mass interior to the hydrogen shell contracts, the temperature at the position of the helium-burning shell eventually begins to increase. It continues to increase until helium is re-ignited under conditions of higher density. Because the calculations of HAYASHI *et al.* (1965) bypassed the earlier unstable helium shell-burning stage studied by SCHWARZSCHILD and HÄRM (1965) it is not obvious that their solutions are relevant.

Calculations by ROSE (1968b) have indicated that in the somewhat idealized case of a $0.6\,M_\odot$ double shell source star with a $0.01\,M_\odot$ hydrogen-rich envelope, the star evolves through a series of relaxation oscillations under non-degenerate conditions until helium burning stops. In this case, the extinction of helium burning occurs before mixing between the hydrogen-rich envelope and the interior regions of the star can take place. However, the re-ignition of helium occurs under semi-degenerate rather than degenerate conditions and does not lead to mass loss. Unfortunately the very small mass of the hydrogen-rich envelope made it impossible for the star to evolve through more than one such cycle, and therefore, helium shell burning did not occur under degenerate conditions for this $0.6\,M_\odot$ star.

The above calculations suggest the following picture for the formation of planetary nebulae. A helium shell-burning red giant star undergoes a series of perhaps 20–30 relaxation oscillations that take place while the helium-burning shell is still non-degenerate. These early oscillations are followed by a smaller number of oscillations that take place under conditions of ever increasing density (see Figure 1). The higher densities inside the burning regions lead to higher rates of nuclear-energy generation. The present picture for the formation of planetary nebulae assumes that the final rates of nuclear-energy generation are sufficiently high so as to lead directly to the ejection of a shell of mass from the red giant. Although the presence of significant mixing between the hydrogen-rich envelope and the interior of the star (SCHWARZ-SCHILD and HÄRM, 1965) would greatly complicate the calculation of the final stages of helium burning, it should be emphasized that one or perhaps several periods characterized by very high rates of nuclear-energy generation are predicted by the calculations. Calculations which hope to clarify the picture of the final phases of helium burning are now in progress. It has not yet been proven that the rates of nuclear-energy generation attained during the final stage of helium burning are sufficiently violent to lead directly to the ejection of a shell of mass such as is observed in a planetary nebula.

At this point in our discussion, it is appropriate to ask under what conditions nuclear burning can lead directly to mass loss. In discussing the physical basis for thermal instability, I compared thermal instability to the theory of thermal explosions. However, the presence of thermal instability in a star does not necessarily mean that a hydrodynamic event such as would be required to produce mass loss will take place. The calculations that have been discussed above indicate that the initial stages of thermal instability when the helium-burning shell is non-degenerate will not lead to a hydrodynamic event in a red giant star. The calculations that have been carried out so far indicate that for a thermal instability to lead to a hydrodynamic event in a red giant, it is necessary that the e-folding time associated with the development of the thermal instability becomes shorter than the fundamental pulsation period of the star. The e-folding time for the development of thermal instability, which can be estimated from Equation (5), is

$$\tau \approx \frac{\frac{3}{2}P/\varrho}{\varepsilon}, \tag{7}$$

where ε is the rate of helium burning evaluated at the mass point with the highest rate of nuclear-energy generation and the internal energy density ($\frac{3}{2}P/\varrho$) is about 2×10^{16} inside the helium-burning shell. The above necessary condition for a hydrodynamic event means that very high rates of nuclear-energy generation are necessary. If it is required that $\tau < 100$ sec, then Equation (7) implies that ε must be about 2×10^{14} ergs/gm-sec.

The calculations indicate that such high rates of nuclear-energy generation can arise only if the helium shell is degenerate during the initial stages of the thermal instability. For such high rates per gm of nuclear-energy generation to be attained, it appears necessary that the total rate of nuclear-energy generation become $\approx 10^{10}$–10^{11} $L_\odot$. Numerical experiments that employ implicit hydrodynamics have been carried out for helium shell-burning models with a pure helium envelope. These numerical experiments indicate that if such high rates of nuclear-energy generation are attained in a blue star, then an efficient ($> 1\%$) conversion of nuclear energy into mechanical energy takes place. It has not yet been shown, however, that such high rates of nuclear-energy generation are attained in the normal evolution of a star. Since the observed kinetic energy of a planetary nebula is about 10^{45} ergs and the total nuclear-energy release in a thermal instability is about 10^{49} ergs, it would appear that it is necessary to convert about one part in 10^4 of the available nuclear energy into kinetic energy if the above mechanism is to lead to the formation of a planetary nebula.

The observations of the nuclei of planetary nebulae provide some evidence that the above picture for their formation may be correct. Calculations have shown that the observed high luminosities and implied short lifetimes for the nuclei of planetary nebulae can be understood if it is assumed that a thermal instability has taken place at about the time the nebula was ejected from the central star (ROSE, 1967). There is some evidence that most stars of low and moderate mass evolve through the planetary nebula stage immediately before entering the white-dwarf state. If the typical mass for the nucleus of a planetary nebula is the same as the typical mass of a white dwarf (i.e. $0.7-0.8 M_\odot$), unstable helium and/or hydrogen burning at the time of the formation of a planetary nebula would appear to be the only plausible mechanism considered up to the present that can explain the short lifetimes indicated by the observations. Figure 1 shows a series of low-amplitude relaxation oscillations for a $0.75 M_\odot$ star that are followed by two oscillations of much larger amplitude. The path of the star in the H-R diagram during the final oscillation follows a track that approximates to the indicated track for the nuclei of planetary nebulae (O'DELL, 1963; SEATON, 1966).

3. Unstable Hydrogen Shells and the Origin of Novae

Astronomers have recognized for some time that most novae are members of close binary systems (STRUVE, 1955). KRAFT (1964) has shown that a number of novae are components of close binary systems that contain a blue component and a red component. The observations suggest that mass accretion is taking place from the red component onto the surface of the blue component.

It is of interest to investigate whether or not thermal instabilities such as have been discussed above may be associated with a nova outburst. In the preceding section, it was concluded that thermal instabilities would probably not lead to hydrodynamic events in red giants unless the nuclear-energy generation took place under degenerate conditions. Even in the latter case, it is not obvious that a hydrodynamic event will result. However, calculations (ROSE, 1967) indicate that thermal instability can lead to pulsational instability in a hydrogen deficient star even if the instability arises under non-degenerate conditions. There are at least two reasons why unstable helium burning is probably not the cause of most nova outbursts. First, the binary nature of most novae is not explained. Second, the recurrence times and thermal relaxation times that are possible if helium burning is the cause of the instability are much too long to explain the corresponding time scales observed for novae. On the other hand, instabilities due to unstable hydrogen shells could arise as a direct consequence of small amounts of mass accretion. Moreover, the required mass in the outer layers of these stars is so small that very short thermal relaxation times are expected.

Although hydrogen-burning shells are generally thermally stable, it has been shown (GIANNONE and WEIGERT, 1967; ROSE, 1968a) that small amounts of mass accretion can lead to thermally unstable hydrogen shell burning. The calculations for a $0.75 M_\odot$ star (ROSE, 1968a) indicate that as mass accretion takes place onto the surface of a hot blue star that is initially burning hydrogen under stable conditions, the hydrogen burning increases and eventually the hydrogen shell becomes thermally unstable. The presence of the thermal instability leads the hot white dwarf into a state such that it becomes pulsationally unstable. It is quite possible, although not yet proven, that pulsational instability in a star of this type can lead to the ejection of a shell of mass such as is observed during a nova outburst. The linear theory of pulsations indicates that approximately 10^{45} ergs may be reasonably expected to go into pulsations for the star under discussion. This amount of energy is comparable to that generally associated with a nova outburst (ARP, 1956). Moreover, the mass of the hydrogen-rich envelope that is above the hydrogen-burning shell is about $2 \times 10^{-4} M_\odot$. This is the approximate amount of mass that is estimated to be ejected during a nova outburst.

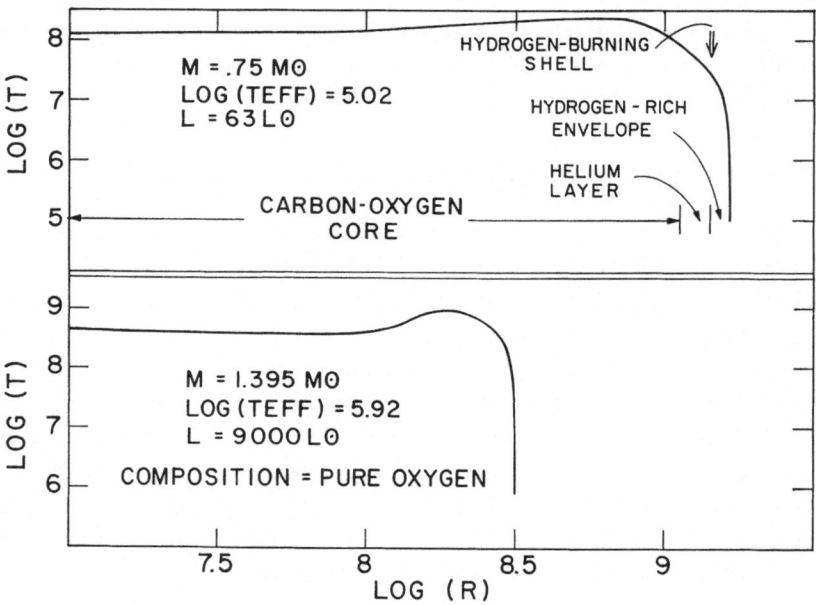

Fig. 3.   The temperature (T) is shown as function of radius (R) for a 0.75 $M_\odot$ model and a 1.395 $M_\odot$ model. The 1.395 $M_\odot$ model has a very high effective temperature (T_{eff}) and a very steep rise in temperature near the surface.

Figure 3 shows the temperature distribution of the outer layers of the initial 0.75 $M_\odot$ model used in carrying out the calculations. The high-temperature sensitivity of the CN cycle means that hydrogen burning will take place in a thin shell whose mass at the onset of thermal instability ($\approx 2 \times 10^{-6} M_\odot$) is a small fraction of the hydrogen-rich envelope. In the previous section, an approximate expression for the e-folding time characterizing a thermal instability was given (Equation (7)) and it was pointed out that this time scale becomes shorter as the rate of nuclear-energy generation increases. For the 0.75 $M_\odot$ hydrogen-burning models under consideration, this time scale becomes as short as $\approx 10^5$ sec. Since the dynamical time scale of the star (i.e. the period of the fundamental mode of oscillation) is much less (≈ 30 sec), the presence of thermal instability does not lead directly to an explosive event. However, the calculations indicate that the star becomes pulsationally unstable for a sufficient time for significant amounts of nuclear energy to be converted into mechanical energy of pulsation.

Figure 3 also shows the temperature distribution of a $1.395 \, M_\odot$ pure oxygen star that is contracting toward the white-dwarf state. The effective temperature of this star is very high (8×10^5 K) and its radius is very small (3×10^8 cm). Although the total luminosity is $\approx 10^4 \, L_\odot$, the amount of energy radiated in the visual region of the spectrum is about $1 \, L_\odot$. The extremely steep temperature gradient near the surface of the model (the mass exterior to the temperature at which hydrogen burning would commence is about $10^{-10} \, M_\odot$) suggests that if mass accretion were taking place onto the surface of this star, the e-folding times associated with the resulting thermal instability might become sufficiently short for the thermal instability to lead directly to an explosive event.

In addition, the recurrence time for explosive events of this type might be very short (i.e. of the order of days). It is perhaps possible that certain X-ray sources such as Scorpius XR1, which is observed to undergo rather sudden outbursts, might be a star of this type. Shock waves resulting from frequent explosions would produce the variable X-ray flares. A more stable ultraviolet flux whose total energy should be considerably greater than that associated with the X-ray emission is predicted. Moreover, the mass of this very hot white dwarf must be rather close to the Chandrasekhar mass limit.

4. Neutrino Emission, Carbon Burning and Mass Loss

It is generally believed that most stars that are initially more massive than the Chandrasekhar mass limit lose sufficient mass that so they are able to evolve into the white-dwarf state. Probably the most striking property of massive stars is their ability to evolve into configurations of very high interior density and temperature. For this reason, it is obvious that those of us who study the interiors of stars should search for instabilities and other physical properties of stars that arise because the star can reach configurations characterized by high interior density and temperature.

Although there is still no direct experimental evidence to prove the existence of a universal Fermi interaction, most theoretical physicists believe that the indirect evidence is sufficiently strong that its existence is highly probable. PONTECORVO (1959) and CHIU and MORRISON (1960) were the first to point out that the neutrino-emission processes that are predicted by a universal Fermi interaction should have a profound influence on the advanced stages of stellar evolution. The influence of neutrino emission on the subsequent evolution of sufficiently massive stars (i.e. $M \geqslant 1 \, M_\odot$) becomes decisive during the helium shell-burning stage of evolution. Evolutionary calculations of models for a $5 \, M_\odot$ star by WEIGERT (1966) have made this point particularly clear. Because the effective mean free path of a neutrino is generally much greater than the radius of the star, neutrino emission removes thermal energy directly from the interior. If the interior is degenerate (as is usually the case during the helium shell-burning stage) the removal of thermal energy from the interior leads to a reduced central temperature, and consequently the onset of carbon burning is delayed.

Evolutionary sequences of models of helium shell-burning stars of $1.45 M_\odot$ and $2.2 M_\odot$ have been computed. The envelopes of these models were assumed to be pure helium. These calculations show that the neutrino emission rates are sufficient to invert the interior temperature distributions of these stars and thereby delay the onset of carbon burning. Moreover, the high neutrino rates, which are concentrated primarily in the semi-degenerate regions of the star, speed up the contraction of the mass of the star interior to the helium-burning shell. When the luminosity of the star has exceeded about $25000 L_\odot$, the star evolves to the red giant branch even though the envelopes are hydrogen deficient. The evolution to the red giant branch (the luminosities of these stars imply that they are supergiants) is a consequence of the discontinuity in molecular weight that exists between the carbon-oxygen cores and helium envelopes of these stars in the same way that the discontinuity in molecular weight between the hydrogen-rich envelopes and helium cores of more conventional stars is responsible for their evolution to the red giant branch. The possible existence of hydrogen-deficient red giants may be of some importance in ascertaining how a star loses mass. Whatever the mechanism for mass loss, the energy requirements are greatly reduced if the star is a red giant. The possible existence of hydrogen-deficient red giants means that helium as well as hydrogen can be removed from a star with a relatively modest expenditure of energy.

The existence of neutrino-emission processes means that for a given mass, a star can evolve into a configuration requiring a higher luminosity than would otherwise be possible. For sufficiently high luminosity, the force exerted on an optically thin, ionized layer surrounding a star by radiation pressure will exceed that due to gravity. It can readily be shown that the condition for radiation pressure to dominate gravity depends only on the ratio L/M ($L/M \gtrsim 3 \times 10^4 L_\odot/M_\odot$ for hydrogen and $L/M \gtrsim \gtrsim 6 \times 10^4 L_\odot/M_\odot$ for helium). The calculations indicate that as the helium-burning shell advances toward the surface and the mass of the carbon-oxygen core increases, the luminosity of the star increases. The computed luminosities become sufficiently high for instability due to radiation pressure to arise when the mass of the core exceeds ≈ 1.1–$1.3 M_\odot$.

In Figure 4, the temperature, density, and radius are shown for a $2.2 M_\odot$ helium shell-burning star as a function of q, the mass fraction interior to a given mass point. The mass of the core (i.e. the mass interior to the helium burning shell) is $\approx 1.3 M_\odot$. The luminosity is $\approx 10^5 L_\odot$. The very high luminosities that characterize the computed models are a consequence of several factors: neutrino emission, which is concentrated in the semi-degenerate region (see Figure 4), the high molecular weight of the core, the convective envelope, and the presence of a nuclear-shell source rather than a nuclear-core source. If neutrino emission takes place in a region of the star where the gas is non-relativistic, the luminosity of the star is increased. This is the case because

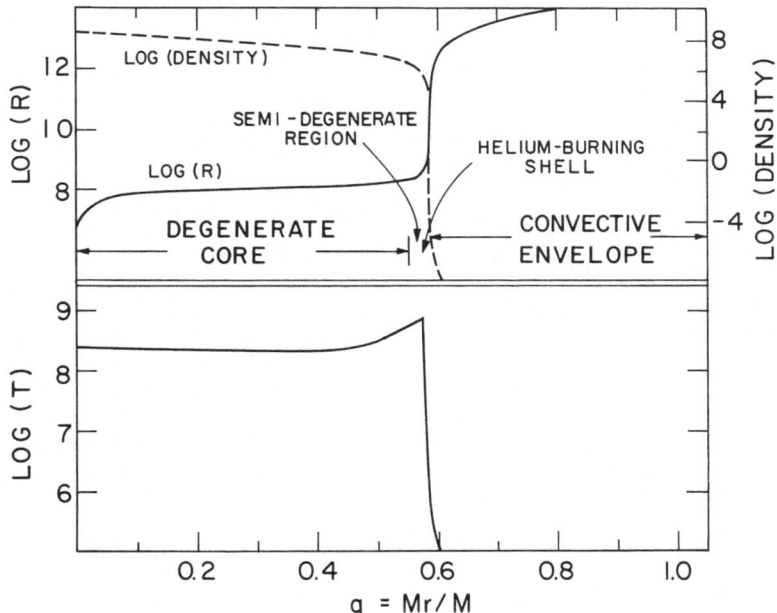

Fig. 4. The temperature (T), radius (R) and density are shown as a function of q, the mass fraction, for a very luminous 2.2 $M_\odot$ helium shell-burning model.

the removal of thermal energy by neutrino emission increases the rate of contraction.

Although the central density of the model described by Figure 4 exceeds 10^8 gm/cm^3, the central temperature has never exceeded that required to ignite carbon. However, the calculations show that in the subsequent evolution of the star, the central temperature increases until carbon burning begins at the center. The onset of carbon burning occurs when the central density is $\approx 2 \times 10^9$ gm/cm^3. The ignition of carbon burning under conditions of extreme degeneracy is expected to lead to an explosive event. Since the available nuclear energy ($\approx 10^{51}$ ergs) exceeds the binding energy of the star ($\approx 2 \times 10^{50}$ ergs) it is possible that the ignition of carbon burning under the calculated conditions could shatter the star.

It should be pointed out that other calculations for carbon burning models of similar mass (RAKAVY et al., 1967; KUTTER and SAVEDOFF, 1968; SUGIMOTO et al., 1968) have indicated that carbon burning can commence under conditions of much lower density (10^5–10^6 gm/cm^3). The difference between the two sets of calculations arises because the present calculations include the helium-burning stage of stellar evolution prior to the onset of carbon burning, and for this reason the lifetime of the star is greatly increased. The calculations referred to above assume that contraction is the only energy source prior to the onset of carbon burning. It should be emphasized that the inclusion of hydrogen burning would serve to strengthen the conclusion that the onset of carbon burning can take place under conditions of extreme degeneracy. A more detailed description of models for helium shell-burning stars that are more massive than the Chandrasekhar mass limit will be published elsewhere (ROSE, 1968c).

References

ARP, H.: 1956, *Astron. J.* **61**, 15.
CHIU, H. and MORRISON, P. 1960, *Phys. Rev. Letters* **5**, 573.
FRANK-KAMENETSKII, D. A.: 1955, *Diffusion and Heat Exchange in Chemical Kinetics*, Princeton University Press, Princeton, N.J.
GIANNONE, P. and WEIGERT, A.: 1967, *Z. Astrophys.* **67**, 41.
HAYASHI, C., HŌSHI, R., and SUGIMOTO, D.: 1965, *Progr. Theoret. Phys.* **34**, 885.
KRAFT, R.: 1964, *Astrophys. J.* **139**, 457.
KUTTER, G. and SAVEDOFF, M.: 1968, Preprint.
O'DELL, C.: 1963, *Astrophys. J.* **138**, 67.
PONTECORVO, G.: 1959, *Soviet Phys. – J.E.T.P.* **9**, 1148.
RAKAVY, G., SHAVIV, G., and ZINAMON, Z.: 1967, *Astrophys. J.* **150**, 131.
ROSE, W. K.: 1966, *Astrophys. J.* **144**, 838.
ROSE, W. K.: 1967, *Astrophys. J.* **150**, 193.
ROSE, W. K.: 1968a, *Astrophys. J.* **152**, 245.
ROSE, W. K.: 1968b, in I.A.U. Symposium 34: *Planetary Nebulae* (ed. by D. E. Osterbrock and C. R. O'Dell), Reidel Publ. Co., Dordrecht, Holland, pp. 390–395.
ROSE, W. K.: 1968c, *Astrophys. J.* (in press).
SCHWARZSCHILD, M. and HÄRM, R.: 1965, *Astrophys. J.* **142**, 855.
SEATON, M.: 1966, *Monthly Notices Roy. Astron. Soc.* **132**, 113.
STRUVE, O.: 1955, *Sky and Tel.* **14**, 275.
SUGIMOTO, D., YAMAMOTO, HOSHI, R., and HAYASHI, C.: 1968, *Progress of Theoretical Physics* **39**, 1432.
WEIGERT, A.: 1966, *Z. Astrophys.* **64**, 395.

Discussion

Dallaporta: In relation to the instability of the burning shells, when an outburst occurs and expels material, is it possible to foresee whether this emission will stabilize the situation, so that no other outbursts will occur, or instability will continue so that a series of similar outbursts is to be expected?

Rose: This is a very interesting question. Unfortunately the calculations that have been completed up to the present time are not sufficiently detailed to make any clear predictions with regard to how many shells of mass should be ejected. However, the calculations do suggest that under some conditions helium shell instabilities may lead to the ejection of more than one shell of matter. O'Dell has pointed out that the observations indicate that for some planetary nebulae more than one shell of matter has been ejected from the central star.

Kutter: You have stated that thermal instabilities in He-burning stars may lead to planetary nebula ejection. Have you been able to estimate from the energy input and the time interval of this input how much mass will be ejected and the speed of ejection?

Rose: The expected ejection velocity would be about the escape velocity from a luminous red giant (i.e., ≈ 20 km/sec).

POSSIBILITY OF MASS LOSS IN THE RED-GIANT STAGE
FROM H-R DIAGRAMS OF GALACTIC CLUSTERS

G. BARBARO, N. DALLAPORTA, and G. FABRIS

Istituto di Fisica dell'Università, Padova, Italy

Abstract. A statistical research on evolved stars beyond hydrogen exhaustion is performed by comparing the H-R diagrams of about 60 open clusters with a set of isochronous curves without mass loss derived from Iben's evolutionary tracks and time scales for Population I stars. Interpreting the difference in magnitude between the theoretical positions thus calculated and the observed ones as due to mass loss, when negative, the results indicate that this loss may be conspicuous only for very massive and red stars. However, a comparison with an analogous work of Lindoff reveals that the uncertainties connected with the bolometric and colour corrections may invalidate by a large amount the conclusions which might be drawn from such research.

1. Although evidence concerning mass loss from stars of different types and especially red giants and supergiants is gradually increasing, still not much is known about the real causes and the quantitative aspects of this phenomenon, so that up to now little can be foretold concerning its bearing on stellar evolution. For this reason, it was considered worthwhile to try to increase our information by some statistical evidence which could follow from comparing the position in the H-R diagrams of red giants belonging to clusters of stars having a known initial mass, with the theoretical position they should have, according to evolutionary tracks calculated with no account of mass loss.

To this aim, a sample of about 40 clusters, already considered by us in a preceding work (BARBARO *et al.*, 1967) (quoted as paper 1), aiming to compare the theoretical and observed evolutionary paths on the H-R diagram, has been amplified to include altogether 62 clusters of different ages for which sufficient data are available. The material thus gathered is presented in Table I.

Since several new theoretical evolutionary tracks have been recently computed, in order not to overcrowd the pattern used for comparison with the data, we have considered only the homogeneous and extended ensemble of trajectories obtained by IBEN (1965, 1966, 1967) from the main sequence up to the post-helium-burning contracting stage for 2.25; 3; 5; 9 and 15 $M_\odot$ stars; to this group we have added Stothers' evolutionary track for a star of 30 $M_\odot$ (STOTHERS, 1963, 1964, 1966), which fits well with the Iben ones and completes the net for large masses. For the special purpose of our research, we have taken account of the tracks only up to the starting-point of He burning for stars with mass lower than 9 $M_\odot$, while for the 15 and 30 $M_\odot$ cases the trajectories have been used for the whole length available. With these limitations, it is then easy to interpolate evolutionary curves with other mass values between the calculated ones.

Comparison between theoretical and experimental positions in the H-R plane requires, as is well known, the transformation from the set of the L-T_e coordinates

M. Hack (ed.), Mass Loss from Stars. All rights reserved

TABLE I

Cluster	t_m	t_{inf}	t_{sup}	References
Group A:				
IC 2944	5.46×10^6			(1)
NGC 4755	7.0×10^6	5.46×10^6	1.0×10^7	(2), (3)
NGC 3293	7.5×10^6	5.46×10^6	1.0×10^7	(4)
NGC 1893	7.5×10^6	5.46×10^6	1.0×10^7	(5), (6)
I Gem.	1.0×10^7	7.5×10^6	1.2×10^7	(7)
h and χ Persei	1.2×10^7	7.0×10^6	1.6×10^7	(8)
NGC 457	1.2×10^7	1.0×10^7	1.5×10^7	(9)
Pup. III	1.2×10^7	1.0×10^7	1.7×10^7	(10)
Ara I	1.25×10^7	1.0×10^7	1.5×10^7	(11)
NGC 581	1.45×10^7	1.0×10^7	1.7×10^7	(12)
Cr 121	2.0×10^7	1.2×10^7	2.8×10^7	(13)
α Per. cl.	2.2×10^7	1.7×10^7	2.8×10^7	(14)
NGC 3766	2.2×10^7	1.7×10^7	3.0×10^7	(15)
Group B:				
NGC 2169	3.7×10^7	1.7×10^7	4.7×10^7	(6)
NGC 2571	4.0×10^7	2.2×10^7	7.0×10^7	(16)
NGC 2422	5.0×10^7	3.7×10^7	7.0×10^7	(17)
NGC 7654	5.0×10^7	4.0×10^7	6.0×10^7	(6), (18)
Tr 18	5.6×10^7	4.6×10^7	7.0×10^7	(19)
NGC 129	6.0×10^7	5.0×10^7	7.0×10^7	(20), (21)
NGC 6405	6.0×10^7	5.0×10^7	7.0×10^7	(22), (23), (24)
NGC 6087	6.0×10^7	5.0×10^7	7.0×10^7	(25), (26)
NGC 2323	6.0×10^7	5.0×10^7	7.0×10^7	(27)
NGC 5617	6.0×10^7	5.0×10^7	7.9×10^7	(28)
NGC 7790	7.0×10^7	5.1×10^7	8.0×10^7	(29)
IC 4725	9.0×10^7	5.0×10^7	1.2×10^8	(30), (31), (32)
NGC 6694	9.4×10^7	7.4×10^7	1.25×10^8	(6)
NGC 3330	1.0×10^8	7.0×10^7	1.25×10^8	(27)
NGC 2546	1.0×10^8	7.0×10^7	1.3×10^8	(19)
NGC 3114	1.0×10^8	7.0×10^7	1.7×10^8	(33)
NGC 6664	1.0×10^8	7.08×10^7	1.7×10^8	(34)
Group C:				
Tr 2	1.15×10^8	7.08×10^7	1.35×10^8	(6)
NGC 2516	1.2×10^8	9.2×10^7	1.5×10^8	(35), (27)
NGC 6124	1.25×10^8	7.8×10^7	1.6×10^8	(36)
NGC 2301	1.25×10^8	1.0×10^8	1.4×10^8	(6), (37)
NGC 2439	1.3×10^8	9.6×10^7	1.45×10^8	(38)
NGC 2451	1.35×10^8	9.6×10^7	1.6×10^8	(39)
NGC 6475	1.5×10^8	1.2×10^8	1.8×10^8	(6), (40)
Ho 17	1.5×10^8	1.2×10^8	2.0×10^8	(41)
NGC 1912	1.5×10^8	1.25×10^8	1.95×10^8	(42)
NGC 2670	1.6×10^8	1.25×10^8	2.5×10^8	(33)
NGC 5316	1.6×10^8	1.25×10^8	2.5×10^8	(43)
NGC 6705	1.7×10^8	1.25×10^8	2.15×10^8	(44)
NGC 2437	1.7×10^8	1.25×10^8	2.5×10^8	(17)
NGC 7063	1.75×10^8	1.25×10^8	2.5×10^8	(6)
NGC 2533	1.75×10^8	1.25×10^8	2.5×10^8	(16)
Mel 105	2.0×10^8	1.5×10^8	2.8×10^8	(15)
NGC 6709	2.0×10^8	1.6×10^8	2.5×10^8	(6)

Table I continued

Cluster	t_m	t_{inf}	t_{sup}	References
NGC 2287	2.05×10^8	1.55×10^8	2.5×10^8	(6), (45)
NGC 6494	2.3×10^8	2.0×10^8	2.8×10^8	(6)
NGC 1528	2.5×10^8	2.0×10^8	2.7×10^8	(6)
NGC 3532	2.5×10^8	2.0×10^8	2.9×10^8	(40)
NGC 4349	2.5×10^8	2.15×10^8	3.0×10^8	(46)
Group D:				
NGC 2447	2.7×10^8	2.5×10^8	3.0×10^8	(38)
NGC 7062	2.8×10^8	1.95×10^8	4.7×10^8	(47)
NGC 1662	2.8×10^8	2.4×10^8	3.0×10^8	(6)
NGC 2548	2.9×10^8	2.5×10^8	3.6×10^8	(48)
NGC 7209	3.4×10^8	2.5×10^8	4.3×10^8	(6)
NGC 2281	3.8×10^8	3.0×10^8	4.9×10^8	(48)
NGC 2360	4.4×10^8	3.0×10^8	5.8×10^8	(49)
NGC 6633	4.7×10^8	3.6×10^8	5.9×10^8	(50)
NGC 1907	5.9×10^8	4.4×10^8	7.0×10^8	(42)
NGC 2324	5.9×10^8	5.1×10^8	8.3×10^8	(49)

t_m = mean value of age determination.
t_{inf} = lower value of age determination.
t_{sup} = upper value of age determination.

References

(1) THACKERAY, A. D. and WESSLINK, A. J.: 1965, *Monthly Notices Roy. Astron. Soc.* **131**, 121.
(2) FEAST, M. W.: 1963, *Monthly Notices Roy. Astron. Soc.* **126**, 11.
(3) ARP, H. C. and VAN SANT, C. T.: 1958, *Astron. J.* **63**, 341.
(4) FEAST, M. W.: 1958, *Monthly Notices Roy. Astron. Soc.* **118**, 618.
(5) SERKOWSKI, K.: 1965, *Astrophys. J.* **141**, 1340.
(6) HOAG, A. A., JOHNSON, H. L., IRIARTE, B., MITCHELL, R. I., HALLAM, K. L., and SHARPLESS, S.: 1961, *Publ. U.S. Naval Observ.* **17**: 7.
(7) HARDIE, R. H., SEYFERT, C. K., and GULLEDGE, I. S.: 1960, *Astrophys. J.* **132**, 361.
(8) WILDEY, R. L.: 1964, *Astrophys. J. Suppl.* **8**, 439.
(9) PESCH, P.: 1959, *Astrophys. J.* **130**, 764.
(10) WESTERLUND, B. E.: 1963, *Monthly Notices Roy. Astron. Soc.* **127**, 71.
(11) WHITEOAK, J. B.: 1962, *Monthly Notices Roy. Astron. Soc.* **125**, 105.
(12) PURGATHOFER, A.: 1961, *Z. Astrophys.* **52**, 22.
(13) FEINSTEIN, A.: 1967, *Astrophys. J.* **149**, 107.
(14) MITCHELL, R. I.: 1960, *Astrophys. J.* **132**, 68.
(15) SHER, D.: 1965, *Monthly Notices Roy. Astron. Soc.* **129**, 237.
(16) LINDOFF, U.: 1968, *Arkiv Astron.* **4**, 587.
(17) SMITH, M. J. and NANDY, K.: *Publ. Roy. Observ. Edinburgh* **3**, 2.
(18) PESCH, P.: 1960, *Astrophys. J.* **132**, 689.
(19) FERNIE, J. D.: 1963, *Observatory* **83**, 33.
(20) ARP, H., SANDAGE, A., and STEPHENS, C.: 1959, *Astrophys. J.* **130**, 80.
(21) LENHAM, A. P. and FRANZ, O. G.: 1961, *Astron J.* **66**, 16.
(22) ROHLFS, K., SCHRICK, K. W., and STOCK, J.: 1959, *Z. Astrophys.* **47**, 15.
(23) EGGEN, O. J.: 1961, *Roy. Observ. Bull.* **27**, 61.
(24) TALBERT, F. D.: 1965, *Publ. Astron. Soc. Pacific* **77**, 19.
(25) FERNIE, J. D.: 1961, *Astrophys. J.* **133**, 64.
(26) LANDOLT, A. U.: 1964, *Astrophys. J. Suppl.* **8**, 329.
(27) BECKER, W.: 1960, *Z. Astrophys.* **49**, 168.

(28) LINDOFF, U.: 1968, *Arkiv Astron.* **4**, 471.
(29) SANDAGE, A.: 1958, *Astrophys. J.* **128**, 150.
(30) LANDOLT, A. U.: 1964, *Astrophys. J. Suppl.* **8**, 352.
(31) SANDAGE, A.: 1960, *Astrophys. J.* **131**, 610.
(32) WAMPLER, J., PESCH, P., HILTNER, W. A.: 1961, *Astrophys. J.* **133**, 895.
(33) LYNGÅ, G.: 1962, *Arkiv Astron.* **3**, 65.
(34) ARP, H. C.: 1958, *Astrophys. J.* **128**, 166.
(35) COX, A. N.: 1955, *Astrophys. J.* **121**, 628.
(36) THE, P. S.: 1965, *Contr. Bosscha Observ.* **33**, 1.
(37) GRUBBISSICH, C. and PURGATHOFER, A.: 1962, *Z. Astrophys.* **54**, 41.
(38) BECKER, W.: 1959, *Z. Astrophys.* **48**, 279.
(39) FEINSTEIN, A.: 1966, *Publ. Astron. Soc. Pacific* **78**, 301.
(40) KOELBLOED, D.: 1959, *Bull. Astron. Inst. Neth.* **14**, 265.
(41) LINDOFF, U.: 1968, *Arkiv Astron.* **4**, 493.
(42) LAVDOVSKII, V. V.: 1965, *Izv. Glav. Astron. Obs. Pulkove* **22**, 138.
(43) RAHIM, M. A.: 1966, *Astron. Nachr.* **289**, 41.
(44) JOHNSON, H. L., SANDAGE, A. R., and WAHLQUIST, H. D.: 1956, *Astrophys. J.* **124**, 81.
(45) COX, A. N.: 1954, *Astrophys. J.* **119**, 188.
(46) LOHMANN, W.: 1961, *Astron. Nachr.* **286**, 105.
(47) FENKART, R. P.: 1965, *Contr. Oss. Padova*, No. 181.
(48) PESCH, P.: 1961, *Astrophys. J.* **134**, 602.
(49) BECKER, W.: 1960, *Z. Astrophys.* **51**, 49.
(50) HILTNER, W. A., IRIARTE, B., and JOHNSON, H. L.: 1958, *Astrophys. J.* **127**, 539.

into the set of the M_v, B-V ones or vice versa; and such a change is always a very controversial problem as many choices exist for these colour and bolometric corrections. One generally never succeeds in obtaining an exact fit between the transformed theoretical main sequence and the experimental one, and the causes of this discrepancy are not plainly understood (BARBARO and FABRIS, 1968). The choice for the corrections is therefore somewhat arbitrary and it will of course influence more or less the different results. This situation, which has frequently been overlooked, must always be kept in mind when judging the results in any investigation comparing evolution theory with data, and we shall discuss later some effects arising from this fact.

In our present case, we have adopted a solution already used by SCHILD (1967), which has the advantage of bringing the two main sequences practically into coincidence. It consists in assuming the colour corrections due to HARRIS (1963) and the bolometric corrections due to JOHNSON (1966); however, for evolved red stars, both colour and bolometric corrections are taken from JOHNSON (1966) as no other complete set is available. The fitting of the two sequences in the M_v, B-V plane thus obtained may be seen in Figure 1.

2. While in paper (1), transformation of the M_v, B-V values of the H-R observational diagrams for the single clusters was operated in order to bring them in the theoretical L-T_e plane, it was found, owing to the increased number of clusters, more convenient to act in the reverse way; that is to transform the theoretical curves once for all in the experimental M_v, B-V plane, and thus compare directly with them the observed diagrams of clusters. The sample of evolutionary theoretical curves in the M_v, B-V plane thus deduced is presented in Figure 1.

From this network, isochronous curves may easily be derived with the help of the evolutionary time scales tabulated by Iben and Stothers, necessary for stars of given mass to reach definite points of the path. Figure 2 presents the network of such isochronous curves.

The age of any experimentally given cluster may now be found by fitting its H-R

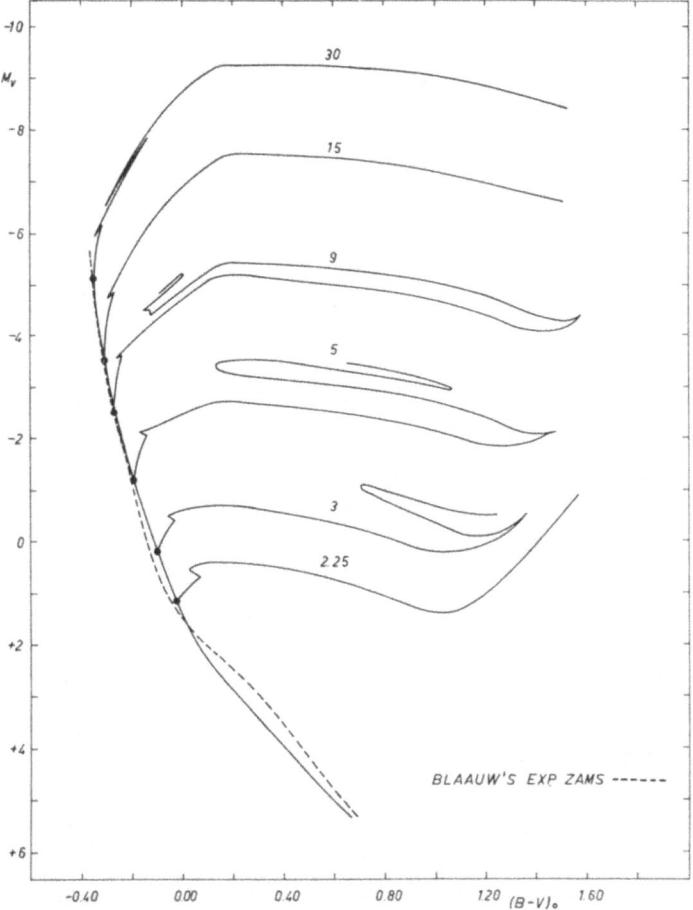

Fig. 1.   Theoretical evolutionary tracks with theoretical and experimental main sequences of zero age.

diagram with one of the possible isochronous curves. The procedure of course is far from being simple, as the stars forming the experimental diagram of a cluster do not appear aligned as the theoretical curves would require, but more or less dispersed around an average behaviour which one should try to guess and to make coincide with an isochronous curve. In many cases we have been rather convinced that it is just a matter of personal appreciation to prefer a given type of fitting to another. We considered therefore that it had not much physical sense to try an exact determination

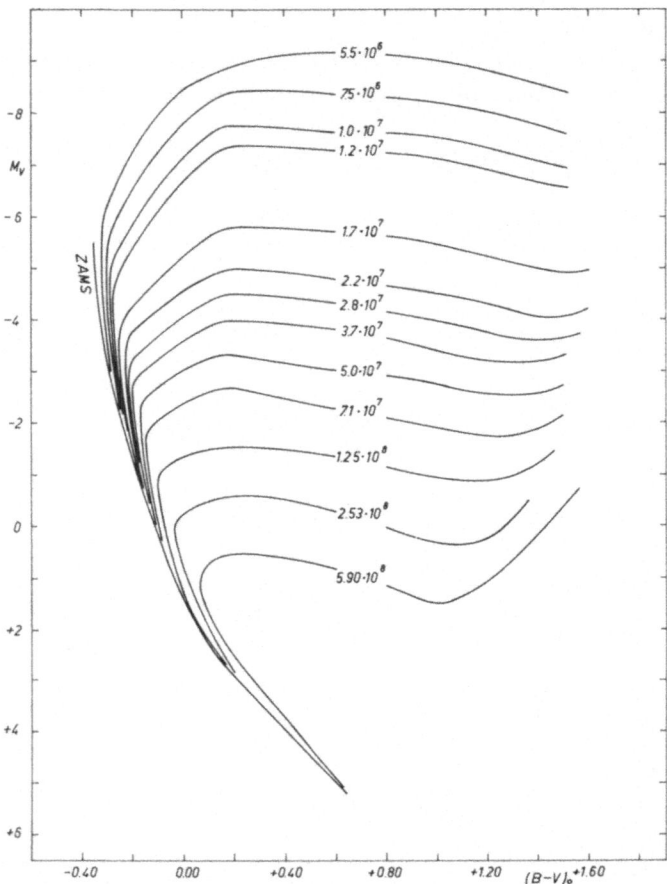

Fig. 2.   Theoretical isochronous curves.

and the only thing allowed by the experimental situation would be to evaluate the most extreme fittings and assume as age for the cluster a mean intermediate value with an error extending up to these outer limits. This has the effect that a given experimental cluster situated according to the stars of its upper main sequence and those just slightly evolving from it above the turning-point, has to be represented in the following portions of the H-R diagram not by a single track, but by a band whose width depends on the error done on the age. A red giant of the cluster inside this band may have evolved without mass loss and is inconclusive for mass-loss evidence; only stars whose representing point lies below this band should be evidence for effective mass loss.

At first sight, it was apparent that the criteria for determining the mean isochronous track corresponding to a given cluster and the uncertainty band around it could not be taken exactly similar for all clusters, but should depend on the age or the mass range of the cluster considered. Generally speaking, young clusters are characterized by a relatively well-defined upper sequence of high-mass stars; there is in fact little background for such sequences as large-mass stars are rare, so spurious stars may be easily discarded and little dispersion in *B-V* is present. The reverse situation is found for the older clusters: the much larger number of medium-mass stars which forms the upper part of their sequence is not easily distinguished from numerous non-cluster similar stars and the difficult task of separating them has been undertaken only in few cases; so their H-R diagram presents a rather confused and strongly dispersed situation.

Owing to these considerations, the clusters have been divided into four groups of different ages: A, B, C, D, according to a first-sight determination, corresponding roughly to the following intervals:

group A: $t < 2.2 \times 10^7$ year $M > 9\ M_\odot$
group B: 2.2×10^7 year $< t < 1 \times 10^8$ year $4.5\ M_\odot < M < 9\ M_\odot$
group C: 1×10^8 year $< t < 2.5 \times 10^8$ year $3\ M_\odot < M < 4.5\ M_\odot$
group D: 2.5×10^8 year $< t < 5.9 \times 10^8$ year $2.25\ M_\odot < M < 3\ M_\odot$,

M being the mass of the evolving stars. No older clusters have been examined for the reasons which will be referred to later.

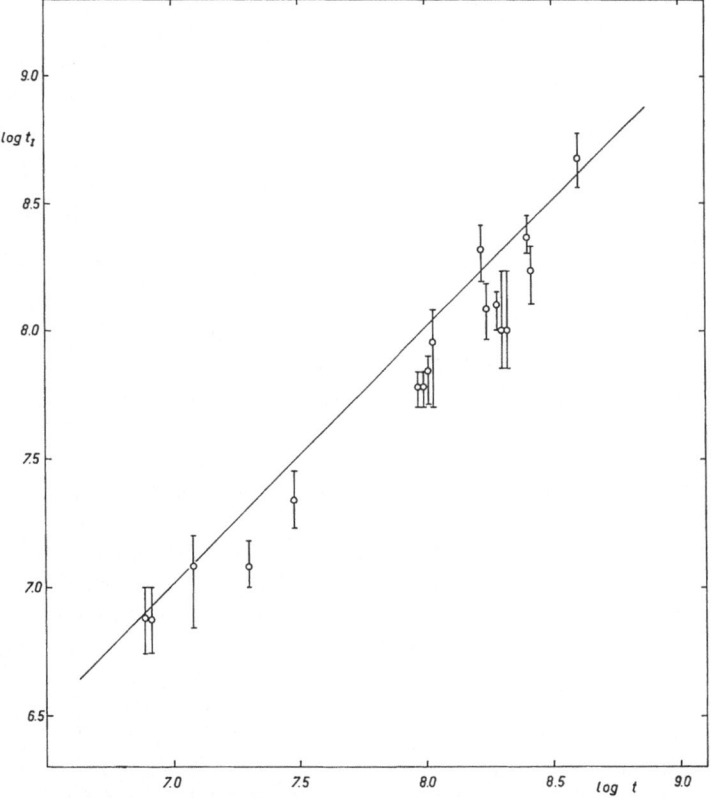

Fig. 3.   Present-age determinations of some clusters plotted vs. the age derived as in Paper (1).
Bars denote the indeterminacy in age.

For the first two groups, A and B, containing in the upper main sequence relatively few stars with moderate dispersion, the uncertainty of the age determination was considered to be measured by the Δt included between the two isochronous curves containing in the interval between them all or most of the stars of the upper end of the sequence. For groups C and D, the terminal aspect of the sequence comprises a large number of strongly dispersed stars; it was felt in this case that most of the dispersion is probably due to background stars, and that the age determination should mostly rely on the more luminous stars of the sequence, regardless of the great number of dispersed stars to the right of it, aiming to appear as evolved ones; so the uncertainty range was confined between two isochronous lines including only the stars of the top of the sequence.

In this way, the results of Table I have been obtained, where the detailed age limits for clusters of the four groups considered are reported.

At this point, one could wonder whether the age uncertainty for the cluster might correspond to a real physical situation, its most obvious sense being the formation time of the cluster. If this were the case, as on the average we should expect that the mean formation time should remain constant in time, we should observe a relative

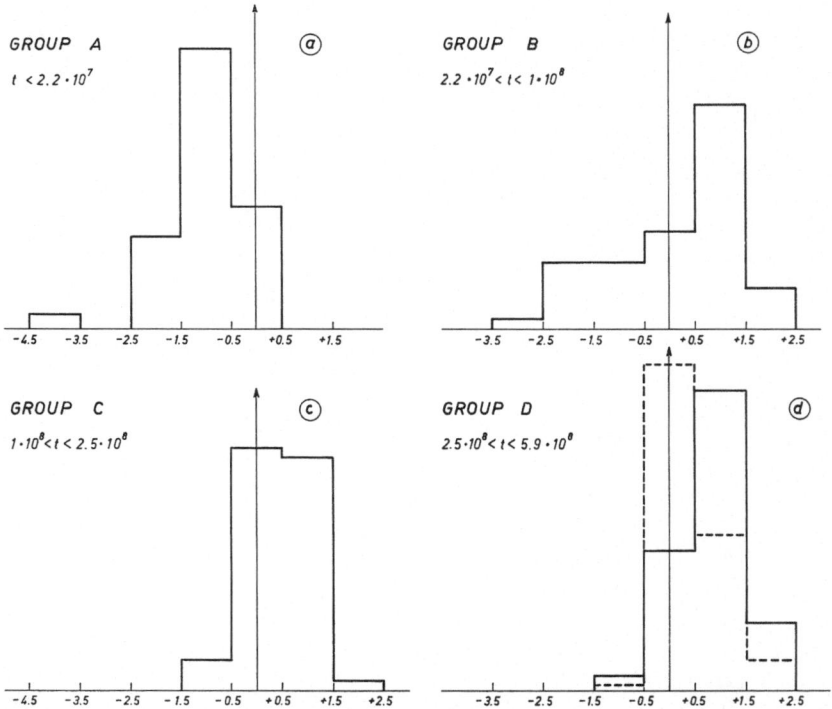

Fig. 4. ΔM_v distribution of stars for different cluster groups. ΔM_v is defined, for each evolved star of a cluster, as the difference between the observed magnitude and the magnitude relative to the mean isochronous curve for that cluster. Dashed line in Figure d refers to the lower limit of age for each cluster.

age uncertainty decreasing as age increases, the most dispersed clusters being the younger ones. Now it is exactly the reverse that we find in Table I, where the age uncertainty results as roughly proportional to the age itself and of the same order of magnitude. This is a clear indication that, should a physical uncertainty be present, it is small, and completely overwhelmed and masked by an instrumental uncertainty, arising from the errors in determining colours and magnitudes, from the spurious stars, and so on, whose average value increases with age, as older clusters, as already stated, are more difficult to analyze and in fact less analyzed than younger ones. This fact seems to set a limit to our capacity of investigating the present material related to objects beyond a certain age value. The results obtained are thus expected to become less and less significant as age increases, so that it appears useless to try, with the present data, to discuss old objects. Figure 3 correlates the age determinations now obtained with those given in paper (1), based on the value deduced from the turning-point; it clearly shows consistency between the two procedures as was in fact to be expected. However, the present paper yields also the uncertainty linked with any determination, and its amount strongly conditions the results in respect to the problem.

3. We shall now present the results obtained by following the procedure just outlined: in order to visualise them better, let us figure the difference of the real position of any red giant in respect to the position of the mean isochronous track corresponding to its cluster for the same B-V value by a difference in magnitude ΔM, with an error corresponding to the width of the band, positives ΔM representing stars above the isochronous track and negative ones, stars lying below it. We thus plot in Figures 4 (a, b, c, d) the number of red giants belonging to the different intervals of magnitude

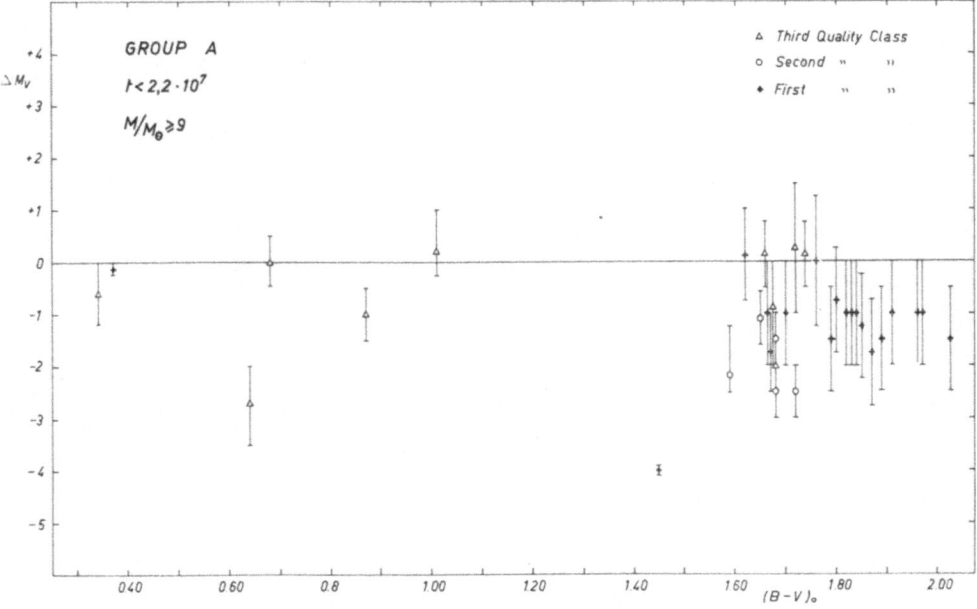

Fig. 5. ΔM_v as function of colour index for group A

ΔM for the four different groups of clusters A, B, C, D already defined, and in Figures 5–8 the mean magnitude differences ΔM of each star with their errors, as a function of their B-V for the same four different cluster groups.

From these diagrams the following conclusions can be drawn:

(A) Considering first Figures 4a and 5, related to group A, there appears to be a continuous distribution of stars from positions lying on the isochronous track to positions well below it, with a maximum frequency for $\Delta M = -1$; this effect may be interpreted as due to mass loss from the star, with a maximum frequency of the order of $\sim 20\%$. Of course, we are not able to say whether the continuous distribution of ΔM observed could depend on a really continuous distribution of masses lost, or the mass loss should be about the same for most stars in this range and the apparent continuous distribution due only to the uncertainty of our determinations. It is moreover evident from Figure 5 that this loss occurs mainly for highly evolved stars in the red supergiant stage, although the few dispersed points corresponding to stars crossing the Hertzsprung gap could well reveal that loss has still begun in earlier stages.

(B) For the second group of clusters B (Figures 4b and 6), while mass loss, although less conspicuous than in the previous case, is still at hand as shown by the number of stars with negative values of ΔM, a maximum for positive $\Delta M = +1$ appears as its characteristic feature. The points above the isochronous track could represent stars evolving on the loops existing in this mass range according to the work of both IBEN (1965, 1966, 1967) and the Göttingen group (HOFMEISTER et al., 1964a, b; KIPPENHAHN et al., 1964). These stars, as shown by Figure 6, possess mostly colour indices lying in the medium range; almost all Cepheids are included in this group and located

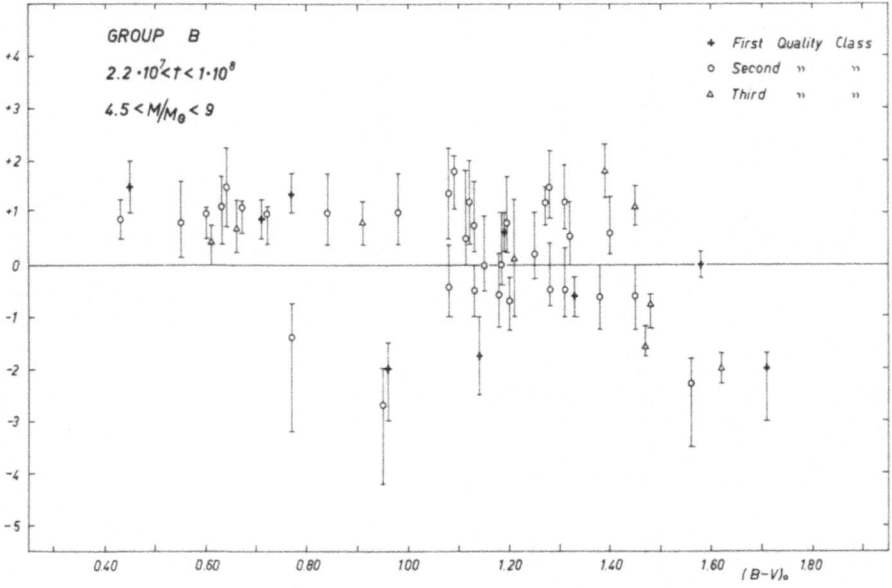

Fig. 6. ΔM_v as function of colour index for group B.

among them, in good agreement with the Göttingen group interpretation of the evolutionary situation of these variables.

(C) When we now turn to the third group of clusters (Figures 4c and 7), stars with negative ΔM are now rare, thus testifying the practical disappearance of mass loss for this mass range. Unexpectedly instead, the $\Delta M = +1$ group is still present and with comparable population to the $\Delta M = 0$ group. Part of it may still be explained, at least for the heavier masses of the group, as due to stars on the loops, which according to Iben do still exist down to 3 $M_\odot$. However, it seems likely that the high number of points in the $\Delta M = +1$ interval is due to some fact connected with the increasing disagreement between theoretical isochronous tracks and experimental clusters: in effect as mass decreases the shapes of the two kinds of curves become increasingly different, so that when fitting is obtained for the upper part of the main sequence and the initial part of the evolutionary track, the observed branches of the cluster

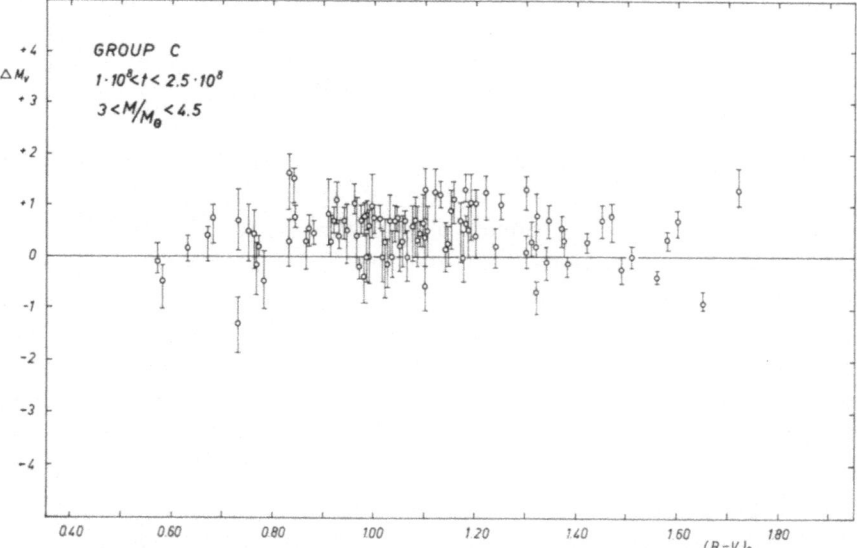

Fig. 7. ΔM_v as function of colour index for group C.

appears to lie systematically higher than the isochronous theoretical curves. The reason for this disagreement, already put forward in paper (1) is not clearly understood up to now; perhaps the exact position and the slope of the red-giant branch, which begins in this interval range to be compelled to follow the Hayashi limit, is rather sensitive to composition, and one should take care, as mass decreases, that in fact older stars are considered while they are still compared with theoretical curves calculated with the same standard composition used for the younger clusters. We are therefore inclined to attribute part of the points lying in the $\Delta M = +1$ range as an apparent effect due to the more rapid increase in luminosity of the giant branch in respect to the theoretical curve.

(D) Similar considerations can be expressed for the fourth group, D, corresponding

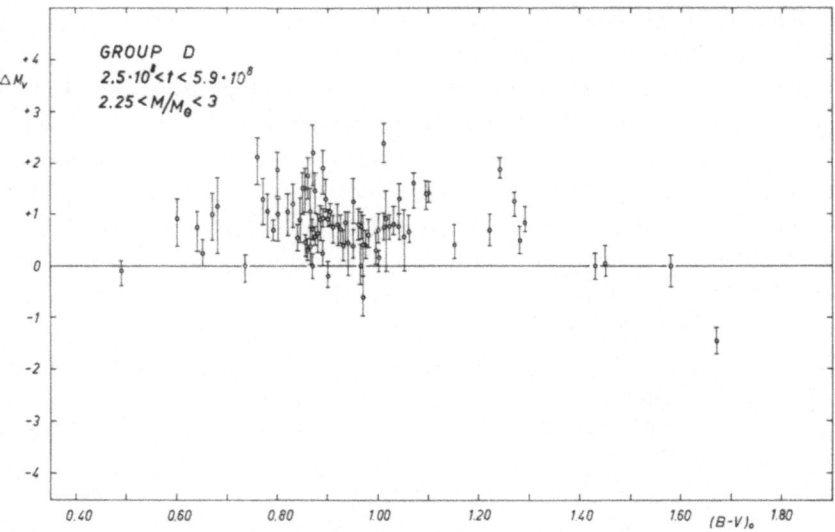

Fig. 8. ΔM_v as function of colour index for group D.

to Figures 4d and 8. Here the $\Delta M = +1$ group is still more conspicuous than the $\Delta M = 0$ one, while in the mass range considered only a few stars could reasonably be supposed to lie on the loops. Therefore most of the points with $\Delta M = +1$ are to be attributed to the observation-theory disagreement between the giant branches. From Figure 8 the tendency of the Hayashi limit to move towards the left as mass decreases is apparent, as most stars are confined to the $0.8 \approx 1.1$ *B-V* interval; stars beyond it being quite few and perhaps stragglers whose real situation is by no means clear. The dashed curve in Figure 4d represents the ΔM values calculated by taking the difference between the star position and the lower limit for the age value instead of the mean value; it is seen that, even thus, the $\Delta M = +1$ group, although weakened, does not disappear.

The disagreement between theoretical and observed giant branches obviously strengthens the reasons already given for avoiding the discussion of old clusters at the present time; we have therefore refrained from trying to analyze and compare with theory still older star groups than those contained in D.

4. Let us now briefly compare our results with those of other authors, and especially those recently obtained by LINDOFF (1968a, b), which have reached us while the present work was still in progress. The main conclusions obtained by Lindoff are somewhat different from ours, as he generally finds higher amounts of mass lost (up to 50% for large-mass stars). We have therefore tried to understand how such differences may arise; and the further considerations we want to discuss here are aimed to clarify our view of the chief reasons responsible for the disagreement in the two investigations.

As a main step, let us compare the age determinations obtained in the two cases.

To this end, we have naturally selected only clusters which have been taken in both with similar photometric data, in order to avoid differences arising only from the diversity of the experimental material used. In Figure 9 the age parameter t_Γ of Lindoff is plotted vs. the age t_1 (with its error) as determined in the present investigation: the full line indicating the locus $t_1 = t_\Gamma$. It is then apparent that, on the average, ages as determined by us are systematically longer than those obtained by Lindoff.

If such is the situation, it is then quite obvious why mass losses are on the average

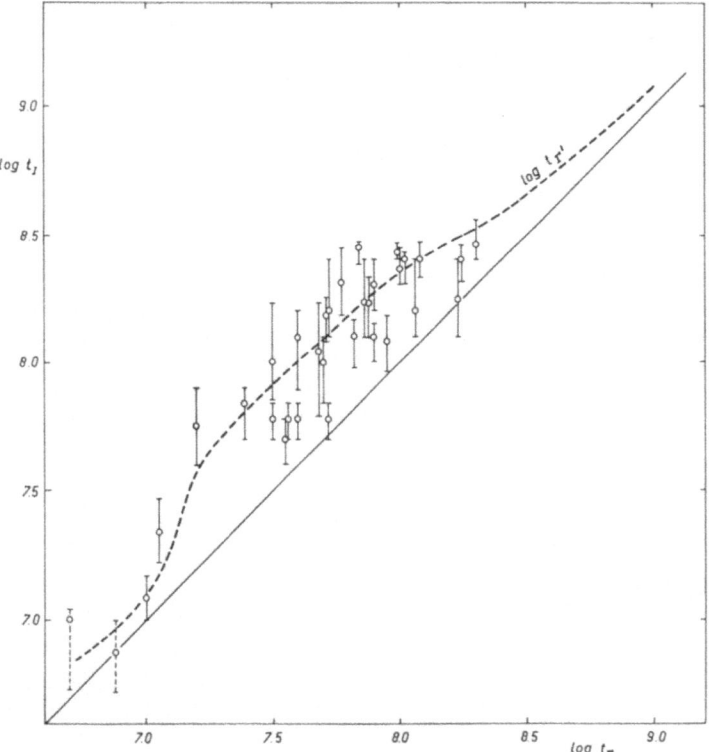

Fig. 9.   Cluster ages derived in the present work compared with the Γ method ages t_Γ of Lindoff. Dashed line shows the relation between $t_{\Gamma'}$ and t_Γ, where $t_{\Gamma'}$ is obtained according to Lindoff's method but using the H + J transformations.

larger in Lindoff's work than in ours. Owing to the systematically shorter ages used by him, his evolutionary theoretical tracks lie higher in the H-R diagram, and therefore his red giants points are more frequently and more deeply underneath them, thus testifying large mass decreases; while in our case, the same points may happen to be included in the indeterminacy band, yielding thus no or less evidence for mass lost. The main point is then confined to understand why age determinations are systematically different in the two researches.

We think that the main cause for the disagreement is not to be looked for in the difference of the methods, which gives only secondary corrections, but mostly in the different choices for the colour and bolometric corrections, as Lindoff uses the Allen ones.

In order to prove this, let us recall that, in Lindoff's work, age determinations are obtained by comparing a typical magnitude Γ, determined for each cluster according to a particular formula, with the corresponding Γ of a set of artificial clusters (in fact the isochronous ones) constructed by transforming the theoretical evolutionary curves; the age then results from the construction procedure of the artificial clusters. In Lindoff's work this construction is based on the use of the ALLEN (1963) colour and bolometric corrections and this yields a particular Γ-t relation (Figure 10, full curve).

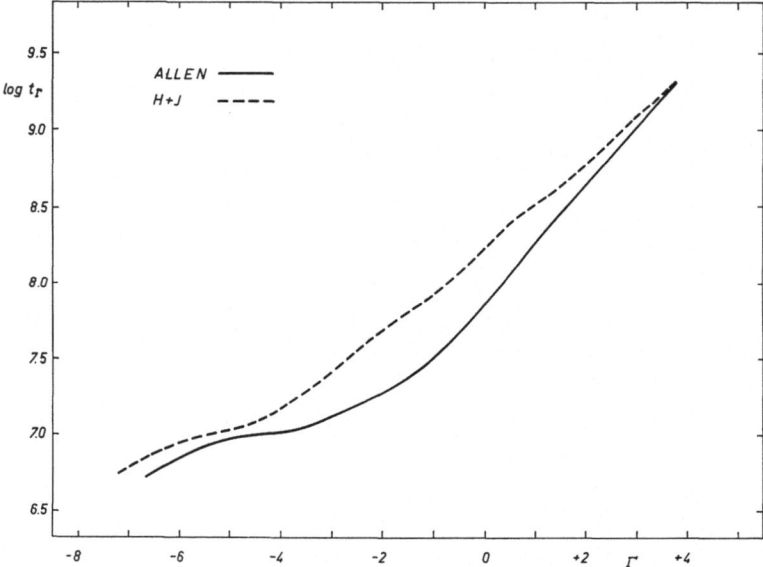

Fig. 10. Relation between the age t_Γ and the magnitude parameter Γ for $N_L = 25$ (see Lindoff). Full line is obtained, as shown by Lindoff, by using Allen corrections, dashed line with H + J transformations.

We have then in principle repeated the same work as Lindoff using instead our Harris-Johnson corrections, and the Γ'-t relation derived is now given in the dashed curve of Figure 10. If we now report on Figure 9 for a given Γ the correlation between the two t_Γ and $t_{\Gamma'}$, we obtain the dashed curve, which appears as a fair mean value for our age determinations of the single clusters. This proves in our view that on the average our method and Lindoff's one lead to similar results concerning age values, and that the systematic differences obtained in the two cases are not due to the method, but to the different choices for the bolometric corrections.

We turn now to consider why, although the average behaviour for age values is similar in the two methods, still there are strong fluctuations for individual clusters. We have identified the following causes for this dispersion.

First, in many cases the material used in the two researches was different, and the different omissions or inclusions of some crucial stars may strongly alter the results.

However, the main causes for the individual disagreement when the material used was the same, are to be looked for in some specific points of the methods themselves, and precisely:

(a) The most brilliant star of the main sequence is omitted in computing the parameter Γ, that is the age of the cluster, in Lindoff's method, while it is considered in ours; this means that Lindoff's method yields the same age for two clusters having all stars exactly similar except the most brilliant one, while ours yields for them different ages.

(b) It may be easily verified that both methods give very similar results in cases in which the main sequence of the cluster is well defined and little dispersion is at hand, while the stronger disagreements occur when the stars show strong fluctuations in colour. In Lindoff's case in fact, no account is taken of the dispersion for the age determination as no dependence on the B-V values of the single stars is considered for calculating Γ; so the same age would be obtained for two clusters having all stars with equal absolute magnitudes but different colour indices; Lindoff's procedure works therefore as if all stars were just aligned on a single sequence. In our method, instead, the age determination is strongly dependent on the dispersion of the colour indices and we mostly rely on it for the age indeterminacy. Moreover, as generally stars are dispersed to the right of the main sequence, this further contributes to lengthen our age determination in respect to Lindoff's one, his ages corresponding mostly to the lower limit in our range of uncertainty.

From the whole discussion, it is clearly apparent that the results obtained in these and similar investigations are strongly and mostly dependent on the choice made for the set of colour and bolometric corrections employed in the comparison between theoretical results and experimental data. In our present case, we have two arguments which could in some sense justify the opportunity of our choice.

(I) Recently a new $(B$-$V)_0$-T_e relation was calculated by ADAMS and MORTON (1968) using model atmospheres with ultraviolet and Balmer-line blanketing. The fluxes obtained in these models are in good agreement with the observed ultraviolet fluxes from hot stars and with the visual energy distribution for A-type stars. The corresponding colour-temperature relation is more similar to the Harris one than to the Johnson one, thus partly justifying our mixing these two sets of transformations.

(II) The bolometric corrections for hot stars of Allen have instead been calculated theoretically without ultraviolet and Balmer-line blanketing and according to somewhat superseded model atmospheres, which yield too large violet and ultraviolet corrections to the spectra of different stars taken with rocket data; the Johnson corrections which we have used, have taken care of these observed reductions.

A further correction which has up to now not been considered should still likely be applied to very cold stars with TiO bands. The bolometric increase necessary to correct the absorption due to such bands amounts, according to SMAK (1964, 1966), from 0 to -4 magnitudes when we go from M0 to M8 spectral types and this effect should further shift on the average the luminosity of the most evolved stars we have considered by about -0.5 to -1 magnitude. Should we take this fact into account, the value for the mass lost obtained should be further reduced.

On the whole, we would stress, as main conclusion to our investigation, the large amount of uncertainty which must be inherent to any research dealing with the confused and unsettled situation related to the colour and bolometric corrections: we would put the accent mostly on the width of the error band in our results related to the age determination of the clusters, as a consequence of the dispersion in the experimental data of any single cluster, and underline that, even if we can probably conclude for some qualitative evidence in favour of mass loss in the advanced red stages of evolution, we are still far away from being able to deduce quantitative values from statistical considerations of the kind of the present one.

Acknowledgments

We wish to express our thanks to Dr. L. Nobili for his help in the analysis of the clusters during the earlier part of this work, and Prof. W. Becker for having kindly provided us with some unpublished data concerning some clusters. We gratefully acknowledge the financial help of the Consiglio Nazionale delle Ricerche.

References

ADAMS, T. F. and MORTON, D. C.: 1968, *Astrophys. J.* **152**, 195.
ALLEN, C. W.: 1963, *Astrophysical Quantities*, The Athlone Press, London.
BARBARO, G. and FABRIS, G.: 1968, *Pubbl. dell'Oss. di Padova*, No. 146.
BARBARO, G., DALLAPORTA, N. and NOBILI, L.: 1967, *Pubbl. dell'Oss. di Padova*, No. 138.
HARRIS, D. L. III: 1963, *Basic Astronomical Data*, Chicago University Press, Chicago, p. 263.
HOFMEISTER, F., KIPPENHAHN, R., and WEIGERT, A.: 1964a, *Z. Astrophys.* **59**, 242.
HOFMEISTER, F., KIPPENHAHN, R., and WEIGERT, A.: 1964b, *Z. Astrophys.* **60**, 57.
IBEN, I.: 1965, *Astrophys. J.* **142**, 1447.
IBEN, I.: 1966, *Astrophys. J.* **143**, 483, 516.
IBEN, I.: 1967, *Astrophys. J.* **147**, 650.
JOHNSON, H. L.: 1966, *Ann. Rev. Astron. Astrophys.* **4**, 193.
KIPPENHAHN, R., THOMAS, H. C., and WEIGERT, A.: 1964, *Z. Astrophys.* **60**, 571.
LINDOFF, U.: 1968, *Arkiv Astron.* **5**, 1; **5**, 2.
SCHILD, R.: 1967, *Astrophys. J.* **148**, 449.
SMAK, J.: 1964, *Astrophys. J. Suppl.* **9**, 141.
SMAK, J.: 1966a, *Acta Astron.* **16**, 1.
SMAK, J.: 1966b, *Ann. Rev. Astron. Astrophys.* **4**, 21.
STOTHERS, R.: 1963, *Astrophys. J.* **138**, 1074.
STOTHERS, R.: 1964, *Astrophys. J.* **140**, 510.
STOTHERS, R.: 1966, *Astrophys. J.* **143**, 91.

Discussion

Underhill: Do similar uncertainties exist in determining the ages of globular clusters from the shapes of the HR diagrams as those which you have found for open clusters? The uncertainties you have found originate largely from difficulties in transforming from ($\log L/L_\odot - \log T_{eff}$) plane to the ($M_v$, B-V) plane according to your findings.

Dallaporta: Owing to the reason I have given, we have discarded from our considerations all old clusters, and therefore we have not considered globular clusters. Maybe some well-studied globular clusters could present a better material than the average of old galactic clusters for the age determination; however, the main sequence available is generally short and this may contribute to increase the error in the fitting. So I am not able to give a definite answer. The uncertainties we have found derive mostly from uncertainties in the ($\log L/L_\odot - \log T_{eff}$)→($M_v - B$-$V$) transformation, but also to the dispersion of the single stars in most of the old galactic clusters probably due to lack of elimination of non-cluster stars.

THE COLOUR-MAGNITUDE DISTRIBUTION OF GIANTS
IN OPEN CLUSTERS

ULF LINDOFF

Lund Observatory, Sweden

(Read by Gunnar Larsson-Leander)

If we want to study the distribution of the giant stars in an HR diagram as a function of age, we must know three parameters: colour, absolute magnitude and age. The

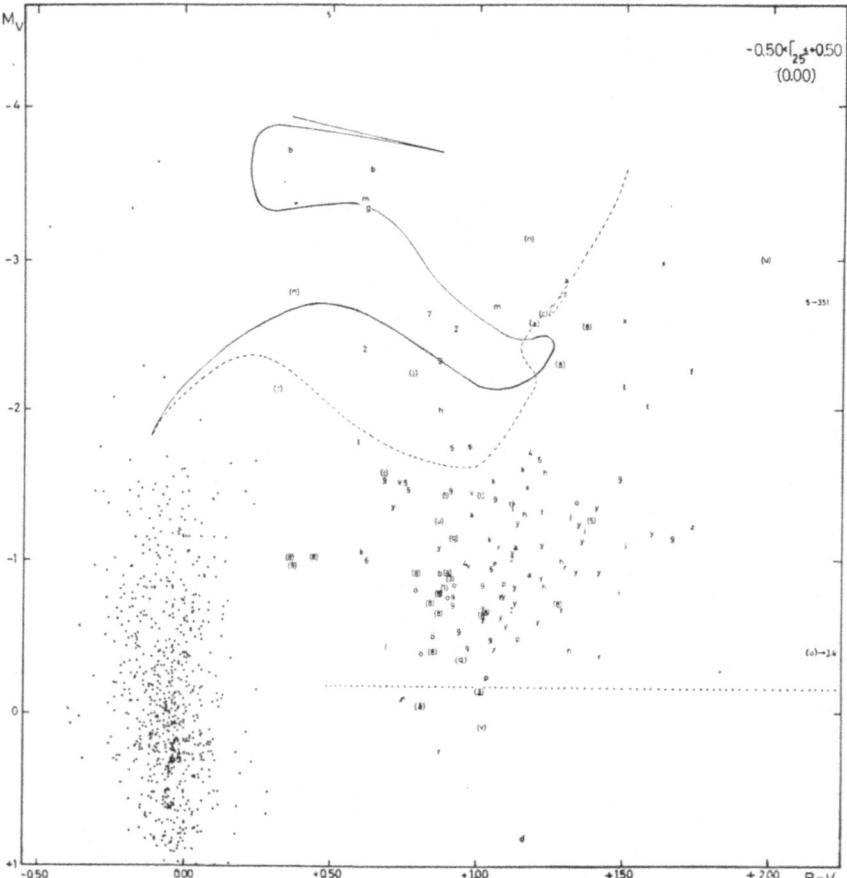

Fig. 1.   The observed colour-magnitude distribution of stars for $\Gamma_{25} = 0.00$, corresponding to the age 6.5×10^7 years. (The material has been collected from the interval $-0.5 < \Gamma_{25} < +0.5$.) The yellow and red giants are denoted by letters or figures, the same symbol for stars belonging to the same cluster. The dotted line shows the limiting magnitude according to Figure 2.   Symbols within brackets indicate objects not very well determined. The main-sequence stars in the same age interval are indicated simply by dots. The absolute magnitudes have been corrected to the group value $\Gamma_{25} = 0.00$, but no correction to the colour has been applied. Hence the colour distribution has been somewhat enlarged. The synthetic giant branches for the same age are shown according to Iben's computations (full curve) and according to Hofmeister's new models (dashed curve).

greatest difficulty is to determine the last parameter, and so far we cannot obtain reliable ages for the field giants. The ages of open clusters, however, must be considered sufficiently well known, so that stars situated in these clusters may be used. Unfortunately, the number of giants in a single cluster is, with occasional exceptions, far too small for useful statistics, and the only possibility to obtain further results is to collect clusters of about the same age and to treat them collectively.

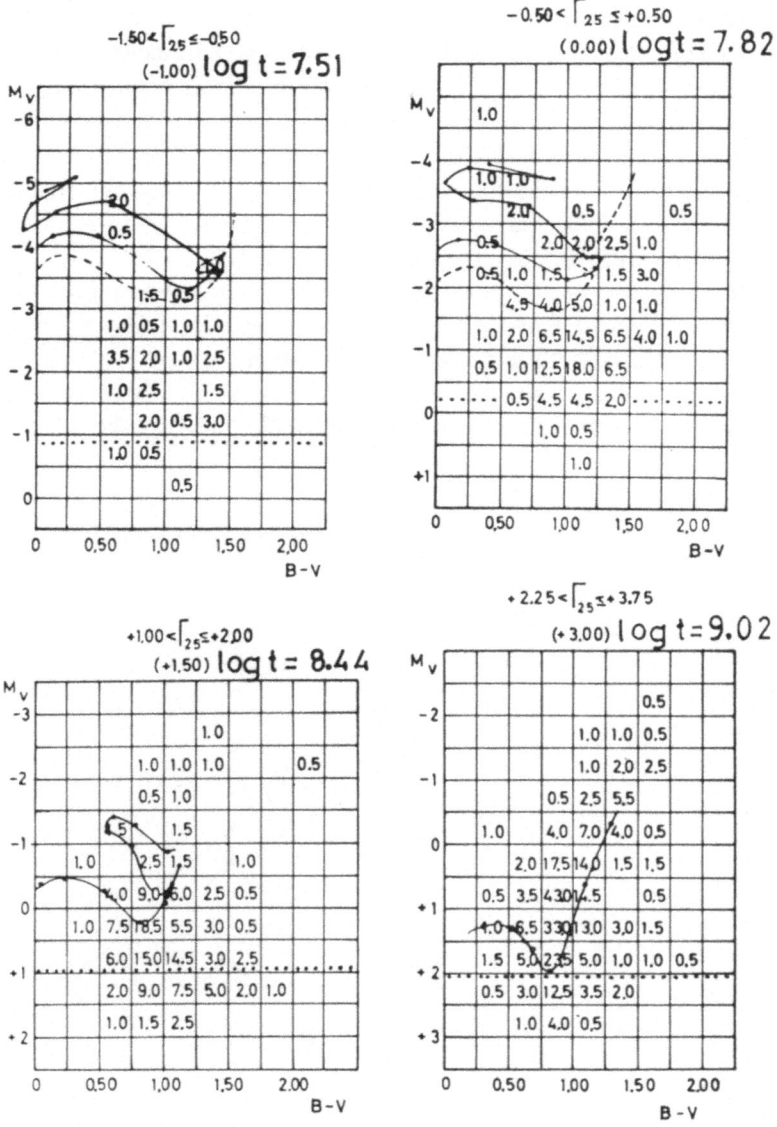

Fig. 2. Colour-magnitude diagram showing the number of giants in squares corresponding to $0^m.5$ in M_v and to $0^m.25$ in B-V for some different values of the age parameter Γ_{25}. Synthetic giant branches have been drawn as full curves according to Iben's models and as dashed curves according to Hofmeister's.

The cluster ages have been determined by the writer (LINDOFF, 1968a) in an objective way from the most recent theoretical computations of stellar evolution. Iben's time scale has been used. Model clusters of various ages have been constructed, and from these the age, or rather an age parameter called Γ (or Γ_{25}, if various star contents are taken into consideration), has been determined for individual clusters. Altogether 180 open clusters have been used.

By means of this age parameter, clusters of about the same age have been collected in groups, and the distribution of their giants have been studied (LINDOFF, 1968b). In Figure 1 of the present paper clusters within the age parameter interval -0.5 to $+0.5$ have been combined and the positions of the giants have been shown in a colour-magnitude diagram. Each symbol represents a certain cluster. Thus, for instance the giants belonging to the cluster M11 are denoted by y.

The full line represents the theoretical giant sequence according to IBEN (1965, 1966a, b, c, 1967) in a model cluster of the same age, 6.5×10^7 years, while the dashed line refers to the computations of EMMI HOFMEISTER (1967). As is seen, there is a deviation between observations and theory in that theory gives brighter giants.

In Figure 2 the number of stars in each square of $0^{m}.5$ in M_v and $0^{m}.25$ in B-V have been plotted for a few age groups. The deviation between theory and observations increases with decreasing age. The explanation of this phenomenon might be mass loss during the giant stage. For stars with 2 $\mathfrak{M}_\odot$ and less the mass loss is in-

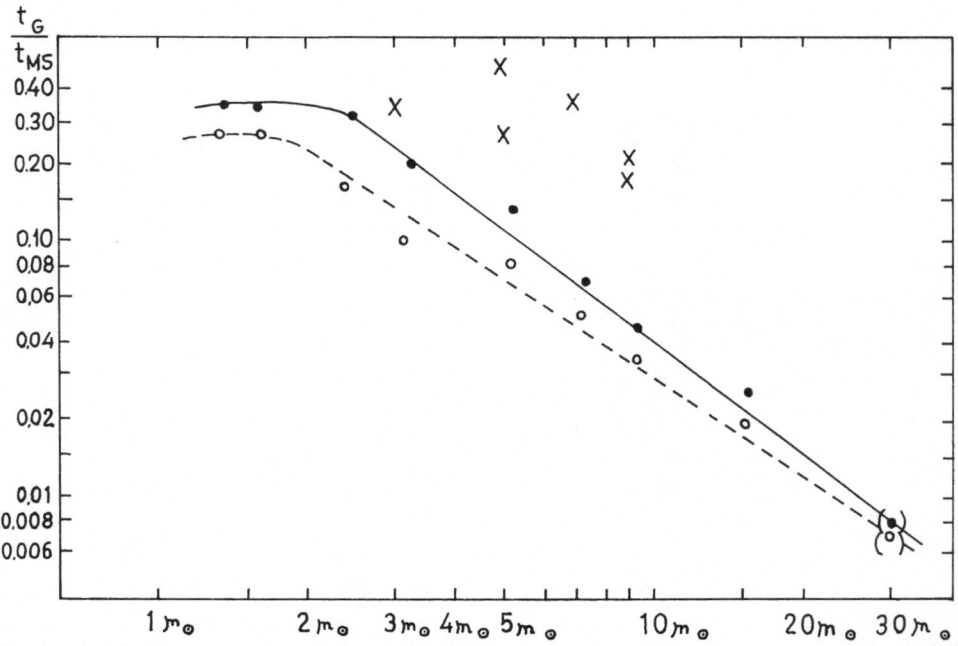

Fig. 3.   The giant lifetime vs. the main-sequence life (derived from clusters), as a function of mass. Open circles and dashed curve refer to results obtained when the initial mass function is taken into consideration. The crosses indicate results from the theories.

significant, but it would amount to 25% at 4 $\mathfrak{M}_\odot$ and to 50% at 10 $\mathfrak{M}_\odot$. Possible errors in the age calibration affect these values only to a small degree.

By studying the number of giants in clusters of various ages it has been possible to estimate the lifetimes of giants of various masses. See Figure 3. The relative lifetime as a giant star seems to be smaller, even considerably smaller, than given by theoretical calculations.

As a secondary result it was found that the masses of Cepheids (before possible mass loss) lies between 6 and 8 $\mathfrak{M}_\odot$ (2–6 × 10^7 years) and that the lifetime as a Cepheid is about 15% of the lifetime as a giant or 1% of the time spent at the main sequence (viz. 3 × 10^5 years for a 7 $\mathfrak{M}_\odot$ Cepheid). This value agrees with theory.

References

HOFMEISTER, E.: 1967, *Z. Astrophys.* **65**, 164.
IBEN, I.: 1965, *Astrophys. J.* **142**, 1447.
IBEN, I.: 1966, *Astrophys. J.* **143**, 483, 505, 516.
IBEN, I.: 1967, *Astrophys. J.* **147**. 624.
LINDOFF, U.: 1968a, *Lund Medd. Ser.* I, **227**.
LINDOFF, U.: 1968b, *Lund Medd. Ser.* I, **228**.

Discussion

McCarthy: I should like to ask for an evaluation of membership criteria used to separate field stars from genuine members in the 180 clusters studied.

Larsson-Leander: The main criteria for membership of yellow giants are MK classes and UBV photometry. Other criteria, as radial velocities and proper motion, are used when available. In many cases the question if a certain star is a physical member or not is very difficult. Dr. Lindoff discussed the problem in his original paper, to which I must refer for further details.

Deutsch: About 40 years ago Prof. Wallenquist sought for evidence bearing on the relative masses of red giants and early-type stars in clusters, by examining the radial density gradients of the two kinds of stars. Has Dr. Lindoff attempted to do this again? If not, would it be useful for another astronomer to try it with Lindoff's clusters?

Larsson-Leander: No, Dr. Lindoff has not made any such attempt. I feel that in the large majority of clusters the number of yellow giants is too small. Of course, for some clusters it might be worthwhile to make an attempt at such an analysis.

SOLAR-WIND MODEL INCLUDING THE EFFECTS
OF ROTATION, MAGNETIC FIELDS, AND
ANISOTROPIC HEAT CONDUCTION

S. GRZĘDZIELSKI

Astronomical Observatory of Warsaw University, Warsaw, Poland

Abstract. A time-independent solar-wind model is considered in the case of spherical symmetry and of radial magnetic field at the sun's surface. The energy equation includes besides the usual terms also the heat conduction and magnetic-energy convection (Poynting vector) terms. The dependence of the thermal conductivity on the magnetic field is taken into account. Numerical integrations of the basic equations were performed under the following assumptions: (i) close to the sun the magnetic field is the dominant azimuthal term and solid-body rotation is enforced; (ii) beyond the Alfvénic point the terms quadratic in B are neglected. The model leads to azimuthal velocity at earth between 0.6 and 2.7 km/sec, to radial velocity at earth between 350 and 500 km/sec, and to angular momentum loss of 5×10^{18} cm²/sec per unit mass of gas leaving the solar equator. The dependence of the solutions on the reduction of the effective thermal conductivity caused by the micro-structures in the solar wind suggests that the conditions at earth may be largely determined by a transition region in the solar wind, in which the conduction régime changes into an almost adiabatic flow.

1. Introduction

The question of the rotation of the expanding solar corona and of the torque exerted upon the sun by the solar wind is closely connected with the problem of the rôle of magnetic fields in coronal expansion. In the early papers (PARKER, 1958, 1961; AXFORD *et al.*, 1963), the magnetic field was considered to float freely on the flow without exerting any force upon the fluid. In recent years many authors discussed the effect of the magnetic stresses upon the gas (PARKER, 1963; PNEUMAN, 1966a; KING and CAROVILLANO, 1966; FERRARO and BHATIA, 1967; WEBER and DAVIS, 1967; TARAFDAR, 1968), paying special heed to the problem of corotation and angular-momentum transfer. Also the combined effects of magnetic fields and viscosity were recently discussed by PNEUMAN (1966b). In all these investigations the magnetic field was included in the equation of motion only. The energy equation was replaced either by assuming constant temperature or by relating pressure and density by a polytropic formula.

The aim of the present paper is to investigate the effects of the magnetic fields and rotation when included both in the equations of motion and energy. In the latter equation the magnetic field may enter via the term describing the convection of magnetic energy (corresponding to the Poynting vector) and via the dependence of the thermal conductivity upon the field. We also limit ourselves to a layer close to the ecliptic plane and so the spherical symmetry is assumed throughout. The effects of viscosity and finite electrical conductivity are neglected. The solutions possess then the same types of singularities as those discussed by WEBER and DAVIS (1967) with an additional singularity lying in infinity. With three singularities present in the solu-

M. Hack (ed.), Mass Loss from Stars. All rights reserved

tions and additional matching conditions at the surface of the sun and in infinity (or at the orbit of earth) the integrations become very awkward and time-consuming. In order to reduce the number of singularities the numerical integrations were performed with the following simplifying assumptions:

(i) Close to the solar surface the magnetic field was supposed to be (almost) radial and strong enough to enforce strict corotation up to a distance $r = r_s$.

(ii) For $r > r_s$ the magnetic field was assumed to be weak enough to permit to neglect the terms quadratic in magnetic-field strength.

Thus, in the second region the coupling between the magnetic field and the motion is due solely to the dependence of the thermal conductivity upon the field, and the gas preserves in that region its angular momentum.

The distance r_s is so far a free parameter of the model. However, the choice of r_s is not arbitrary since $r = r_s$ has to divide the region where the magnetic stresses are dominant (region I) from the region where the Reynolds stresses of the fluid prevail (region II). The above condition corresponds to such a value of r_s for which the Alfvén Mach number $N = (4\pi\rho)^{1/2} v_r/B_r$ is of the order of unity (v_r, B_r and ρ denoting radial velocity, radial component of the magnetic field, and density, respectively). This circumstance enables us to reduce the number of singularities to one (GRZĘDZIELSKI, 1968) and also to obtain an estimate of the angular momentum carried away by the gas. The numerical results obtained indicate that despite the simplifications adopted in the treatment, the model represents reasonably well the magnitude of the influence of the magnetic field upon the gross features of the solar wind.

2. The Basic Equations of the Problem

Let us consider a time-independent expansion of the corona of a star of mass M rotating with angular velocity Ω. We shall assume the gas to be fully ionized and subjected to the gravitational pull of the star as well as to the forces of the kinetic and magnetic pressures. To simplify the discussion we adopt a simple geometry of the flow, the same as in a paper by WEBER and DAVIS (1967), which assumes that close to the equatorial plane (plane of ecliptic) the flow is spherically symmetric: $\partial/\partial\varphi = \partial/\partial\theta = 0$.

Thus, the present description is relevant to a narrow cone around the equatorial plane to which, by the way, refer the existing experimental measurements of the solar wind by the space probes. The model of the magnetic field one may think of corresponds to a hedgehog, with the lines of force being more and more bent as the distance from the sun increases. Since no φ-dependence is retained no allowance is made for the possible effects of the sector structure discovered recently (NESS, 1967).

We shall also assume the transport of energy in the corona to be determined by heat convection, heat conduction and flow of electromagnetic energy associated with the magnetic field. No radiative cooling is taken into account. In the case of our sun this last restriction may be not applicable to the lower corona, within $0.4\ R_\odot$ from the limb. Farther up, however, it seems fully justified. Also no deposition of energy by shock wave dissipation is taken into consideration.

The flux of heat due to thermal conduction is equal to $\mathbf{Q} = -\sum_j K_{ij}\, \partial T/\partial x_j$ where K_{ij} denotes the thermal conductivity tensor and T is the kinetic temperature supposed to be equal for ions and electrons. Under the present assumptions only the diagonal terms matter in the tensor and

$$\sum_j K_{ij}\, \partial T/\partial x_j = (K_\parallel \cos^2 \chi + K_\perp \sin^2 \chi)\, dT/dr\,,$$

where χ denotes the angle between the radius vector and the direction of the line of force. Since the value of the ratio of electron gyro-to-collision frequencies exceeds 10^6 for $B \simeq 1$ gauss in the lower corona and reaches 10^9 for conditions prevailing at the distance of earth, only the thermal conductivity parallel to the magnetic field is important (unless χ is very close to $\pi/2$). Assuming

$$K_\parallel = 6 \times 10^{-7} \cdot T^{5/2} \; \text{erg cm}^{-1}\, \text{sec}^{-1}\, \text{deg}^{-1}\,,$$

one obtains

$$Q = -K_\parallel \cos^2 \chi\, dT/dr = -(K_\parallel\, dT/dr)/(1 + B_\varphi^2/B_r^2)\,.$$

The Poynting vector in the case of infinite electrical conductivity may be written in the form
$$\mathbf{P} = (4\pi)^{-1}\mathbf{B} \times (\mathbf{v} \times \mathbf{B})\,.$$

The equations of continuity of matter and, respectively, of the magnetic lines of force yield

$$r^2 \rho v_r = F/4\pi = \text{const} \tag{1}$$

and

$$r^2 B_r = r_0^2 B_0 = \Gamma = \text{const}\,, \tag{2}$$

where $F/4\pi$ is the total flux of matter per second and per steradian, and B_0 is the radial component of the magnetic field at a chosen reference distance r_0. We shall identify r_0 with the base of the corona in the following.

The Faraday law gives

$$r(v_\varphi B_r - v_r B_\varphi) = \Delta = \text{const} = \Omega\Gamma\,, \tag{3}$$

if one assumes that the magnetic lines of force are rooted into the star and stationary in a frame rotating with the star.

The rth component of the equation of motion is

$$\rho v_r\, dv_r/dr - \rho v_\varphi^2/r + dp/dr + B_\varphi/4\pi \cdot d/dr(rB_\varphi) + \rho GM/r^2 = 0 \tag{4}$$

and the φth component gives after integration

$$Frv_\varphi - \Gamma rB_\varphi = C = \text{const}\,, \tag{5}$$

G being the constant of gravitation.

The first integral of the energy equation may be written as

$$F\left[c_p T + (v_r^2 + v_\varphi^2)/2\right] - rB_\varphi \Omega\Gamma - 4\pi r^2 K_\parallel\, dT/dr/(1 + B_\varphi^2/B_r^2)$$
$$- FGM/r = E\,, \tag{6}$$

where $E/4\pi$ denotes the total flux of energy per second and per steradian. The set of equations is completed by the equation of state

$$p = k\rho T/\mu m_H, \tag{7}$$

where μ is the mean mass of a free particle in atomic units and m_H denotes the hydrogen mass.

In this way we are left with seven equations for seven functions: p, ρ, T, v_r, v_φ, B_r, B_φ. The distance r from the centre of the sun is the independent variable. Let us now suppose we know from observations or direct measurements the numerical values of the integration constants F, Γ, hence $\Delta = \Omega\Gamma$, C and E. We then may eliminate the variables v_φ, B_r, B_φ, p, ρ by means of Equations (1–3) and (5–6), and obtain a set of two non-linear differential equations of the first order for v_r and T. These equations are linear in dv_r/dr and dT/dr, and one may look for the singularities in the solutions by exploring the zero points of the denominators in the corresponding expressions for dv_r/dr and dT/dr. The energy equation, as containing a transport coefficient, does not generate any singularity, and therefore the singularities are determined by the remaining equations only. Hence, the topological properties of the solutions for finite r are essentially the same as those discussed by WEBER and DAVIS (1967). The point $r = \infty$ represents another singular point in the solutions. This additional singularity is related to the existence of the $d/dr\,[K(T)dT/dr]$ term in the energy equation with heat conduction, and hence it was absent in the case discussed by Weber and Davis. From the point of view of that new singularity the solution with $T=0$ for $r=\infty$ represents a critical solution attainable only when the values of the integration constants Γ, Δ, C, F, E satisfy certain additional conditions. This question will be discussed in more detail elsewhere.

The position of the Alfvénic point for which $N=1$ (the principal singularity) is determined by the conditions

$$r_A^2 F\Omega = \Gamma^2\Omega/v_A = C, \tag{8}$$

where r_A and v_A denote the values of radius and radial velocity at the principal singularity. Expressing v_φ in terms of r_A and v_A one may easily verify that for large N, i.e. for large distances from the sun, the angular momentum carried away by unit mass of gas becomes

$$(rv_\varphi)_\infty = r_A^2\Omega(1 - v_A/v_\infty) = \Omega(r_A^2 - \Gamma^2/Fv_\infty).$$

When the heat-conduction term and the Poynting-vector term can be neglected in the energy equation for $r=\infty$, as seems plausible in the case of solar wind, one has $v_\infty^2 = 2E/F$ and the angular-momentum formula may be written in the form

$$(rv_\varphi)_\infty = \Omega\,[r_A^2 - \Gamma^2/(2EF)^{1/2}]. \tag{9}$$

3. The Numerical Results

Numerical integrations were carried out separately for regions I and II and the corresponding solutions were joined at $r=r_s=r_A$ with proper continuity conditions satisfied.

The details of the numerical integrations are discussed at length in another paper (GRZĘDZIELSKI, 1968). Here we shall present only some physical arguments in favour of the adopted procedure.

The juncture of the solutions I and II (corresponding to regions I and II, respectively) takes place at $r = r_A$ at which $N = 1$ and

$$v_r^2 = \frac{B_r^2}{4\pi\rho} = \frac{\Gamma^4}{F^2 r^4} = \frac{B_0^4 r_0^8}{F^2 r^4}. \tag{10}$$

For a given value of Γ, i.e. for a given magnitude of the radial component of the

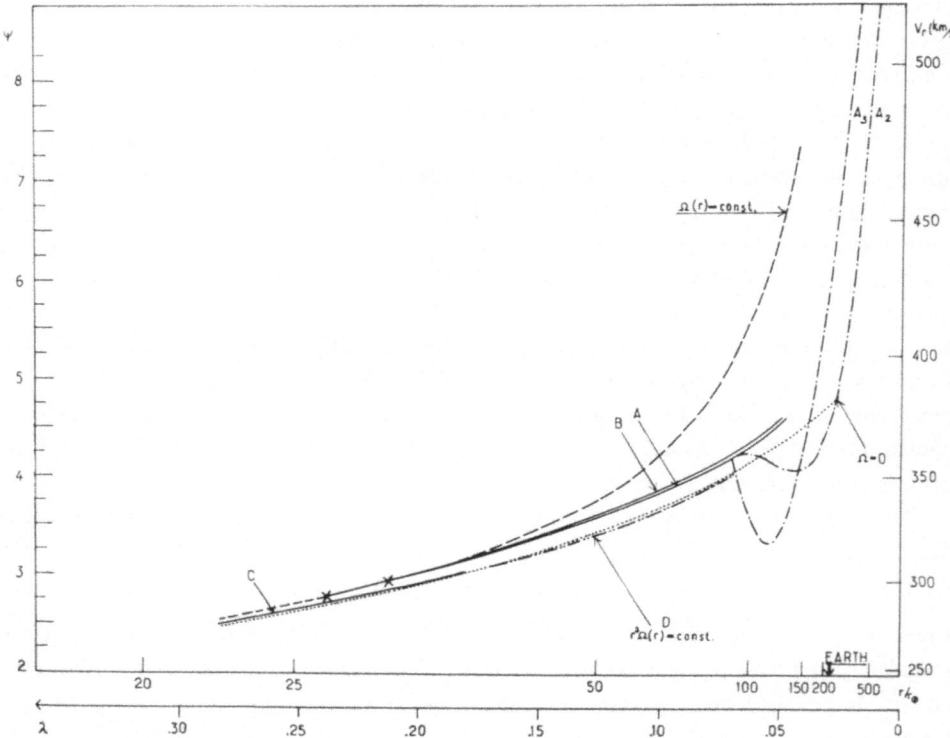

Fig. 1. The dependence of the radial velocity v_r on the distance r expressed in solar radii. Solution type I corresponding to solid-body rotation is shown by a dashed line ($\Omega(r) = $ const). The dotted line ($\Omega = 0$) shows the solution corresponding to non-rotating sun and hence to a purely radial magnetic field. Curves B and C (solid lines) represent the solutions type II attached to solution I at $r_s = 29.6\ R_\odot$ and at $r_s = 10.77\ R_\odot$. The position of the principal singularity (Alfvénic point) is indicated by $N = 1$. At this point ($r_s = r_A = 26.45\ R_\odot$) a solution type II is attached (solid line, labelled 'A'). The solid-body solution up to the Alfvénic point followed by the constant angular momentum solution (curve A) form the required solution of the problem as stated in the present paper. Curves B and C also representing constant angular-momentum solutions show the influence of the assumed corotation radius upon the run of solutions (different corotation radius implying different angular momentum and different total energy flux). Curve D (double dot dash) corresponds to solution type II beginning right at the base of corona. To the A-curve the asymptotically adiabatic solutions corresponding to parabolic conduction decrease (curve labelled 'A₂') and to linear conduction decrease (curve labelled 'A₃') are attached at $r = r_d = 89.3\ R_\odot$. λ is defined in the text and ψ is the radial kinetic energy per unit mass (arbitrary units).

magnetic field the position of the Alfvénic point depends on the run of $v_r(r)$, which is not known *a priori*. Fortunately, as the numerical integrations show, the dependence of v_r on r is such as to allow us to circumvent this difficulty.

It is obvious on physical grounds that for given initial conditions any possible solution $v_r(r)$ will run between two limiting solutions corresponding to cases:

(a) when $\Omega(r) = $ const everywhere in space, i.e. when region I extends from the base of the corona up to infinity. The $v_r(r)$ solution corresponding to this case (solution type I) is shown by the curve labelled '$\Omega(r) = $ const' in Figure 1 (dashed line).

(b) when $\Omega(r)r^2 = \Omega(r_0)r_0^2 = $ const everywhere in space, i.e. when region II extends from the base of the corona up to infinity. The $v_r(r)$ solution corresponding to that case (solution type II) is shown by the curve labelled 'D' in Figure 1 (dash dot dot).

Therefore, one may expect that the coordinate r_A of the Alfvénic point and the value v_A of the radial velocity at that point will lie between the corresponding values determined for cases (a) and (b). The difference between these cases amounts in actual computations to about 0.5% in r_A and to 2% in v_A. Thus we hope the solutions corresponding to cases (a) and (b) provide a fair estimate of the possible range of values of r_A and v_A for given initial conditions. In actual computations the values of r_A and v_A were determined from solution I using condition (10).

Special heed has to be paid when carrying on the integrations for large values of r. There exist strong physical arguments against integrating the equations of the problem with the $d/dr(T^{5/2}\,dT/dr)$ term up to $r = \infty$. The use of the conductivity $K_\parallel$ is justified only when the scale of the magnetic field is much greater than the mean collisional free path of the heat carriers. That is the plausible case in the corona. However, at large distances from the sun, of the order of tens or hundred solar radii the scale of the magnetic field may be much smaller than the mean free path which is of the order of 1 AU. There is a wealth of indirect astronomical evidence based on radio-source occultations by the corona that large density fluctuations (streamers) exist, whose dimensions are far less than the Coulomb mean free path for the electrons. Clearly, this implies magnetic irregularities of a small scale, of the order of 10^5 km or even less.

Direct magnetic-field measurements are available for the region between earth and Venus (NESS, 1967), The rms magnetic field is of the order of tens of gammas exceeding as a rule the average value of the field. We may expect the mean free path determining the transport of heat to be either of the order of the gyroradius or of the order of the longitudinal scale of the magnetic-field irregularities, depending on the local geometry of the field. This may reduce the effective thermal conductivity by so many orders of magnitude that heat conduction may be neglected altogether. It goes without saying that the transition from one régime to the other takes place gradually. Thus, one might expect the only solution physically acceptable is that which becomes adiabatic for $r = \infty$ (asymptotically adiabatic).

An adiabatic solution cannot be attached to a conduction solution at a given r unless one is prepared to accept a discontinuity either in the energy flux or in the thermodynamical variables. The way chosen in the present paper consists in assuming a gradual damping of the conduction term in the energy equation starting from a given distance r_d somewhere in region II, so as to obtain a pure adiabatic solution for $r = \infty$. Two cases were discussed:

(1) Linear damping with conduction term $\propto \lambda/\lambda_d$;

(2) Parabolic damping with conduction term $\propto 1 - (\lambda - \lambda_d)^2/\lambda_d^2$, where $\lambda_d/\lambda = r/r_d$.

The numerical data about the interplanetary medium were taken following NESS (1967), who reviewed the experimental results obtained by space probes operating between 0.7 and 1.5 AU. Although the day-to-day data vary enormously, the scatter of individual measurements is due to real cosmic noise, and one may expect that the average values do in fact refer to some 'average conditions'. In Table I we reproduce some of the average data as reported by Ness.

TABLE I

Average data for the neighbourhood of earth

Flux (ions cm^{-2} sec^{-1})	3×10^8
Radial velocity (km sec^{-1})	400–500
Density (ions cm^{-3})	5
% He to H by number	4
Temperature (K)	2×10^5
Magnetic-field strength (gamma)	6
Magnetic-field direction at earth (angle χ)	45°, 225°

TABLE II

Initial values adopted in numerical integrations

Quantity	Cgs units
F	9.32×10^{11}
E (at sun)	2.21×10^{27}
B_r (at earth)	4.24×10^{-5}
Ω (at sun)	2.79×10^{-6}
T_0 (at sun)	2.16×10^6 K
v_0 (at sun)	9.26×10^5
$r_0 = 1.09\, R_\odot$	7.58×10^{10}

The principal data serving as initial conditions for the integrations are listed in Table II. The chemical composition assumed was 90% H and 10% He by number which yields $\mu = 0.616$. The total flux of energy (E) is given at sun's surface since E at earth includes the work of the magnetic constraints and is slightly greater, depending on the case considered (in any case the increase is less than 1%). The adopted base of the corona corresponds to 1.09 $R_\odot$ from the centre of the Sun. The flux of ions at earth corresponding to the adopted value of F amounts to 1.54×10^8 ions (H + He) per cm^2 and per second.

The dependence of v_r on the distance is shown in Figure 1. The surface of the sun is at the left and the distance scale is labelled both in terms of r and λ. The base of the corona (r_0) corresponds to $\lambda = 5.77$, and the position of earth to 215 $R_\odot$ ($\lambda = 0.0293$). All curves start at $r_0 = 1.09$ $R_\odot$ with the same values of v_0, T_0 and E. The dependence of the temperature on r is shown in Figure 2.

The comparison of curve 'D' (dash double dot) with the curve labelled '$\Omega = 0$' (dots) in Figure 1, shows the influence upon the motion of the dependence of the thermal conductivity on the direction of the field. Curve 'D' describes the case when no magnetic torque is present and angular momentum is preserved starting right

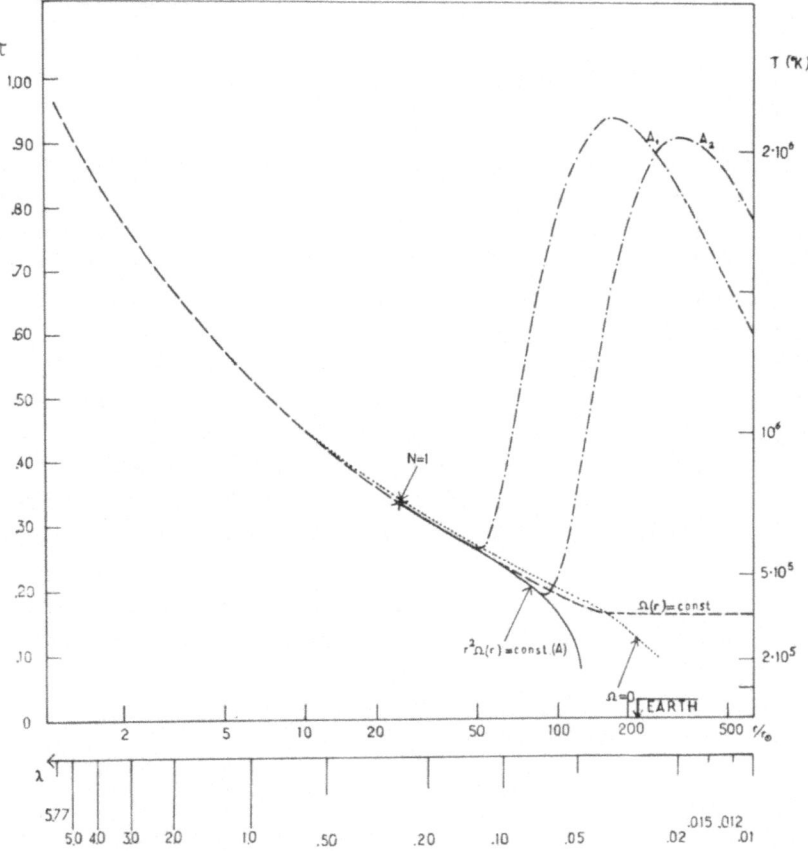

Fig. 2. The dependence of the temperature on the distance r expressed in solar radii. The meaning of symbols as in Figure 1. The A_1-curve describes the asymptotically adiabatic solution corresponding to parabolic conduction decrease starting at $r = r_d = 48.2$ $R_\odot$ and attached to the best solution type II curve (curve A). τ denotes a dimensionless temperature (arbitrary units).

from the base of the corona. Curve '$\Omega = 0$' corresponds to a non-rotating sun and thus to a purely radial field. Because of the slowness of solar rotation the difference is very small indeed.

The solution corresponding to solid-body rotation (solution type I) is shown in Figures 1 and 2 by dashed lines and labelled '$\Omega(r) = $const.' Using that curve the Alfvénic point was found from Equation (10). Its position ($N = 1$) is indicated by an arrow.

At this point a solution of type II is attached labelled 'A' (solid line). The principal singularity determined from the curve '$\Omega(r) = $const' is

$r_A = 26.45\ R_\odot$, $v_A = 292$ km/sec, $T_A = 7.37 \times 10^5$ K.

If it were determined from solution 'D' one would have

$r_A = 26.33\ R_\odot$, $v_A = 287$ km/sec, $T_A = 7.48 \times 10^5$ K.

In order to investigate the dependence of the solutions on the assumed position of juncture ($r = r_s = r_A$), solutions of type II were attached to solution I for

$r_s = 29.6\ R_\odot$, curve 'B', solid line in Figure 1,
$r_s = 10.7\ R_\odot$, curve 'C', solid line in Figure 1.

The influence of the assumed reduction in thermal conductivity upon the run of the solutions is shown by the curves labelled 'A_1, A_2, A_3' (dash-dot lines). The linear decrease in the conduction term corresponds to the A_3-curve and the parabolic decrease to the A_2- and A_1-curves (Figures 1 and 2). In both cases drawn in Figure 1 the distance at which the conduction damping begins corresponds to 89.3 $R_\odot$. For the case A_1 this distance is 48.2 $R_\odot$.

In Figure 3 the density profile in the corona as obtained from solution I (solid-body rotation) is reproduced. This is compared with the observational determinations of density profiles by different authors and with the conduction solution found by NOBLE and SCARF (1963). The discrepancy between the computed values and the observed ones close to the sun's surface is a common feature of many theoretical models. It may be due to the deposition of energy in that region (wave-energy dissipation).

Knowing r_A and v_A, one can estimate the angular momentum per unit mass carried away by the solar wind (Equation 9). For the adopted initial values one gets

$$(rv_\varphi)_\infty = 5.45 \times 10^{18}\ \mathrm{cm^2/sec}.$$

The torque exerted on sun by the solar wind is then

$$\alpha \cdot F \cdot (rv_\varphi)_\infty = \alpha \cdot 5.1 \times 10^{30}\ \mathrm{g\ cm^2/sec^2},$$

where $\alpha < 1$ is a factor resulting from the (unknown) θ-dependence of the angular momentum. If one applies the $\sin\theta$ factor to the angular momentum at $90° - \theta$ heliographic latitude, then $\alpha = 0.39$. For the total angular momentum of the sun of the order of 1.6×10^{48} g cm^2/sec, the typical time scale for angular momentum loss is about 4×10^{10} years.

For given initial conditions the exact value of v_r at earth depends on the kind of asymptotically adiabatic solution adopted. For the cases shown in Figure 1, v_r at earth is between 350 and 500 km/sec, i.e. in agreement with the velocities observed (Table I). This leads to the azimuthal velocity at earth

$$v_\varphi = 2.7\ \mathrm{km/sec\ for}\ v_r = 500\ \mathrm{km/sec},$$
$$= 0.6\ \mathrm{km/sec\ for}\ v_r = 350\ \mathrm{km/sec},$$

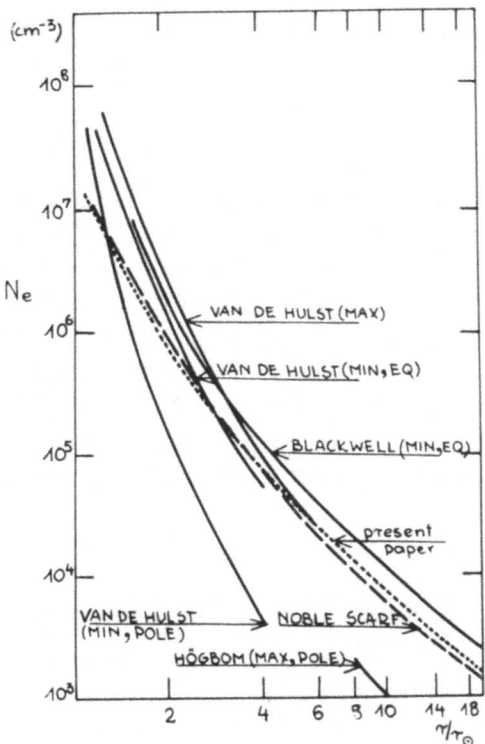

Fig. 3. The density profile in the corona obtained in the present paper (dashed line) as compared with observational determinations by different authors and with the solution found by NOBLE and SCARF (1963).

compared with the 9 ± 3 km/sec azimuthal velocity inferred by BRANDT (1967) from the observations of the comet tails.

The point that deserves most attention is the question of a gradual transition from the conduction régime close to the sun to an approximately adiabatic flow at large distances. The necessity of such a transition seems unavoidable when one takes into account the highly chaotic magnetic fields near earth and hence the very short effective mean free path. However, it is not possible to tell at present how abrupt or how smooth this transition is and at what distance the role of heat conduction begins to decline. In the present paper this change in the energy transport was described formally by the linear and parabolic damping formulae and thus no weight should be attached to the exact values of the velocity and temperature at earth's orbit implied by these formulae (curves A_1, A_2, A_3 in Figures 1 and 2). For instance the temperature at earth (Figure 2) is much higher than measured. This may easily be removed by shifting towards the sun the point r_d at which the damping of conduction begins (cf. curves A_2 and A_1 in Figure 2) or by assuming a different damping formula. The important thing is that if the magnetic field becomes disordered somewhere between the sun and earth, then the temperature observed at earth might be in no direct relation with the coronal temperature but might depend primarily on the structure of the region in which the magnetic noise becomes important. This is due to the circumstance that for reasonable initial conditions the transport of heat by conduction represents an important fraction of the total energy flux even at large distances from the sun (e.g., 0.707 at $r=48.3\ R_\odot$ and 0.682 at $r=89.3\ R_\odot$ for solution A). Hence, any substantial decrease in the conduction flux requires a drastic re-modelling of the flow as the temperature and velocity waves in Figures 1 and 2 show.

The above considerations suggest that the damping of heat conduction may be of paramount importance when one tries to establish a close correspondence between the values observed at earth and at the base of the corona. It seems therefore that one cannot construct a satisfactory model of the solar wind without a proper description of the micro-structures (micro-instabilities?) determining the effective mean free path of the thermal particles. One may imagine also some of the flux and temperature variations observed at earth to be caused not so much by direct energy release on the sun but by thermal conductivity changes occurring in the interplanetary space.

References

AXFORD, W. I., DESSLER, A. J., and GOTTLIEB, B.: 1963, *Astrophys. J.* **137**, 1268.
BRANDT, J. C.: 1967, *Astrophys. J.* **147**, 201.
FERRARO, V. C. A. and BHATIA, V. B.: 1967, *Astrophys. J.* **147**, 220.
GRZĘDZIELSKI, S.: 1968, *Acta Astron.* **18**, 479.
KING, J. H. and CAROVILLANO, R. L.: 1966, *Astrophys. J.* **146**, 372.
NESS, N. F.: 1967, Goddard Space Flight Center Rep. X-612-67-293, June.
NOBLE, L. M. and SCARF, F. L.: 1963, *Astrophys. J.* **138**, 1169.
PARKER, E. N.: 1958, *Astrophys. J.* **128**, 664.
PARKER, E. N.: 1961, *Astrophys. J.* **133**, 1014.
PARKER, E. N.: 1963, *Interplanetary Dynamical Processes*, New York.
PNEUMAN, G. W.: 1966a, *Astrophys. J.* **145**, 242.
PNEUMAN, G. W.: 1966b, *Astrophys. J.* **145**, 800.
TARAFDAR, S. P.: 1968, *Ann. Astrophys.* **31**, 5.
WEBER, E. J. and DAVIS, L.: 1967, *Astrophys. J.* **148**, 217.

Discussion

Nariai: I presume that you used the data around the earth in deriving the time scale of decelerating the solar rotation. Then, could you tell me what is the reason for the difference between your results and those of Brandt and Weber and Davis?

Grzędzielski: It is because I used the energy equation, while Weber and Davis substituted it with a polytropic law.

Nariai: I did not understand well why the temperature reversal occurs. Could you explain it again?

Grzędzielski: In general a conduction model is characterized by a large fraction of energy (say, of the order of 50%) being still transported by conduction at the earth's distance. This is the case when one takes $K \propto T^{5/2}$. Suppose now that because of large magnetic fluctuations the conductivity begins to decrease more or less rapidly in the region $r > r_d$. That means that the conduction term in the energy equation decreases, more or less rapidly. Since the total energy flux has to be conserved, the decrease of the conduction term has to be compensated by a corresponding increase of other terms (flux of enthalpy and flux of mechanical energy). This results in a more steep velocity increase for $r > r_d$ (cf. Figures 1 and 2) and in enthalpy (or temperature) wave visible in Figure 3.

MASS LOSS FROM CORONAE AND ITS EFFECT
UPON STELLAR ROTATION

KYOJI NARIAI*

NASA-Goddard Space Flight Center, Greenbelt, Md., U.S.A.

Abstract. The acoustic energy-generation rate from the convective zone was calculated for various models. Results show that chromosphere and corona can be expected around stars with temperature lower than 8000 K at the main sequence, and lower than 6500 K at $\log g = 2$.

When a star is rotating rapidly, mass loss from its corona is large, and can be an effective mechanism of braking the stellar rotation. If this mechanism is effective, we can explain the slow rotation of stars later than F2 to be the result of the loss of the angular momentum through a stellar wind that is effective in their main sequence phase. Stars with mass $M > 1.5\ M_\odot$ lose mass through a stellar wind during their contraction phase. The mass-loss rate is larger than the solar value because of the larger energy input into the chromosphere-corona system and because of the smaller gravitational potential at the surface. T Tauri stars may be the observational counterparts for such stars. As the duration of contraction phase is very short (less than 10^7 years), the braking mechanism works only in the presence of a strong magnetic field (Ap) or in the presence of a companion (Am).

1. Energy Supply to the Stellar Corona

Chromospheres and coronae are characterized by their high values of the ratio of the electron temperature to the radiation temperature and by their small optical depths, both measures being extreme in the coronae. Here the radiative process works only as a cooling mechanism. The dissipation of mechanical energy suggested by BIERMANN (1948), SCHWARZSCHILD (1948), and SCHATZMAN (1949) seems to be adequate for heating the chromosphere and the corona. The rate of the acoustic energy generation in the solar convection zone with LIGHTHILL's (1952) equation gives roughly the same value as the rate of energy loss from the solar chromosphere and corona by conduction and radiation (OSTERBROCK, 1961). Therefore we may say that a star has a corona if the turbulent motions in the convective region are violent enough to produce the mechanical energy required to maintain the chromosphere-corona.

2. Atmospheric Models

In order to see the dependence of the convective zone of T_e and g, we have calculated a grid of model atmospheres with BÖHM-VITENSE's (1958) mixing-length theory as modified by HENYEY et al. (1965), who took into account SPIEGEL's (1957) suggestion concerning the temperature distribution in the turbulent element.

A stellar atmosphere is convectively stable when

$$\left(\frac{\mathrm{d}\ln T}{\mathrm{d}\ln p}\right)_{\text{ad}} > \left(\frac{\mathrm{d}\ln T}{\mathrm{d}\ln p}\right)_{\text{rad}}. \tag{1}$$

* NRC-NASA Resident Research Associate.

Near the surface, the atmosphere is stable and is in radiative equilibrium. The temperature and pressure distribution can be obtained by integrating the hydrostatic equation.

$$dp = - \rho g dh = g/\bar{\kappa} \, d\bar{\tau} \tag{2}$$

with the help of the temperature dependence on the optical depth,

$$T^4 = \tfrac{3}{4}T_e^4 \left(\bar{\tau} + q(\bar{\tau})\right). \tag{3}$$

In the above equations, p represents the pressure, T the temperature, ρ the density, g the gravity, h the height, $\bar{\kappa}$ the mean absorption coefficient, $\bar{\tau}$ the mean optical depths, and T_e the effective temperature. In the present study, we have used an analytical expression by SWAMY (1966),

$$q(\tau) = 1.39 - 0.815 \, e^{-2.54\tau} - 0.025 \, e^{-30\tau}, \tag{4}$$

which is based upon the observation of limb-darkening and the line profiles of the sun (MITCHELL, 1959). The use of this expression for a wide range of values of the effective temperature T_e and the surface gravity g does not guarantee us correct models, but nevertheless might be permitted as providing a first approximation.

In the convective region, the equations are

$$\pi F_{\text{rad}} = \frac{16\sigma T^4}{3\bar{\kappa}\rho H} \, \nabla, \tag{5}$$

$$\pi F = \frac{16\sigma T^4}{3\bar{\kappa}\rho H} \, \nabla_{\text{rad}}, \tag{6}$$

$$\pi F_{\text{conv}} = \tfrac{1}{2}C_p\rho\bar{v} \, l/H \, T(\nabla - \nabla'), \tag{7}$$

$$\bar{v}^2 = gl^2/4H \, (\nabla - \nabla'), \tag{8}$$

$$\pi F = \sigma T_e^4 = \pi F_{\text{conv}} + \pi F_{\text{rad}}, \tag{9}$$

where πF, πF_{rad}, and πF_{conv} represent the total, the radiative, and the convective flux, respectively, σ the Stefan-Boltzmann constant, H the scale height, C_p the specific heat for constant pressure per 1 g, l the mixing length, $\bar{v}$ the average velocity of the turbulent element, and ∇ the logarithmic gradient $(d\ln T/d\ln p)$. Among the logarithmic gradients, ∇_{rad} is a gradient which would be required if the total flux is carried in the form of radiation, ∇ the average gradient, ∇' the gradient for individual turbulent element, ∇_{ad} the adiabatic gradient. A convenient interpolation formula for the efficiency factor Γ which can be used both in optically thin and thick cases has been given by HENYEY et al. (1965):

$$\Gamma = \frac{\nabla - \nabla'}{\nabla' - \nabla_{\text{ad}}} = \frac{c_p\rho T\bar{v}}{8\sigma T^4\theta}, \tag{10}$$

where

$$\theta = \omega/(1 + y\omega^2), \tag{11}$$

$$\omega = \bar{\kappa}\rho l, \tag{12}$$

$$y = 3/4\pi^2. \tag{13}$$

TABLE I

Elements used in calculating models

Element	Relative number	X_1(eV)	X_2(eV)
Hydrogen	1.0	13.595	
Helium	0.15	24.581	54.403
CNO	10^{-3}	13.614	35.108
Heavy elements	10^{-4}	7.870	16.180

(For details, see references.) Model atmospheres were constructed with the composition shown in Table I. References concerning the absorption coefficients are H: KARZAS and LATTER (1961), H⁻: OHMURA and OHMURA (1960) using approximation formulae given by GINGERICH (1961), He: HUANG (1948), GOLDBERG (1939), and UENO (1954), He⁺: UENO (1954), He⁻: McDOWELL et al. (1966), and electron scattering: e.g. ALLEN (1962).

3. Acoustic Energy from the Convective Zone

Quadrupole noise generation from isotropic turbulence has been studied by LIGHTHILL (1952, 1954) and PROUDMAN (1952), and applied to the solar convection zone by OSTERBROCK (1961). The rate of the acoustic noise generation per unit volume derived by LIGHTHILL (1952) is

$$j_1 = \frac{\alpha \rho \bar{v}^8}{l V_s^5}, \tag{14}$$

where α is a numerical constant which depends only weakly upon the form of the spectrum of the turbulence and has the value 38 for the Heisenberg spectrum (PROUDMAN, 1952), l is the scale length of the turbulence, and V_s is the velocity of sound. The upward flux of energy in acoustic waves is

$$\pi F_m = \tfrac{1}{2} \int j_1 \, \mathrm{d}h. \tag{15}$$

Table II shows the results for models with various values of T_e and g when the mixing length is $l = H$. As j_1 is proportional to $\bar{v}^8$, the flux is very much dependent upon the assumed value of the mixing length. Calculation for several cases show that the result should be multiplied by a factor of 10 if we take $l = 2H$. Therefore, the results in Table II should be understood as indicating not the absolute value of the acoustic energy flux but the active region in the T_e-g plane on a relative scale. KUPERUS (1965) has estimated the acoustic energy flux in a very approximate way using the scale height and the maximum velocity given by BÖHM-VITENSE (1958). Comparisons between his and the present results show that a model calculation, at least, is necessary for a proper discussions of the history of coronae.

TABLE II

Mechanical energy flux for models with $l = H$

$\log g$	T_e	$\log \pi F_m$	$\log g$	T_e	$\log \pi F_m$
2.0	5000	8.05	4.0	5000	7.23
	6000	8.27		6000	7.72
	6500	7.72		7000	7.94
	7000	2.60		7500	7.38
	8000	-10		8000	4.67
3.0	5000	7.67	5.0	6000	7.36
	6000	8.18		7000	7.60
	6500	8.07		8000	7.44
	7000	7.47		8500	5.91
	7500	4.23			
	8000	-2.47			

4. Stellar Rotation and Mass Loss from Coronae

BRANDT (1966), WEBER and DAVIS (1967) and MODISETTE (1967) studied the torque exerted by the solar wind upon the sun through its magnetic field. According to them, the characteristic time of the deceleration of the solar rotation is 7×10^9 years, which is comparable to the age of the sun. MESTEL (1968) treated the same problem in a more elaborate way but this result was 20 times larger than the ones by Brandt and Weber and Davis. The difference between Mestel and the others may have been caused by the use of different boundary conditions, namely the coronal temperature and density in MESTEL's (1968) case, and the solar-wind data in BRANDT's (1966) and WEBER and DAVIS' (1967) work. The characteristic time is expressed as follows by WEBER and DAVIS (1967):

$$\frac{1}{\tau} = \frac{2r_a^2}{3I} \frac{\mathrm{d}M}{\mathrm{d}t}, \tag{16}$$

where I is the moment of inertia, $\mathrm{d}M/\mathrm{d}t$ the rate of mass loss, and r_a is the distance at which radial gas velocity u becomes equal to the radial Alfvénic velocity

$$\frac{4\pi\rho u^2}{B_r^2} = 1 \quad (\text{at } r = r_a). \tag{17}$$

Equation (16) indicates that the magnetic field can exert torque to the gas flow up to $r = r_a$, therefore the matter which flows along the magnetic line of force gains energy as well as angular momentum until it reaches $r = r_a$. In the solar wind, r_a lies between 15 and 50 $R_\odot$ according to WEBER and DAVIS (1967).

The rate of mass loss is almost independent of the value of r_a if it is larger than r_c, which is the distance where the flow is trans-sonic. In the case of zero rotation (PARKER 1958),

$$\frac{GMm_H}{r_c} = \frac{kT}{2}. \tag{18}$$

The above implies that the change of the flow from subsonic to supersonic occurs at the distance where the gravitational force becomes so small that it can no longer hold the gas against the expansion caused by the thermal energy. In analogy to equation (18), we may consider that the following relation holds in a rigidly rotating corona;

$$m_H \left(\frac{GM}{r_c} - r_c^2 \Omega^2 \right) = \frac{kT}{2}. \tag{19}$$

Let us define two distances $r_{c,1}$ and $r_{c,2}$, where

$$\frac{GM}{r_{c,1}^2} = r_{c,1} \Omega^2, \tag{20}$$

and

$$\frac{GMm_H}{r_{c,2}} = \frac{kT}{2}. \tag{21}$$

Then

$$r_c \sim r_{c,1} \tag{22}$$

when Ω is large (stage I), and

$$r_c \sim r_{c,2} \tag{23}$$

when Ω is small (stage II).

For the coronae of solar-type stars, the transition from stage I to stage II occurs at around $\Omega \sim 2 \times 10^{-5}$, which corresponds to about 15 km/sec of rotational velocity at the surface.

If the critical distance is small, the escape of gas from the gravitational potential occurs where the gas is dense. Therefore we might expect a larger rate of mass loss when a star is rotating faster. The increase in mass loss causes a slight decrease in r_a, but the product $r_a^2 \, dM/dt$ increases. Thus we may expect a faster decay of stellar rotation when the rotational velocity is greater than 15 km/sec.

5. The Evolutionary Point of View

In Figure 1, evolutionary tracks of stars from the contraction phase to the main sequence (IBEN, 1965) have been drawn on the T_e-g plane, together with a plot of the mechanical energy flux calculated in Section 3.

Stars with mass $M < 1.5 \, M_\odot$ remain in the active region from the contraction to the main-sequence phase. From the discussions in the previous section we may conclude that the characteristic time for decelerating the rotation was smaller than the lifetime of the star by a factor of more than, say, 10, when the star was rotating rapidly. WALKER (1956) has pointed out that in the young cluster NGC 2264, the rotational velocities of stars of the spectral class later than F8 are large, and suggested that the deceleration time might be larger than the lifetime of the cluster (3×10^6 years). KRAFT (1967) studied the rotation of stars in galactic clusters and of field stars and derived 4×10^8 years for the time scale for reduction of the rotational

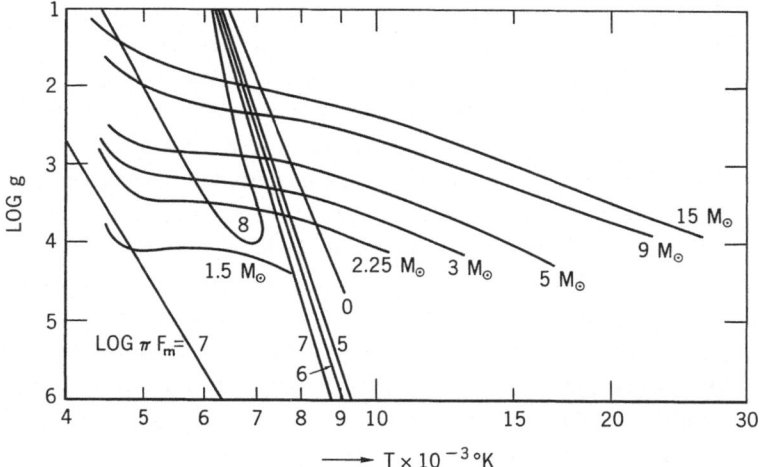

Fig. 1.   Acoustic energy-generation rate and evolutionary tracks.

velocities by a factor of 2. This period of fast rotation and large mass loss might also be the period of strong Ca II H-K emission, with strong mass loss changing the chromospheric structure. Then, the above interpretation is in good agreement with WILSON's (1966) conclusion that strong H-K emission is found in stars near the zero-age main sequence.

Stars with mass $M > 1.5\ M_\odot$ have active coronae only during a part of the contraction phase, which is usually less than 10^7 years. In ordinary cases these stars cannot lose their angular momentum because the time scale is smaller compared to the characteristic time of deceleration. This is in very good correlation with the observational fact that the stellar rotational velocity changes abruptly at spectral type F2, from more than 100 km/sec for the hotter stars to less than 20 km/sec for the cooler stars within a mass range of 5% of the mass of a F2 star (WILSON, 1966).

T Tauri stars might be interpreted to be stars with $M > 1.5\ M_\odot$ passing the active phase. If a strong magnetic field is present as in Ap stars, even massive stars can lose their angular momentum in such a short period. Duplicity helps the mass loss to become greater (NARIAI, 1967). Am stars may be interpreted to be massive stars that had strong mass loss and have been decelerated during the contraction phase because of duplicity. (T Tauri, Ap, and Am stars will be discussed in the next two sections.)

6. T Tauri Stars

T Tauri stars show emission in the hydrogen and H-K lines, show displaced absorption line indicating violent mass loss, have large rotational velocities ranging from 20 to 65 km/sec, have higher luminosities than stars of the same spectral type at the main sequence, and are associated with young clusters whose ages are $\lesssim 10^7$ years. (For details, see the review paper by HERBIG, 1962.) These last two facts indicate that these stars are in the contraction stage.

In the active convection region of contracting stars (see Figure 1), there is usually more acoustic energy generation and less gravitational energy per unit mass at the surface than is to be found in the main-sequence stars. These two conditions together with rapid rotation provide favorable conditions for mass loss from the corona. KUHI's (1964) analysis showing that the more massive stars lose mass at larger rates is consistent with the present interpretation. KUHI (1964) derived masses from luminosities and colors with the help of IBEN's (1965) evolutionary track for the contraction phase. Therefore, 'more massive' means 'more luminous' here, and can be interpreted as meaning 'having less gravitational energy at the surface'. The mechanical energy transferred into the corona is spent either in (1) accelerating the flow, or (2) pushing the gas out of the gravitational potential of the star, or (3) causing radiation loss through X-rays, or (4) causing the conduction flow back to the chromosphere. When the gravity is sufficiently high, the first two processes are negligible and the temperature of the corona can be obtained by a procedure used by KUPERUS (1965) as a function of the acoustic energy-generation rate. But when the gravity is not so large, as in the case of T Tauri stars, the temperature does not attain so high a value as KUPERUS (1965) suggests, because the energy is consumed in the first two processes. The above discussions are drawn from UCHIDA (1967) and UNNO (1967). What we can expect from the above discussions and the calculations in Section 3 is that more luminous stars that have lesser gravity but have larger acoustic energy generation must show larger mass loss, which is in good agreement with KUHI's (1964) observation.

7. Peculiar A Stars

The discussions in Section 5 showed that a star with mass $M > 1.5\ M_\odot$ cannot lose its angular momentum during the T Tauri phase if the magnetic field is of the order of the solar value (1 gauss at the surface). However, there are stars called peculiar A-type stars because they show peculiar abundance anomalies. These stars have strong magnetic fields ranging from several hundred up to 34 kilogauss in the extreme case. The surface temperatures of these stars correspond to those of normal stars in the spectral range B5 ~ F0. The implication is that the masses of these stars are greater than 1.5 $M_\odot$. The rotational velocities of these stars are remarkably smaller than those of normal stars of the same spectral class. (For details, see CAMERON, 1967). Since the Alfvén velocity is proportional to the magnetic-field intensity, r_a is larger in Ap stars than in ordinary stars. Therefore the characteristic time of deceleration of the stellar rotation may become less than the contraction time for Ap stars. As the strong magnetic field and rapid rotation both tend to suppress the turbulent motion, the whole discussion in this section may prove to be invalid when the problem is solved rigorously, because we are applying the rate of acoustic energy generation under conditions of zero (weak) magnetic field to the deceleration of rotation through a stellar wind with a strong magnetic field. Nevertheless, the following comment may be worth mentioning. In the gravitational field, heavier elements tend to concentrate toward the center, while lighter elements rise toward the surface, according to the

Boltzmann's law. The reason we do not see the separation of elements in the spectrum of stars is that the characteristic time of diffusion is longer than the stellar lifetimes. (See EDDINGTON, 1926.) The diffusion time is expressed as

$$\tau_{diff} = x^2/D,\tag{24}$$

where x is the characteristic length and D, the diffusion coefficient, is given by

$$D = \tfrac{1}{3}\lambda V_s \sim 10^8 \; T^{2.5}/N,\tag{25}$$

where λ is the mean free path and V_s the velocity of sound. In the corona, however, the characteristic diffusion time is of the order of from a few hours to several days because the density is so low and the temperature is so high. Therefore, if the mixing time is more than the diffusion time, we may expect that heavier elements will be relatively less abundant in the upper corona. Then we may also expect that the same situation will occur in the stellar wind, which is governed by the physical state of the upper corona. Thus, the remaining corona becomes richer in the heavy elements. If we take the particle point-of-view in interpreting the stellar wind, the situation becomes more extreme. In this interpretation, the stellar wind is composed of the particles which do not collide with other particles since they have gained escape velocity. Then, as the heavier elements have smaller thermal velocities, the height at which the escape occurs effectively for heavier elements is higher than that for lighter elements. Since larger height implies smaller density, the enrichment of heavier particles in the remaining corona occurs even without element separation caused by the gravitational field.

Keeping this fact in mind, let us proceed now to the next part of the discussion. Figures 2 and 3 show the acoustic energy generation rate and the thickness of the convection zone in g cm^{-2} as a function of time, respectively, for a star with 2.25 $M_\odot$. A gap in the thickness at 1.75×10^6 years is due to the separation of the convection zone into two parts, the upper one being due to H ionization and the lower one to He II ionization. Acoustic energy generated after this separation is 1.5×10^{21} erg cm^{-2}. A

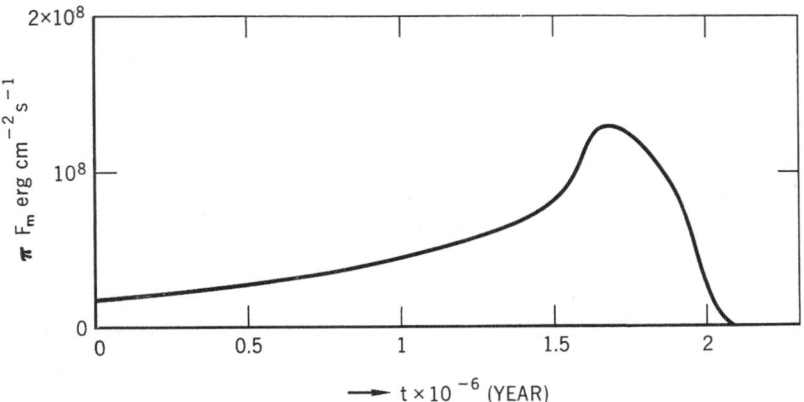

Fig. 2.   Change of the mechanical energy flux with time for $M = 2.25 \; M_\odot$.

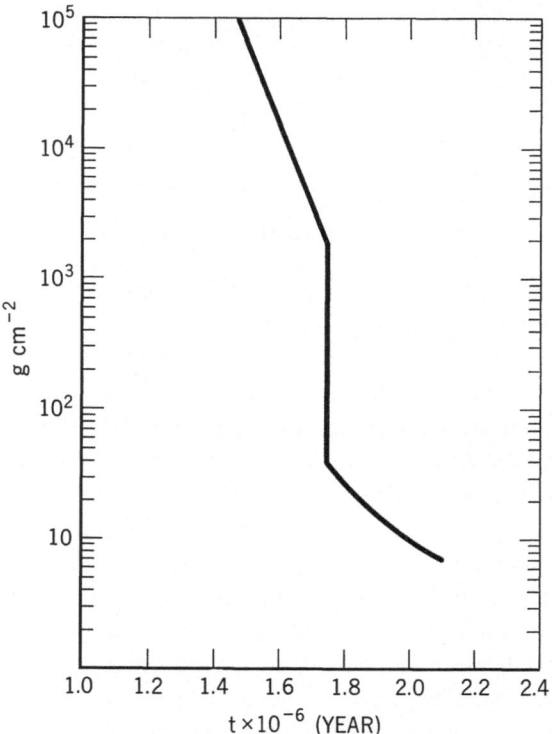

Fig. 3.   Mass contained in the convection zone for $M = 2.25\ M_\odot$.

part of this energy is spent in pushing the material out of the gravitational potential of the star. The ratio of the energy required for pumping the matter out and the energy generated in the convection zone is about 10^{-2} for the sun. This factor might be larger for stars with mass $M > 1.5\ M_\odot$ at T Tauri stage because these stars have smaller values of g. Stronger magnetic fields in Ap stars also assist the transfer of energy into the corona without its being consumed in the chromosphere. If we assume 0.1 for that ratio, then the total amount of material which passes through the convection zone is about 10^3 times larger than the material contained in the convection zone. Thus we can expect enrichment of heavier elements in the visible thin layer in stars with mass $M > 1.5 M_\odot$. The diffusion time at the bottom of this concentrated region is of the order of 10^{12} years. Therefore, unless a mechanism other than diffusion is at work, we would observe metal-rich stellar surfaces. Then the spectral difference between Ap and normal stars could be interpreted as due to (1) the small circulatory motion because of slow rotation, of Ap stars, and (2) the suppression of the circulatory motion by the magnetic fields in Ap stars.

Almost the same discussion can be applied to Am stars. The difference between Ap and Am cases is that the existence of a companion may raise the mass loss rate from coronae of Am stars, which may cause the differences in the degrees of concentration of metals.

Acknowledgements

The author's thanks are due to Dr. R. C. Cameron and Dr. M. P. Nakada and many other colleagues of the Goddard Space Flight Center for useful discussion and encouragement. The author holds an NRC-NASA Resident Research Associateship.

References

ALLEN, C. W.: 1962, *Astrophysical Quantities*, The Athlone Press.
BIERMANN, L.: 1948, *Z. Astrophys.* **25**, 161.
BÖHM-VITENSE, E.: 1958, *Z. Astrophys.* **46**, 108.
BRANDT, J. C.: 1966, *Astrophys. J.* **144**, 1221.
CAMERON, R. C. (ed.): 1967, *The Magnetic and Related Stars*, Mono Books Corp., Baltimore.
EDDINGTON, A. S.: 1926, *The Internal Constitution of Stars*, Cambridge University Press, Cambridge.
GINGERICH, O.: 1961, *Astrophys. J.* **134**, 653.
GOLDBERG, L.: 1939, *Astrophys. J.* **90**, 414.
HENYEY, L., VARDYA, M. S., and BODENHEIMER, P.: 1965, *Astrophys. J.* **142**, 841.
HERBIG, G.: 1962, *Adv. Astron. Astrophys.* **1**, 47.
HUANG, S-S.: 1948, *Astrophys. J.* **108**, 354.
IBEN, I.: 1965, *Astrophys. J.* **141**, 993.
KARZAS, W. J. and LATTER, R.: 1961, *Astrophys. J. Suppl.* **6**, 167.
KRAFT, R. P.: 1967, *Astrophys. J.* **150**, 551.
KUHI, L. V.: 1964, *Astrophys. J.* **140**, 1409.
KUPERUS, M.: 1965, *Rech. Astron. Obs. Utrecht* **17** (1).
LIGHTHILL, M. J.: 1952, *Proc. Roy Soc.* **A211**, 564.
LIGHTHILL, M. J.: 1954, *Proc. Roy. Soc.* **A222**, 1.
MCDOWELL, M. R. C., WILLIAMSON, J. H., and MYERSCOUGH, V. P.: 1966, *Astrophys. J.* **144**, 831.
MESTEL, L.: 1968, *Monthly Notices Roy. Astron. Soc.* **138**, 359.
MITCHELL, W. E.: 1959, *Astrophys. J.* **129**, 369.
MODISETTE, J. L.: 1967, *J. Geophys. Res.* **72**, 1521.
NARIAI, K.: 1967, *Publ. Astron. Soc. Japan* **19**, 564.
OHMURA, T. and OHMURA, H.: 1960, *Astrophys. J.* **131**, 8.
OSTERBROCK, D. E.: 1961, *Astrophys. J.* **134**, 347.
PROUDMAN, I.: 1952, *Proc. Roy. Soc.* **A214**, 119.
SCHATZMAN, E.: 1949, *Ann. Astrophys.* **12**, 203.
SCHWARZSCHILD, M.: 1948, *Astrophys. J.* **107**, 1.
SPIEGEL, E. A.: 1957, *Astrophys. J.* **126**, 202.
SWAMY, K. S. KRISHNA: 1966, *Astrophys. J.* **145**, 174.
UCHIDA, Y.: 1967, *Proc. Tokyo Symp. UV and IR Radiation from Celestial Bodies* (in Japanese), p. 60.
UENO, S.: 1954, *Contr. Inst. Astrophys. Univ. Kyoto*, No. 42.
UNNO, W.: 1967, *Proc. Tokyo Symp. UV and IR Radiation from Celestial Bodies* (in Japanese), p. 71.
WALKER, M. F.: 1956, *Astrophys. J. Suppl.* **2**, 365.
WEBER, E. J. and DAVIS, L., Jr.: 1967, *Astrophys. J.* **148**, 217.
WILSON, O. C.: 1966, *Astrophys. J.* **144**, 695.

CORONAE AROUND HELIUM STARS AND X-RAY SOURCES

KYOJI NARIAI*

NASA-Goddard Space Flight Center Greenbelt, Md., U.S.A.

Abstract. Calculations of the acoustic energy generation for helium-rich composition show that the maximum acoustic energy generation is located around 12000 K at $\log g = 4$ and 15000 K at $\log g = 6$. The author's suggestion in his last paper that a helium star υ Sgr may have a corona seems to be justified. X-ray from a corona around a helium star is strongest when the physical parameters of the star are $\log g \sim 6$ and $T_e \sim 15000$ K. But the total energy flux is too small to account for the observed X-ray sources.

The star υ Sgr is a single-lined spectroscopic binary with a helium-rich envelope (for refs., see NARIAI, 1967). This star exhibits a displaced absorption line of Hα for about one-quarter of the orbital period, the center of the phase of this phenomenon being the conjunction time. The present author has interpreted this phenomenon to be the result of the flow from the corona through the Lagrangian point (NARIAI, 1967). The gas flow is subsonic inside the Lagrangian surface, becomes trans-sonic at the Lagrangian point, and then expands nearly adiabatically getting cooler and acquiring the translational energy.

Calculations similar to those of Section 3 of another paper (NARIAI, 1968) have been made for a helium-rich composition in order to see where the maximum activity occurs in helium stars.

$$q(\tau) = 0.7104 - 0.1331 \, e^{-3.4488\tau},\tag{1}$$

which represents the gray atmosphere very well (LABS, 1950) was used in the temperature-depth relation

$$T^4 = \tfrac{3}{4} T_e^4 (\tau + q(\tau)).\tag{2}$$

Acoustic energy generation rates for helium to hydrogen ratio 400:1 are presented in Figure 1. The excitation temperature for hydrogen, helium, nitrogen, etc., by HACK and PASINETTI (1963) – 12800 K – indicates that this star has a violent convection zone due to helium I ionization, and consequently, a strong corona. A large value of the turbulent velocity ($\xi_t = 9$ km/sec for He, $\xi_t = 18$ km/sec for H) also provides evidence for the existence of the convection zone.

The change of the acoustic energy-generation rate with the helium-to-hydrogen ratio was calculated and is shown in Figure 2 for $T_e = 12000$ K and $\log g = 4$. Practically speaking, stars at this temperature have coronae if N_{He}/N_H is more than 5.

In the solar corona, the gravitational energy at the bottom is about 10 times larger than the thermal energy. But the energy lost in X-ray radiation from the solar corona is about one-tenth of the energy spent in accelerating the gas and pushing the gas

* NRC-NASA Resident Research Associate.

M. Hack (ed.), Mass Loss from Stars. All rights reserved

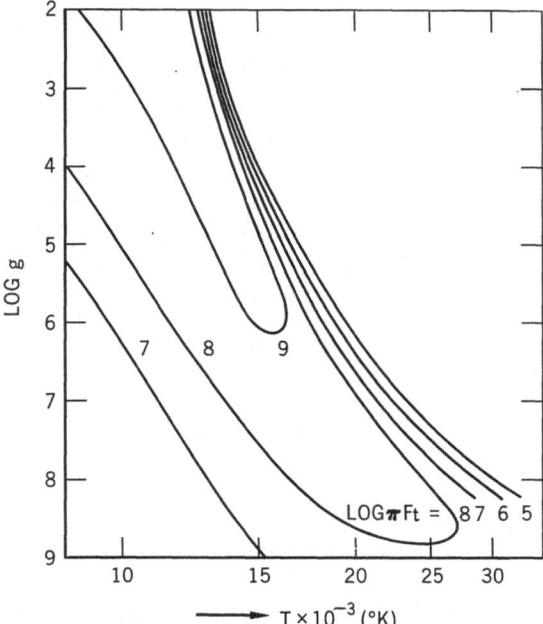

Fig. 1.   Acoustic energy-generation rate for helium-rich atmospheres. $N_{He}/N_H = 1000/2.5$.

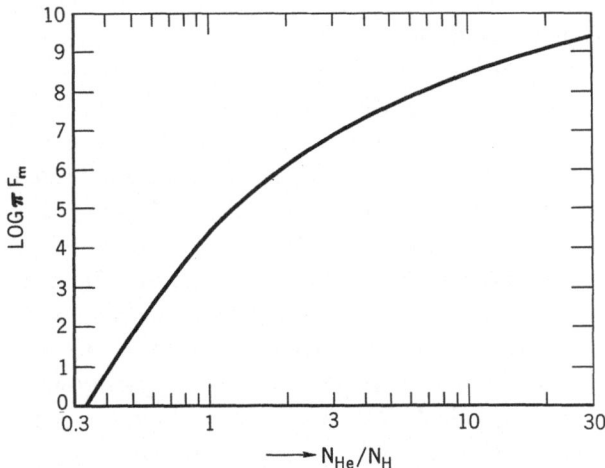

Fig. 2.   Change of acoustic generation rate with helium-to-hydrogen ratio at $T_e = 12000\,\text{K}$ and $\log g = 4$.

out of the gravitational potential of the sun in the solar wind. This means that a higher ratio of gravitational to thermal energy is necessary in order to change the mechanical energy into X-ray radiation. The most promising case would be a helium-rich star with $\log g \leqslant 6$ and $T_e \sim 15000$ K. A star with a helium-rich composition and with mass $M > 10\,M_\odot$ can take such values in its contraction phase toward the helium main sequence. Assuming an optimistic value of 0.1 for the ratio of the

X-ray flux to the energy generation rate in the convection zone, we obtain X-ray radiation of 10^{31} erg sec^{-1}. But this value still seems to be too small to account for the observed X-ray sources.

Acknowledgement

The author wishes to express his thanks to Rev. F. J. Heyden, Director of the Georgetown University Observatory for his generosity to the author in his using the library of the observatory. The author holds the NRC-NASA Resident Research Associateship.

References

HACK, M. and PASINETTI, L. E.: 1963, *Contr. Milano Merate Obs.* No. 215.
LABS, D.: 1950, *Z. Astrophys.* **27**, 153.
NARIAI, K.: 1967, *Publ. Astron. Soc. Japan* **19**, 564.
NARIAI, K.: 1968, *Astrophys. Space Sci.* **3**, 150.

Discussion

Underhill: You have computed the amount of acoustic energy generated in the convection zone of stars, but does this energy escape and move in sufficient quantity to an outer part of the stellar atmosphere and create there a corona? I believe that in the case of the sun it is confirmed that acoustic energy is a very likely source for heating the corona, but no satisfactory theory exists for transporting the energy from the convection zone to the corona. Is this so?

Nariai: If the mechanical energy is generated in the form of acoustic energy first, and then transferred into corona somehow, then the problem you pointed out becomes the theoretical problem for solar corona and my arguments are safe because the solar case is used as a scaling measure. But if the mechanical energy is initially generated in the form of gravity wave, then the results might be a little different.

Kutter: In a paper (in *Solar Physics*) about one year ago, Stein presented formulae for the generation of acoustic as well as gravity (or internal) waves. Under certain circumstances the gravity waves are more efficient in transferring energy to the corona than acoustic waves. Have you considered the effect of gravity waves?

Nariai: In the present calculations, I did not include the generation of gravity waves. Such calculations should really be done in finer works later.

MASS LOSS FROM FAST ROTATING STARS

L. NOBILI and L. SECCO

Istituto di Fisica dell'Università, Padova, Italy

Abstract. Evolutionary models of a rotationally unstable $7\,M_\odot$ star are computed; preliminary results during the H-burning stages are obtained and discussed.

These are very preliminary results concerning mass loss from fast rotating stars.

With the program related to the integration of the stellar equilibrium equations for stars of extreme population I ($X=0.602$, $Z=0.044$) kindly lent to us by Prof. Kippenhahn, we have considered the rotational instability of a $7\,M_\odot$ star, during the first stages of its post-main-sequence evolution.

As a first approach, we have treated a spherical and rigidly rotating structure, owing to the obvious difficulties of more complicated models, without examining the very involved and up to now open problem of the possible mixings due to meridional circulations.

Such a structure becomes rotationally unstable as soon as the ratio of the centrifugal to the gravitational force at the surface is $\geqslant 1$, that is when:

$$\lambda = \frac{\omega^2 R}{GM/R^2} \geqslant 1, \tag{1}$$

where ω, R and M are the angular velocity, the radius and the total mass respectively. Writing the total angular momentum in the form

$$I = kR^2 M\omega$$

with

$$k = \frac{2}{3R^2 M} \int_0^M r^2 \, \mathrm{d}M,$$

the expression (1) may be written as follows:

$$\left(\frac{I^2}{GM^3}\right) \cdot \frac{1}{k^2 R} \geqslant 1.$$

In absence of mass loss the bracketed term is obviously a constant; if we therefore assume an initial angular momentum for the star, the behaviour of $1/k^2 R$ as a function of age determines the stages of rotational instability. Such a dependence of $1/k^2 R$ from age for a stable star of $7\,M_\odot$ has been calculated and results as shown in Figure 1a (solid line), while the corresponding H-R diagram of the stable star is reported in Figure 2.

We may note that in the portion AB of its evolutionary track the star is burning

M. Hack (ed.), Mass Loss from Stars. All rights reserved

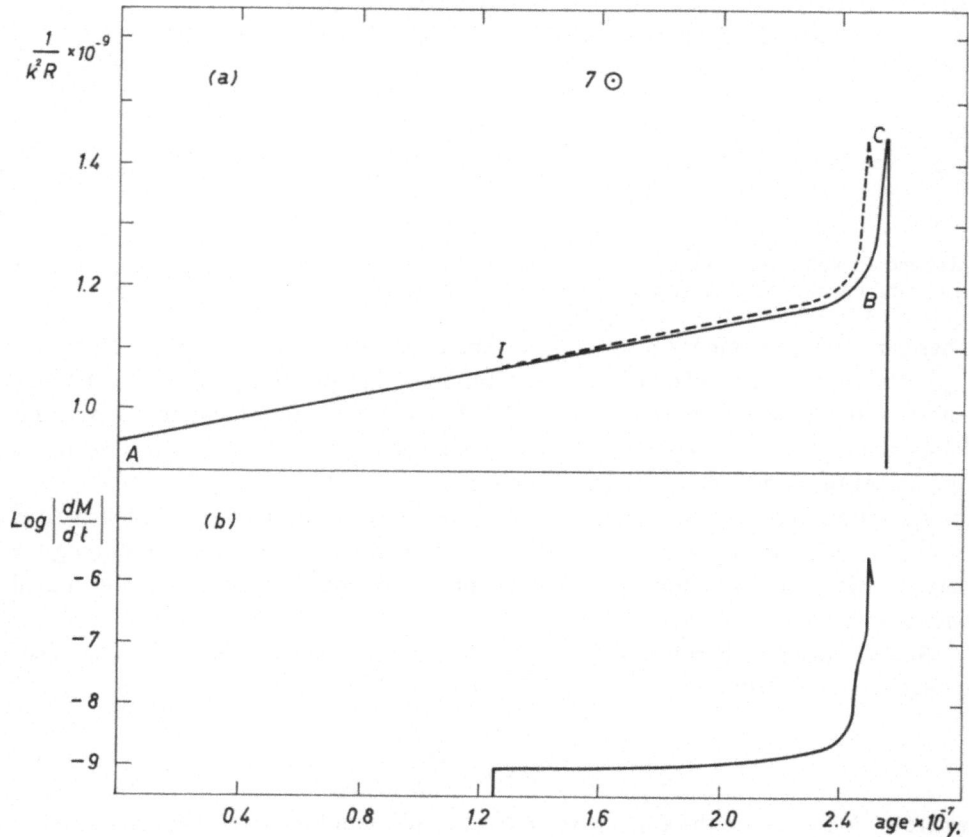

Fig. 1(a). The time-dependence of the function $1/k^2R$ for a $7\ M_\odot$ star without mass loss (solid line); the modified curve by mass loss due to rotation is given by the dashed line. – (b) Mass-loss rate (in solar masses per year) for the $7\ M_\odot$ star with rotational instability and mass loss.

its hydrogen and is expanding, while in the portion BC it goes through the stage of hydrogen exhaustion and gravitational contraction; the behaviour of k explains the results of Figure 1a, as in portion AB the function k decreases but the radius increases in such a way as to make the function $1/k^2R$ increase slowly, while in the portion BC both k and R decrease, so the function $1/k^2R$ increases quickly.

We see also, if the rotating star is initially on the main sequence with a value of λ <1, that it is possible, owing to the progressive increase of $1/k^2R$ during the stages from A to C, for λ to become $\geqslant 1$: thus the star becomes unstable and is expected to lose mass from its equator.

As soon as the mass loss begins, the instability is maintained if

$$\frac{\Delta\lambda}{\lambda} = 2\frac{\Delta I}{I} - 3\frac{\Delta M}{M} + \frac{\Delta F}{F} \geqslant 0, \tag{2}$$

where $F=1/k^2R$, the quantities ΔI, ΔF being evaluated between two consecutive models differing in mass by an amount $\Delta M(<0)$ and in age by Δt.

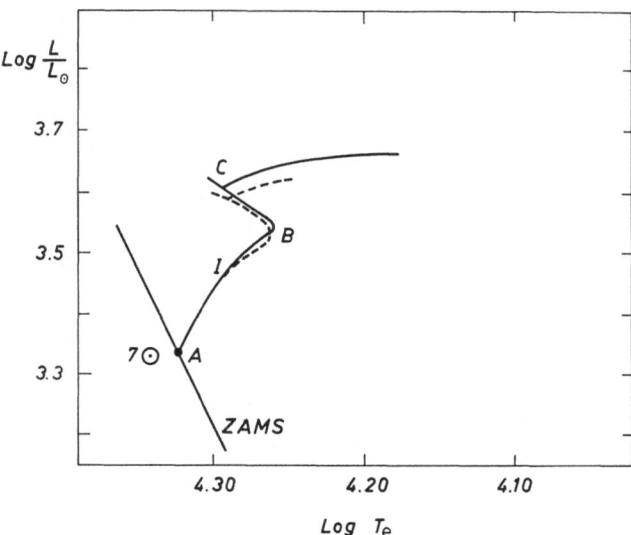

Fig. 2. Evolutionary paths in the H-R diagram for the stable (solid line) and rotationally unstable star (dashed line) of the text.

As the star loses mass from its equator, we have

$$\Delta I = \omega R^2 \, \Delta M$$

and we may write (2) in the form

$$\frac{\Delta \lambda}{\lambda} = \left(\frac{2}{k} - 3 \right) \frac{\Delta M}{M} + \frac{\Delta F}{F} \geqslant 0. \tag{3}$$

The amount of mass lost by the star has been computed in the following way. Let us start from a configuration in which condition (3) with equal sign has been exactly reached. This means that the star fills a radius corresponding to condition (1) with equal sign. Then for a given ΔM lost we calculate the corresponding ΔF adequate to maintain condition (3) with equal sign, and this leads us to a new value of the instability radius. One then calculates according to the evolution process the time interval Δt running between the two configurations considered and $\Delta M/\Delta t$ yields the mass-loss rate. This procedure is of course always possible as long as F increases (ΔF being >0 and $\Delta M < 0$, (3) with equal sign may always be satisfied); but when F decreases, owing to the negative values of both ΔM and ΔF, (3) cannot be satisfied any more. The star has thus reached its stability and mass loss has to stop. The behaviour of $1/k^2 R$ during the mass-loss rate is not modified in a sensitive manner in respect to its behaviour in the non mass-loss case: it is represented in Figure 1a (dashed line).

In the first region immediately following the beginning of mass loss (point I in Figures 1a and 2) the loss rate is very small (about $10^{-9} M_\odot$ per year), but it increases during the stages of hydrogen exhaustion to values up to 10^{-7} and $10^{-6} M_\odot$ per year (Figure 1b).

Nevertheless, the total amount of mass lost is by no means conspicuous on account of the fact that during the slow rate stage the lifetime of the star is long, while, on the contrary, in the quick loss rate stage its lifetime is short. Owing to the smallness of the effect, the evolutionary track is only slightly modified in respect to the track of constant mass (dashed track in Figure 2).

The results obtained do not seem inconsistent with the data related to mass loss from Be stars, and to some qualitative theoretical provisions by several authors (CRAMPIN and HOYLE, 1960; SCHMIDT-KALER, 1964; IBEN, 1967), interpreting them as due to rotation.

The work is still in progress in order to follow the evolution further and extend the considerations of rotation to several different masses.

Acknowledgements

We want to thank Prof. N. Dallaporta for having suggested to us the idea of this work and for several useful discussions. We gratefully acknowledge the financial support of the Consiglio Nazionale delle Ricerche.

References

CRAMPIN, J. and HOYLE, F.: 1960, *Monthly Notices Roy. Astron. Soc.* **120**, 33.
IBEN, I.: 1967, *Ann. Rev. Astron. Astrophys.* **5**, 571.
SCHMIDT-KALER: 1964, *Bonn. Veröffentl.* **70**, 1.

Discussion

Lauterborn: (1) Which value is taken for R in the equation $\Delta I = \omega R^2 \Delta M$? (2) Did the calculations appreciate the changes of star evolution as a result of mass loss? That is: did you calculate models? (3) What was the amount of angular momentum fed into the star at the beginning?

Nobili: (1) The value we have taken is not the total radius of the star, as could appear from the formula $\Delta I = \omega R^2 \Delta M$, but a mean value, calculated in the outer unstable shells according to our model distribution, which may vary between 0.8 R_{tot} to 1 R_{tot} according to the amount of ΔM used in any of the steps.

(2) Yes, we have calculated an evolutionary sequence. From models unstable against rotation we have taken matter off by an amount depending upon the extent in mass of the unstable shells.

(3) The initial value of angular momentum corresponds to an equatorial velocity of about 500 km/sec.

Nariai: Why can the mass loss be larger after the main-sequence stage?

Nobili: As the mass concentration towards the centre becomes larger after the main sequence stage, the momentum of inertia becomes smaller, and thus, as mentioned, the function $1/k^2 R$ is always increasing. Therefore, even if the radius becomes larger, we can expect rotational instability at the surface.

ANGULAR MOMENTUM CONSIDERATIONS
AND MASS LOSS

I. P. WILLIAMS

The University of Reading, England

Abstract. The conservation of energy and angular momentum in a contracting star that is ejecting material is considered. It is found that in cases likely to be of interest, either an upper limit to the mass loss exists when the material escapes to infinity, or, when the primary aim is to remove angular momentum, a lower limit to the rate of mass loss exists.

1. Introduction

It is evident that for all stars following a fairly normal process of evolution, some restrictions on the mass loss must exist. By considering energy sources and requirements in a star that is ejecting matter, WILLIAMS (1967) has obtained an upper bound to this mass loss. It was shown that the maximum rate of losing mass, m, is given by

$$m = \lambda LR/GM, \tag{1}$$

where L, R and M denote the luminosity, radius and mass of the star respectively, G is the gravitational constant, and λ a numerical constant whose value is less than unity.

However, if a star is rotating it is clear that conservation of angular momentum leads to additional constraints and possibly further restrictions. In this communication we derive the equations for the conservation of energy and momentum for a star contracting on to the main sequence and from there obtain restrictions on the mass loss. We consider a star ejecting material, but not transferring any more angular momentum once ejection has occurred.

One simple aspect of these problems, already discussed by PORFIRIEV (1967), is that the total angular momentum removed from a star during its contraction phase is equal to the average amount per unit mass removed times the mass loss. By expressing this in mathematical form Porfiriev obtains a lower limit to the amount of matter removed in terms of its initial and final radius and rotation and its initial mass. This contains rather more variables than are usually known, and it does not give the mass loss at a given epoch in terms of convenient parameters. We therefore re-investigated the problem, considering all the mathematical cases that arise. Some of these cases have very little physical significance but for completeness have been included.

2. Basic Equations

At the epoch of interest, let the star be of mass M, radius R and rotating with an angular velocity ω. After an interval of time dt, let the corresponding quantities be $M+dM$, $R+dR$ and $\omega+d\omega$. The rate of mass loss, m, is therefore given by $-dM/dt$. Assume

M. Hack (ed.), Mass Loss from Stars. All rights reserved

that k is a constant throughout the interval of interest so that the radius of gyration is Rk. Similarly let Rk_1 be the radius of gyration of the ejected material, k_1 being also assumed to be constant throughout the interval dt. The angular momentum carried away by the ejected matter is therefore

$$R^2 k_1^2 \omega (- \mathrm{d}M).$$ (2)

Conservation of angular momentum for the system consisting of the star and the ejected matter gives

$$Mk^2 R^2 \omega = k^2 (M + \mathrm{d}M)(R + \mathrm{d}R)^2 (\omega + \mathrm{d}\omega) - R^2 k_1^2 \omega \, \mathrm{d}M,$$ (3)

or to the first order in small quantities,

$$R^2 \omega \, \mathrm{d}M (k_1^2 - k^2) = Mk^2 R^2 \, \mathrm{d}\omega + 2 Mk^2 R\omega \, \mathrm{d}R.$$ (4)

It is not only angular momentum that is conserved for the two-body system. Energy, if radiation is included, is also conserved, and we now consider this. For a system that is changing slowly enough for $\mathrm{d}^2 I/\mathrm{d}t^2 = 0$, where I is the total moment of inertia, then a virial theorem is valid, its form being

$$2T + \Sigma \mathbf{r}_i \cdot \mathbf{F}_i = 0,$$

where T is the total kinetic energy, $\mathbf{r}_i$ denotes the position of the ith particle, and $\mathbf{F}_i$ the force acting on it. When mutual gravitational attraction is the only force acting $\Sigma \mathbf{r}_i \cdot \mathbf{F}_i = \Omega$ the gravitational potential energy, and for a star of polytropic index n,

$$\Omega = - \frac{3}{5 - n} \frac{GM^2}{R}.$$ (5)

If the star is rotating, then consider a rotating set of axes rotating with the star. The kinetic energy in such a system will remain unaltered and will be given by

$$T = \tfrac{3}{2}(\gamma - 1) U,$$ (6)

where U is the total internal energy and γ the ratio of specific heats. However, an additional term will be present in $\Sigma \mathbf{r}_i \cdot \mathbf{F}_i$, for $\mathbf{F}_i$ will now consist of an additional part, because of rotation, such that $\Sigma \mathbf{r}_i \cdot \mathbf{F}_i = \Sigma_i M_i R_i^2 \omega^2 + \Omega$ (provided a unique ω exists for the body) where R_i is the projection of $\mathbf{r}_i$ on the plane defined by $\boldsymbol{\omega}$. The virial theorem therefore gives

$$2T + \Omega + \Sigma M_i R_i^2 \omega^2 = 2T + \Omega + Mk^2 R^2 \omega^2 = 0.$$ (7)

This result can equally be reached by considering a non-rotating frame with the total kinetic energy consisting of two parts, T and $\tfrac{1}{2}Mk^2 R^2 \omega^2$, but the potential energy now invariant. The total energy, E, of the system is given by

$$E = U + \tfrac{1}{2}Mk^2 R^2 \omega + \Omega.$$ (8)

From Equations (7) and (8) we derive

$$E = (4 - 3\gamma) U - \tfrac{1}{2} M k^2 R^2 \omega = \Omega \frac{(3\gamma - 4)}{3(\gamma - 1)} + \frac{(3\gamma - 5)}{6(3\gamma - 1)} M k^2 R^2 \omega. \tag{9}$$

For most stars of interest to us, $\gamma = \tfrac{5}{3}$, and we have

$$E = \Omega/2 ,$$

precisely as if no rotation had been present, Alternatively the work which follows can, with minor modifications, apply to a star with any given value for γ provided the rotational energy is small. The amount of energy lost to the star is $-dE$, that is

$$\frac{1}{2} d \left(\frac{3}{5 - n} \frac{GM^2}{R} \right) = \frac{3G}{2(5 - n)} \left\{ \frac{2M \, dM}{R} - \frac{M^2 \, dR}{R^2} \right\}. \tag{10}$$

The energy lost to the star is that radiated away plus that given to the ejected matter, that is

$$L \, dt + P(- dM), \tag{11}$$

where P is the energy per unit mass given to the ejected matter. If total escape from the star of the matter occurs, then $P \geqslant 0$.

Combining (10) and (11), we have

$$\frac{3G}{2(5 - n)} \left\{ \frac{2M \, dM}{R} - \frac{M^2 \, dR}{R^2} \right\} = L \, dt - P \, dM . \tag{12}$$

Eliminating dR between this equation and Equation (4), we obtain

$$\frac{3G}{2(5 - n)} \left\{ \left(\frac{5k^2 - k_1^2}{2k^2} \right) \frac{M}{R} \, dM + \frac{M^2}{2R\omega} \, d\omega \right\} = L \, dt - P \, dM \tag{13}$$

or

$$\frac{3GM^2}{4(5 - n) R\omega} \frac{d\omega}{dt} = L + \left(\frac{- dM}{dt} \right) \left\{ P + \frac{3GM(5k^2 - k_1^2)}{4(5 - n) k^2 R} \right\}. \tag{14}$$

This is the basic equation which governs the mass loss. As we originally had only two equations connecting the three variables dM/dt, dR/dt and $d\omega/dt$ we cannot hope to eliminate further variables to obtain dM/dt unless some other physical principle can be found. There does not seem to be such a principle readily available and so all the information to be gathered must be obtained from Equation (14) by discussing cases that arise. It becomes slightly easier to do this by modifying the equation as follows. We can write $\omega^2 = \beta(GM/R^3)$ for any star, where β is a dimensionless parameter expressing how close to rotational instability the star is. From this expression, on substituting from Equation (4) for dR, we obtain

$$\frac{1}{\omega} \frac{d\omega}{dt} = \frac{5k^2 - 3k_1^2}{k^2} \frac{1}{M} \frac{dM}{dt} + \frac{2}{\beta} \frac{d\beta}{dt}. \tag{15}$$

Substitution of this expression into Equation (14) gives

$$\frac{3GM^2}{2(5-n)R\beta}\frac{d\beta}{dt} = L + \left(\frac{-dM}{dt}\right)\left\{P + \frac{3MG}{2R(5-n)}\left(\frac{5k^2-2k_1^2}{k^2}\right)\right\}. \tag{16}$$

We now discuss as many cases as possible arising out of this equation.

3. Cases of Interest

An obvious distinction can be made between $d\beta/dt$ negative and $d\beta/dt$ positive. In one case the star is becoming more stable, while in the other it approaches rotational instability.

Case 1: $d\beta/dt > 0$. The star is not therefore ejecting matter to overcome instability, and this case might represent some stellar wind type phenomenon. Numerical investigation shows that this cause is likely to be applicable initially for most stars. Equation (16) now requires that

$$L + \left(\frac{-dM}{dt}\right)\left\{P + \frac{3MG}{2R(5-n)}\left(\frac{5k^2-2k_1^2}{k^2}\right)\right\} \geqslant 0. \tag{17}$$

We obtain two sub-cases from this that are of interest.

Case 1a: $P = 0$, in which case the ejected material just escapes to infinity as expected for a stellar wind type model. Inequality (17) is now automatically satisfied if $5k^2 > 2k_1^2$. As the maximum value for k_1^2 is 1, the condition is automatically satisfied if $k^2 > \frac{2}{5}$. Fortunately, for all bodies other than a uniform spherical sphere, k is appreciably less than this and so this situation has no real physical interest.

If $2k_1^2 > 5k^2$ then in order to satisfy the inequality we must have

$$L > \left(\frac{-dM}{dt}\right)\left\{\frac{3MG}{2(5-n)R}\frac{(2k_1^2-5k^2)}{k^2}\right\}.$$

This gives an upper bound for the mass loss $(-dM/dt)$ such that

$$m = \frac{-dM}{dt} < \frac{2(5-n)k^2}{3(2k_1^2-5k^2)}\frac{RL}{GM} = \lambda'\frac{RL}{GM}. \tag{18}$$

This case, which nearest resembles the solar wind type of mass loss, therefore gives the same type of upper bound as was derived by another method, a pleasing verification of this upper bound.

Case 1b: $P < 0$. The ejected material is now captured around the star. Therefore, let $P = -Q$, where Q is positive, and as there must be initial motion, $Q < GM/R$. Inequality (17) now requires that

$$L > \left(\frac{-dM}{dt}\right)\left\{Q + \frac{3MG}{2(5-n)R}\left(\frac{2k_1^2-5k^2}{k^2}\right)\right\},$$

with $0 < Q < GM/R$. If $2k_1^2 > 5k^2$ again we have an upper bound (which is not attained) such that

$$m = \frac{-\,\mathrm{d}M}{\mathrm{d}t} < \frac{2\,(5-n)\,k^2}{3\,(2k_1^2 - 5k^2)} \frac{RL}{GM} = \lambda' \frac{RL}{GM}, \tag{19}$$

which is precisely the same as (18).

Now again if $5k^2 > 2k_1^2$ (as unlikely now as in case 1a), then we must have

$$L > \left(\frac{-\,\mathrm{d}M}{\mathrm{d}t}\right)\left\{Q - \frac{3MG}{2\,(5-n)\,R}\left(\frac{5k^2 - 2k_1^2}{k^2}\right)\right\}.$$

A restriction can now exist on $-\mathrm{d}M/\mathrm{d}t$ only if

$$Q > \frac{3MG}{2\,(5-n)\,R}\,\frac{5k^2 - 2k_1^2}{k^2},$$

which, apart from a mathematical infinity as the two terms tend to equality, will
again lead to an upper bound of the form

$$\lambda RL/GM \,.$$

Case 2: $\mathrm{d}\beta/\mathrm{d}t < 0$. Physically this will be the case where mass loss occurs primarily
to remove excess angular momentum. Equation (16) now demands that

$$L + \left(\frac{-\,\mathrm{d}M}{\mathrm{d}t}\right)\left\{P + \frac{3MG}{2R\,(5-n)}\left(\frac{5k^2 - 2k_1^2}{k^2}\right)\right\} < 0\,. \tag{20}$$

If the material just escapes to infinity, so that $P=0$, (20) can only be satisfied if
$2k_1^2 > 5k^2$ and

$$L < \left(\frac{-\,\mathrm{d}M}{\mathrm{d}t}\right)\frac{3MG}{2R\,(5-n)}\left(\frac{2k_1^2 - 5k_1^2}{k^2}\right)$$

or

$$m = \frac{-\,\mathrm{d}M}{\mathrm{d}t} > \lambda'' \frac{RL}{GM}\,. \tag{21}$$

If the material reaches infinity within a finite velocity, the lower bound to the mass
loss becomes higher. The smallest lower bound is obtained when P is as small as possible, namely $P = -GM/R$ and inequality (20) now requires that

$$L < \left\{-\frac{\mathrm{d}M}{\mathrm{d}t}\right\}\left\{\frac{GM}{R} + \frac{3MG}{2R\,(5-n)}\left(\frac{2k_1^2 - 5k^2}{k^2}\right)\right\}$$

or

$$m = \frac{-\,\mathrm{d}M}{\mathrm{d}t} > \frac{RL}{GM}\,\frac{2\,(5-n)\,k^2}{6k_1^2 - (5+2n)\,k^2}\,,$$

which demands that $6k_1^2 > (5+2n)k^2$. With the usual value for n of around $\frac{3}{2}$ this last
requirement is $k_1^2 > \frac{4}{3}k^2$, and as $k_1^2 > \frac{2}{3}$, while $k^2 < \frac{1}{2}$ for all stars, we see that all stars

can avoid instability through mass loss, provided the material does not move very far from the star and that the mass loss is in excess of a lower bound of the form

$$m = \frac{-\,\mathrm{d}M}{\mathrm{d}t} > \lambda'' \, \frac{RL}{GM},$$

with λ'' of the order of unity. We note that this is a very high rate of mass loss and could be only sustained for a short period of time, such as envisaged by HOYLE (1960) in his theory for the origin of the solar system. As mentioned by Hoyle, if the removal of angular momentum has to be maintained for a reasonable period, there must be some coupling mechanism between the ejected material and the star.

4. Conclusions

We have shown that any star can avoid rotational instability by ejecting mass, but that the rate of mass loss in order to do this has to be in excess of a lower bound. It has also been shown that most stars can eject matter in such a way that their contraction still will eventually lead them to rotational instability, the ejected matter now being lost to the star completely, but now an upper limit to this mass loss exists, being of the same form as that already given by WILLIAMS (1967).

We can therefore give a tentative outline to the type of mass loss that occurs in pre-main sequence stars. Initially they lose mass by a stellar wind type process, there being an upper limit to the amount they lose in this time. During this stage the star is gradually approaching rotational instability and, because of the time scale involved, will not have altered substantially in mass. When rotational instability becomes imminent, the rate of mass loss increases in order to remove angular momentum, the actual rate now having to be in excess of a minimum value. During this phase the rate of carrying away energy by the mass loss is in excess of the luminosity, and so we would expect the luminosity to be rather ill-defined, and the star's position not confined to well in the HR diagram.

References

HOYLE, F.: 1960, *Quart. J. Roy. Astron. Soc.* **1**, 28.
PORFIRIEV, V. V.: 1967, Paper presented at IAU Meeting, Prague.
WILLIAMS, I. P.: 1967, *Monthly Notices Roy. Astron. Soc.* **136**, 341.

Discussion

Cremin: By putting $\gamma = \frac{5}{3}$, thus making the term containing ω equal to 0, are you not using an energy equation equivalent to that of the non-rotating case? Is it not therefore to be expected that you will get an upper bound to the rate of mass loss similar to that obtained, by yourself, in a previous investigation?

Williams: While it is true that by using $\gamma = \frac{5}{3}$ the total energy takes the same form as it would in the non-rotating case, in this investigation we still have the equation of conservation of angular momentum as a second constraint. The energy considerations used in the previous investigation are also slightly different to those incorporated in the energy equation used here.

MASS LOSS FROM BINARY STARS. OBSERVATIONS

MASS LOSS FROM CLOSE BINARIES, 1941–68

FRANK BRADSHAW WOOD

University of Florida, Gainesville, Fla., U.S.A.

Abstract. The history of studies of two types of mass loss – particle ejection and gradual loss caused by evolutionary expansion – is discussed briefly. Difficulties are encountered when we try to compare the true shapes of close binaries with theoretical models. In particular, the evidence at present indicates the W UMa systems are not 'contact' binaries as has been generally assumed, although the results of narrow-band observations or theoretical developments in rectification may change this picture. The question is raised of the possible importance of pre-main-sequence evolution in understanding present peculiarities. Finally, the importance of observing systems such as ζ Aur for possible secular changes is discussed.

The first portion of these introductory remarks is chiefly for those who are not specialists in close double stars and is intended to serve in part as background for the following papers on specific topics. However, in addition to summarizing some of the work of recent years, I intend to consider certain aspects of it rather critically and to indicate areas where further work seems to me to be most urgently needed.

At the Prague joint discussion on evolution and close double stars held in 1967, I summarized some of the 'ancient' history of ideas concerning mass loss and the application of the concept of 'zero-velocity' surfaces to close binaries going back to Kuiper's work on β Lyrae (KUIPER, 1941) and my own ideas of the connection between mass loss and period changes (WOOD, 1950). There is no need to repeat this here. I am now concerned with the later developments, which have led to the current state of our thinking about close double stars. These have been influenced predominantly by two factors. The first, of course, is the improvement in our knowledge of the evolution of single stars. The second is the concept of the zero-velocity surface which at certain stages of development alters drastically the evolution of a close double star as compared to a single star of the same mass. This difference of evolution is intimately associated with mass loss.

Considerations of mass loss from a member of a double-star system have fallen into two general patterns. In one, the loss is considered to be violent in nature. At one extreme the physical processes involved may be similar to those responsible for the solar wind and at the other to those responsible for novae. The other pattern considers a less violent type of mass loss with matter usually considered as passing through the well-known inner Lagrangian point of the zero-velocity surface. As far as our present understanding goes, it seems likely that both types of mass loss occur. The first may be responsible for erratic period changes, flares, and temporary spectral changes while the latter may be more pertinent to larger evolutionary changes.

A number of authors have computed trajectories of ejected particles making various assumptions as to the location and velocity of ejection. An excellent summary of the most recent of these and the effect on the orbital elements of the system has been given by PIOTROWSKI (1967). On certain assumptions, the mass lost by one component is transferred to the other; on others, it is lost to the system entirely. In reality probably both processes occur; in the present state of our knowledge it is difficult to estimate the relative amount of each.

A serious drawback in these computations, as various authors have pointed out, is the assumption of individual, completely independent particles. If, as some writers have computed, the mean free path of the atoms is several orders of magnitude smaller than the size of the system, then a hydrodynamical approach is needed. An initial attempt has been made by PRENDERGAST (1960). His work indicates the complexity of the problem but, by using a number of simplifying assumptions, he obtained figures which reproduce in a general way many features of the physical models used to interpret the observations. Future work along these lines will be very difficult but highly desirable.

The second type of mass loss is that generally discussed when considering changes in the stellar radius connected with the evolution of the star. The initial suggestion of the role of the zero-velocity surfaces in this context is due to CRAWFORD (1955); the first detailed computations were made by MORTON (1960). Various authors have been active in recent years in making detailed calculations. The most consistently active groups have been those headed by Kippenhahn, by Paczyński, and by Plavec. The following summary is taken from PACZYŃSKI (1967a).

"Stage 1. Both components of the binary system are on the main sequence. They are smaller than their Roche lobes, and evolve independently of each other. The component initially more massive (primary) evolves faster and approaches the Roche limit first.

Stage 2. Rapid mass transfer between the components occurs on a thermal time scale. At the end of this stage the mass ratio is reversed, and thermal equilibrium is restored in the primary (now less massive component), that fills the Roche lobe.

Stage 3. Slow mass transfer between the components proceeds on a nuclear time scale. Primary fills the Roche lobe.

Stage 4. During the subsequent evolution the radius of the lighter component decreases. Now this star is below the Roche limit."

KIPPENHAHN and WEIGERT (1967) have distinguished between two different cases at the time the mass loss begins, and this distinction has been preserved by most of the workers in the field. In the first, the more massive component attains the Roche limit when the hydrogen burning is confined to the core and the star is in the slow main-sequence expansion. In the second case – binaries with longer periods – the primary reaches this limit during the rapid expansion interval when the hydrogen burning is in a shell and the star is presumably on its way to the red giant stage.

Many individual problems have been investigated in recent years. Details of mass exchange for systems of different assumed masses have been computed. Perhaps the most striking characteristic of these is that significant changes in the properties of the star are computed to occur in intervals of less than 10^3 years and that relatively large amounts of mass transfer appear to occur in such short time intervals. It is impossible to summarize all the interesting papers of the past 3 years but PLAVEC's

(1967a, b) work on Algol systems and PACZYŃSKI's (1967b) suggestion concerning the formation of Wolf-Rayet stars should certainly be mentioned.

One caution should be added. Tacit assumption is made in all these calculations that when a star approaches the Roche limit, it actually assumes the shape of the zero-velocity surface defining this limit. We do not actually know this to be the case. This surface represents a limit which the star cannot exceed, but it is computed on simplified assumptions and we do not know that the real star actually assumes precisely this shape. In principle, this could be checked by the careful analysis of well-observed light curves, but when one looks into the theoretical difficulties of determining precisely the shape of each star from such an analysis, the difficulties are such that an entirely new treatment is needed for the rectification of the light curves of systems in this state. Computations made on the best available evidence 20 years ago (WOOD, 1950) indicated that the real stars approached the Roche limit at the end of the shorter equatorial axis rather than at the inner Lagrangian point, but it is clear that these calculations should be repeated using modern developments in theory and more recent observational data. To make the matter more complex, it is not unreasonable to expect that as the star approaches the limiting surface, mass loss from the prominence-like ejections referred to earlier will become more important and the real conditions on the stellar surface at this time must be chaotic indeed. (For example, would anyone like to discuss the corona of a close binary?)

From this general discussion, I turn now to an area in which a serious breakdown in communication between observers and theoretical workers has occurred and I think it important to emphasize this. This is in the case of the W Ursae Majoris stars. It is extremely common now in the literature to find these referred to as 'contact binaries' – that is, systems in which both components have somehow reached their Roche limits. It is extremely difficult to find sound documentation supporting this assertion. At least one review paper has started with the statement that 'It is well known' that this is the case, but makes no reference to the basis of such knowledge. The papers that do make reference, cite papers by KOPAL (1955, 1956). The table in the first of these is exceedingly uncritical, even for the time. One set of results cited as from a photo-electric light curve is found, by looking up the source, to be an extremely sketchy one based on photographic estimates. Elsewhere, photographic estimates are treated with the same respect as photoelectric observations. No effort is made to assess the precision of the solutions or the coverage of the critical points of the light curve. At a symposium of this nature, I do not want to take time for detailed criticism but to repeat that, at the best, this was a somewhat weak source and it is now well more than a decade out of date. The second paper is even more curious. It follows an earlier paper in which a purely mathematical investigation was made of the sizes of Roche surfaces for given mass ratios. The statement is then made that "assumption is made (in the absence of any evidence to the contrary) that the two stars form a contact pair". With this assumption thus made, the elements determined by using on the real stars the mathematical relationships derived from the imaginary surfaces naturally show them all to be contact binaries – a curious example of what is called reasoning in a circle. As an illustration of the results, in one system the inclination is computed as 61°; later work shows clearly a long total phase with an inclination very close to 90°.

Again, I do not want to take time here in a detailed discussion of a paper of this type. What I do want to say is that even if there were no evidence to the contrary 12 years ago – and there was – there is a great deal of evidence to the contrary today. Of the well-observed W UMa systems, whose light curves are suitable for reasonably precise analysis, not a single one is a 'contact binary' in the sense now used. There are now about a dozen of these – more if one chooses to be a little less critical of the determinancy of the solution – and in none of them do both components approach the zero-velocity surface. This is not the place for detailed documentation, but I can point out that in his invited paper at the December 1967 meetings of the American Astronomical Society, R. H. Koch made essentially these same remarks – i.e., that for systems which show complete eclipses, the evidence is conclusive that both component stars do not fill their surrounding lobes. Far earlier, nearly 20 years ago, STRUVE (1950) pointed out that if both components were really in contact, the wings of the absorption lines would touch at elongation and that this was not true for W UMa.

Let me be clearly understood. I am not saying – yet, at least – that contact systems do not or cannot exist. I am saying that according to the best modern evidence, few if any W UMa type stars are contact systems and any study based on this assumption is at the best a very rough approximation. Unfortunately, I cannot give a table of photometric elements to which theoreticians can turn with confidence as a check on their models. I hope that one will be prepared within the next year, possibly under the sponsorship of Commission 42 of the IAU, but until then there is no recourse other than critical consideration of the individual papers.

As a note of caution, let me repeat my earlier statement but with different emphasis. The W UMa stars are not contact binaries, *according to the best present evidence*. There are two reasons for thus emphasizing what ordinarily would be taken for granted. One is that preliminary studies by R. H. Koch have shown that in some cases light curves observed through narrow-band filters peaked at λ 4275 and λ 4165 (KOCH, 1968) differ significantly from published ones using broad-band filters. From inspection alone, it appears these may yield quite different value for the relative radii. Caution must be used for various reasons. It is not known whether these differences between light curves from broad-band and narrow-band filters exist for all systems or only a few; it is not known whether they continue systematically from season to season; other interpretations than the one mentioned are possible; and finally, until complete and careful rectifications have been carried out, we cannot be sure that different solutions will really result. This should be a fruitful field for observation in the next few years. A second reason for caution is that several astronomers – Mauder at Bamberg is one example – are working on theoretical developments, which may be far more suitable for treatment of W UMa systems than methods currently in use.

I turn now to quite a different question. The majority of computations of recent years have begun with both components of the binary on the main sequence, and have considered changes in radius and luminosity as the system evolves. The obvious reason is that in the evolution of single stars, these are the best understood phases. But the components somehow must reach the main sequence and this presents a problem. PLAVEC (1967a) has raised the question of why few if any close binaries are detected in this stage and has answered it by pointing out the shortness of the contraction phase as compared to the subsequent life of the star. But nevertheless, according to our present ideas, this is a phase through which each component of the system must pass. (It is true that ideas of fission have been with us at least since the time of Jeans, but I think it fair to say they have not yet explained the formation of binary systems. However, recent papers by ROXBURGH (1965, 1966a, b) should not be ignored. BATTEN (1967) has given an interesting review of the whole problem of the origin of binary stars, including earlier suggestions that more than one mechanism may be needed.) If we begin at the zero-age main sequence and in imagination go backward in time, we soon find both components exceeding the critical surfaces, and this raises the question of how any close double star system can form at all. We can picture an intermediate state in which the more massive component has already reached the main sequence but the less massive is still contracting. A large amount of matter of what ordinarily might contract into the less massive component may be captured by the heavier or lost to the system. If, as some theoreticians working on single stars have assumed, the heavier elements are more concentrated during this stage, the material thus lost would be chiefly hydrogen and we might have a metal-rich star. If we extend further these speculations – I do not think they can be called more than that – we might consider at an earlier era, when both components were very large, that hydrogen was lost from each and that both stars should be metal-rich. (Obviously this refers to second- or third-generation stars.) Can this be the explanation of metallic-line stars? When these ideas first occurred to me in a somewhat different connection (WOOD, 1964) I was soon tempted to abandon any such thoughts by the belief that only the primary component of metallic-line binary showed metallicity. However, a letter from Y. Kondo dated February, 1968, lists three systems (50 Dra, η Vir, and HR 6611) in which both components have been found to exhibit characteristics of a metallic-line star. He further adds that at that time there appeared to be no known cases where the secondary component of an Am binary system has been found to have a normal spectral type. He adds, however, that it is rather difficult to determine the existence of such secondary components.

The whole matter could be carried on at some length. ABT (1961, 1965) has suggested that all metallic-line stars are members of binary systems. ABT (1965) further found that all close binaries with spectra A4–F2 (IV–V) with periods 1–100 days were metallic-line stars. These limits of course raise questions. Are binaries with periods over 100 days well enough separated to evolve normally in the contraction stages? Do more massive stars than A4 pass through them so rapidly that there is no appreciable mass loss? In the joint discussion on Evolution and Close Double Stars held at the 1967 IAU meetings, P. S. Conti suggested on statistical grounds that the Am stars were not a product of post-main-sequence evolution. The whole matter is not at all clear and probably will not become so until the evolution of single stars in this stage is better understood. Nevertheless, I think we should keep in mind the idea that some observed features of close double stars may be the result of pre- and not post-main-sequence evolution.

The whole matter of the age of these systems is one which deserves attention. It has long been recognized that eclipsing stars are extremely rare, if indeed they exist at all, in globular clusters. A list of those in open clusters, provided recently by K. C. Leung, indicates that most of the eclipsing stars occur in very young clusters. Considering again the chaotic conditions that must exist at least at the surfaces of these stars, we can wonder whether a system can remain in this state over an extended interval. JASCHEK and JASCHEK (1959) have published an interesting paper based on spectrographic data, and more work along this line is in order.

One note of caution should be sounded. Most data on clusters include very few observations of any individual star, and Algol systems in particular might exist in normal frequency and yet be undetected. Still it is strange that the few that have been found are in the young clusters.

As a final topic, I want to call attention briefly to stars relatively late in their evolutionary stages and in particular to the system ζ Aurigae. I do this in part because of the computations referred to earlier which indicate that at certain stages changes in the radius of an evolving star may occur at a rate which a few years ago would scarcely have been thought possible. This adds particular interest to recent observations of ζ Aurigae. KIYOKAWA (1967) has collected the observed durations of totality during the past 30 years and has shown that they indicate – although they do not definitely establish – a secular increase. One interpretation of this is a rate of increase of the diameter of the K component of 0.18 $R_\odot$/year. This is roughly equivalent to 0.1 radius of the giant star per century and tends to lend credence to computations showing appreciable mass exchange in this interval. Precise observations in the 1971–72 eclipse may make this evidence much stronger.

The foregoing remarks have touched on only a few of the exciting topics concerning close double stars. Suffice it to say that we are a long way from the era when the only reason for observing these systems was thought by many to derive stellar masses, radii, and densities and when a light curve or velocity curve was once well observed, it was labelled 'definitive' and the general feeling was that this system no longer needed observation. Fortunately, a few wise astronomers – R. S. Dugan, for example – realized that the real situation was much more complex, although no one dreamed of the exciting developments we are finding today.

References

ABT, H.: 1961, *Astrophys. J. Suppl. Ser.* **6**, 37.
ABT, H.: 1965, *Astrophys. J. Suppl. Ser.* **11**, 429.
BATTEN, A. H.: 1967, *Rev. Astron. Astrophys.* **5**, 25.
CRAWFORD, J. A.: 1955, *Astrophys. J.* **121**, 71.
JASCHEK, C. and JASCHEK, M.: 1959, *Z. Astrophys.* **48**, 263.
KIPPENHAHN, R. and WEIGERT, A.: 1967, *Z. Astrophys.* **65**, 251.
KIYOKAWA, M.: 1967, *Publ. Astron. Soc. Japan* **19**, 209.

Koch, R. H.: 1968, Abstracts of Papers to be Presented at the 126th Meeting of the Am. Astr. Soc., p. 71.
Kopal, Z.: 1955, Ann. Astrophys. 18, 379.
Kopal, Z. and Shapley, M. B.: 1956, Jodrell Bank Ann. 1, 141.
Kuiper, G. P.: 1941, Astrophys. J. 93, 133.
Morton, D. C.: 1960, Astrophys. J. 132, 146.
Paczyński, B.: 1967a, Commun. Obs. roy. de Belgique, Series B, No. 17.
Paczyński, B.: 1967b, Acta Astron. 17, 355.
Piotrowski, S. L.: 1967, Commun. Obs. roy. de Belgique, Series B. No, 17, 133.
Plavec, M.: 1967a, Bull. Astron. Inst. Col. 18(5), 253.
Plavec, M.: 1967b, Joint Discussion, 'Close Binaries and Stellar Evolution', XIII General Assembly, IAU Prague.
Prendergast, K. V.: 1960, Astrophys. J. 132, 162.
Roxburgh, I. W.: 1965, Nature 208, 65.
Roxburgh, I. W.: 1966a, Astrophys. J. 143, 111.
Roxburgh, I. W.: 1966b, Astron. J. 71, 133.
Struve, O.: 1950, Stellar Evolution, p. 175.
Wood, F. B.: 1950, Astrophys. J. 112, 196.
Wood, F. B.: 1964, Vistas in Astronomy 5, 119.

Discussion

Field: I reduced the data on semi-detached systems with undersize subgiant secondaries provided by Kopal using the assumption that the primary was in pre-main-sequence contraction. It seemed possible this was correct for four or so of the systems – notably TV Cass, KO Aquilae (see paper in *AJ* by I. W. Roxburgh).

Wood: This is a very interesting result and I hope it will be published soon. I think work on stars in these stages will be important. Certainly, a number of features should eventually be explained by the history of the star's evolution during the contracting stages.

Hall: I examined the galactic distribution of the stars in Kopal's list of detached Algol binaries and found their distribution indistinguishable from that of the semi-detached Algol binaries. Furthermore it was not nearly flat enough to be characteristic of contracting stars, which according to Roxburgh, would be approximately 10^6 years old. This will appear in the *Publ. Astron. Soc. Pacific*.

Wood: This is an interesting approach and I hope it will be carried further. Unfortunately, that particular catalogue contains a large number of errors; for more results, it is necessary to consider critically the original publications. If the proposed IAU catalogue is prepared, it should aid in many studies of this kind.

Sahade: I agree with Brad Wood in what he said about the W Ursae Majoris stars but perhaps mention should be made that Struve used the term 'contact binaries' to describe them because he thought that they have a common envelope. In a sense one could still use the denomination 'contact binaries' to describe those systems where the Roche lobes are either filled by the stars themselves or by matter surrounding smaller stars.

Wood: This is a good suggestion, if further investigation shows there really are no contact binaries in the sense now used. I spoke today only of W UMa stars. This would be a very useful application of the term.

GENERAL REVIEW OF OBSERVATIONAL SPECTROSCOPIC EVIDENCE FOR MASS LOSS IN CLOSE BINARIES

JORGE SAHADE*

*Observatorio Astronómico, Universidad Nacional de La Plata,
La Plata, Argentina*

The evidence for mass loss in close binaries provided by spectroscopic observations is of two types; there is evidence for the existence of gaseous streams from one of the components and also for the existence of envelopes that surround a number of systems.

I. Evidence for the existence of gaseous streams from one of the components:
   (1) Direct evidence for the existence of gaseous streams, indicating transfer of matter, is given by extra lines that appear in the spectrum either in absorption or in emission.
   (a) In β Lyrae immediately before and after the midpoint of principal eclipse the spectrum displays a set of so-called 'satellite' lines, which have been interpreted as evidence for the existence of two gaseous streams, one going towards the primary component ('satellite' lines before the midpoint of principal eclipse) and the other one towards the secondary star in the system ('satellite' lines after the midpoint of principal eclipse).
   (b) A number of Algol-type systems display extra absorption lines in the phase interval from about $0.75 P$ towards the next conjunction. These extra lines do not necessarily appear as blends to all the spectral features but only to those of some elements, primarily H, and give rise to the well-known Barr effect in the distribution of the ω's in spectroscopic binaries because of the 'peaked' velocity curves that are produced; they are interpreted as arising from the gaseous stream from the secondary component towards the advancing hemisphere of the primary when seen projected upon it. The matter in these streams form the gaseous rings that surround the primaries in many systems (RZ Scuti, SX Cassiopeiae, etc.), and presumably is responsible for the drastic drop, in one-half of the cycle, in the intensity of the absorption lines of the component towards which the stream moves, in systems such as 29 Canis Majoris.
   (c) The gaseous streams give rise in some systems to relatively narrow emission features that appear superimposed to the broader emissions in the spectrum: such is the case in HD 47129 and in the Wolf-Rayet binary V 444 Cygni. These narrow emissions suggest a velocity distribution that provides information as to the direction and velocity of the stream. The stream velocities are of the order of 700 km/sec in the case of V 444 Cygni and 400 km/sec in the case of HD 47129 while it is of the order of 200 km/sec in systems like U Sagittae.

* Member of the Carrera del Investigador Científico, Consejo Nacional de Investigaciones Científicas y Técnicas, Argentina.

(d) In systems such as β Persei the spectrum displays emission lines during quadratures that indicate that they arise from matter which is concentrated between the two stars presumably in the vicinity of the Lagrangian point L_1. A similar conclusion was reached in the case of UX Monocerotis and may be true also in SX and RX Cassiopeiae, although the presence of emission at quadratures can only be inferred from the weakening of the H absorptions.

II. Evidence for the existence of envelopes that surround close binary systems:

The existence of outer envelopes is indicated also by lines that appear in absorption or in emission.

(a) Several systems display sets of absorption lines that show the effect of diluted radiation and suggest the existence of an expanding tenuous outer envelope. In β Lyrae the expansion velocity of the outer shell is of the order of -200 km/sec, in HD 47129, of -700 km/sec, and in γ_2 Velorum, of -1300 km/sec. It is interesting to point out that at least the strongest line from the outer shell, namely He I 3888, is multiple in β Lyrae and in γ_2 Velorum, in the latter case the behavior of the structure being probably connected with the phase in the cycle of the orbital motion.

(b) Other systems – e.g. HD 698, U Coronae Borealis, AO Cassiopeiae – display broad emissions specially at Hα the behavior of which – radial velocity-wise – suggests that they are produced in a gaseous envelope that surrounds the entire system.

(c) The presence of emission arising from forbidden transitions are common in close binaries that are also eruptive variables, but otherwise they are not. W Serpentis is one of the few exceptions and displays very weakly the strongest predicted forbidden transitions of Fe II.

III. There is also indirect evidence for mass loss from components in close binary systems.

(a) There are systems, such as S Velorum and UX Monocerotis, where the radial velocities from consecutive plates show large and erratic variations that have been attributed to 'prominence activity' in a scale much larger than the scale that is prevalent in the sun.

(b) In a number of systems the systemic velocities as derived from the lines of each component are different and this fact can be interpreted in terms of ejection of matter from the component that yields a more negative value of γ. Systems in this group are HD 47129, V 448 Cygni, α Virginis, AO Cassiopeiae, β Scorpii. I exclude the Wolf-Rayet binaries where the reported differences in the γ's from different lines may arise from problems connected with the assigned wavelength and/or with the line profiles.

(c) The spectral behavior in objects like β Lyrae and SX Cassiopeiae suggests that matter is streaming out of the system from the outer Lagrangian points in the restricted solution of the three-body problem.

IV. There is no need to make specific reference to the close binaries that are also eruptive variables like novae, symbiotic stars, ultrashort period variables, etc., where the spectroscopic evidence for mass loss is well known.

V. Our knowledge of the amount of mass being lost by the systems is very meager. Miss Underhill has mentioned at this Symposium that the estimates for the mass loss in Wolf-Rayet objects put the figure at about 10^{-5} $\mathfrak{M}_\odot$/year; in the case of novae outbursts the figure currently accepted is that of 10^{-5} $\mathfrak{M}_\odot$, and in the case of β Lyrae Struve figured it at 10^{-5} the mass of the primary component per year.

VI. Theoretical computations were made several years ago to ascertain the trajectories that particles ejected by one of the components of a close binary would follow under certain initial conditions and the gravitational field of the other component, the original idea being to try to reproduce the gaseous rings that surrounds the early-type component in many Algol systems. More recently the emphasis has been shifted to try to explain configurations such as the Algol-type and other systems; and work in this direction has been carried by Kippenhahn, Paczyński and Plavec and the respective groups. These computations start with specific mass distributions of the components and consider the evolution of the more massive star, the assumption being made that all the mass lost by this component goes to the companion and increases its mass. Although I must admit that the configuration of the Algol systems seems to be nicely explained as far as present mass distribution is concerned, I do not feel that the basic assumption is always true nor do I forget that so far no attempt has been made to ascertain whether the spectra of the theoretical 'present' configuration would match the spectra that are actually observed.

The average amount of mass that is lost per year by the originally more massive component is, according to the computations, of the order of 10^{-4} or 10^{-3} of the original mass, depending on the case considered.

Discussion

Kitamura: Do you think that, in those systems with emission lines or shell spectra which you mentioned, at least one of the components fills the Roche limit generally?

Sahade: It seems to be so. My feeling is that those components that the emission lines are connected with, are smaller than their Roche lobes, but the space in between is filled with matter.

Morton: What is the direction of gas streams in binary systems with subgiant secondaries? Is it from the larger to the smaller star?

Sahade: From the larger to the smaller.

Hack: In υ Sgr there is evidence of two streams, one going from the brighter to the fainter star, and one in the opposite direction. (There are two Hα absorptions observable during the eclipse of the primary star: one is violet-shifted at -300 km/sec, and the weaker one is red-shifted at $+150$ km/sec.)

SPECTROSCOPIC STUDY OF THE ECLIPSING
SYSTEM R CANIS MAJORIS

MASATOSHI KITAMURA

Tokyo Astronomical Observatory, University of Tokyo, Mitaka, Tokyo, Japan

Abstract. Variation of the residual intensities of metallic and hydrogen lines of R CMa with phase is presented from measurement of its spectrograms obtained with dispersions of 10.3 Å/mm and 4.1 Å/mm at the Okayama Astrophysical Observatory. It is found from variation of the residual intensities of these lines, with the exception of the Ca-K line, that the duration of the eclipse is longer than expected from the photometric elements. The fractional loss of light of the eclipsed component at mid-eclipse has been derived from the ratio between residual intensities of the lines at mid-minimum and outside eclipse. The value so deduced is found to be larger than expected from the light-curve analysis. The behaviour of the residual intensities of the K line around the mid-minimum provides some evidence of emission at the line center. The possible existence of gaseous matter surrounding the components is discussed.

1. Introduction

The close binary system R CMa is well-known to be unusual in that both components appear to have very small masses compared to those expected from their luminosities or spectra (KOPAL, 1959), if we interpret the observational aspects as the result of binary motion of two stellar bodies. Though various interpretations have been given (SMAK, 1961; SAHADE, 1963; etc.), the system still remains one of the most baffling close binary stars.

The spectral type of the primary component was classified as F1V by FRINGANT (1956), but the spectrum of the secondary component has never been observed. The U-B and B-V colors of the star outside eclipse were determined by KITAMURA (1967) to be U-B=0.00 and B-V=0.33, which are quite in agreement with those for normal main-sequence stars of spectral type F1.

Photoelectric studies of this system have been done by WOOD (1946), by KOCH (1960), by KITAMURA and TAKAHASHI (1962), and quite recently by Sato (unpublished). Photoelectric light curves obtained by these workers reveal that the light variation is very asymmetric not only at the minima (KITAMURA, 1960) but also at the maxima (KOCH, 1960). It was reported that some transient humps were observed after the primary minimum (PICKERING, 1904; WENDELL, 1909). Similar humps have also been seen in some photoelectric light curves. Figure 1 is a schematic illustration of the asymmetric light variations around the minima which were taken from KOCH's (1962) blue observations.

Spectrographic observations of this system have been done by JORDAN (1916), by SITTERLY (1940), and by STRUVE and SMITH (1950). Only the spectrum of the brighter component is ever visible on the spectrograms and the mass-function is very small. The scatter in their velocity curves is always large. According to STRUVE and SMITH (1950), there were found to be no systematic differences between velocities measured from lines near the violet end as compared to lines near the red end of the spectrum.

M. Hack (ed.), Mass Loss from Stars. All rights reserved

2. Spectroscopic Observation

R CMa was observed with the coudé spectrograph of the 74-inch reflector at the Okayama Astrophysical Observatory in February 1966 and January 1967. Thirty-five spectrograms on Eastman 103a-0 emulsion were obtained, of which 20 good ones were used for the present study.

Of the spectrograms studied, 19 were taken with the $f/4$ Schmidt camera having a focal length of 40 cm and with the second order and a grating of 1200 grooves per

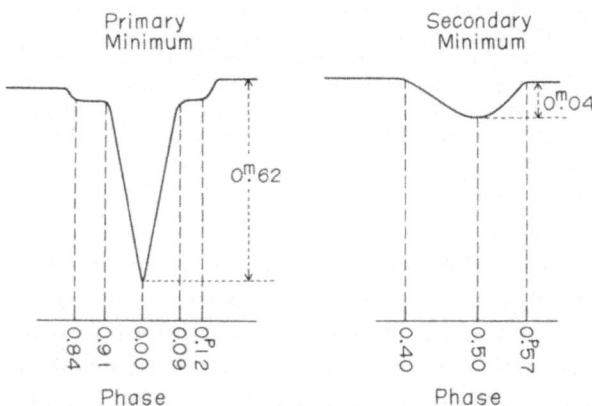

Fig. 1.   Schematic example of asymmetric light variations of R CMa at the minima (taken from blue observations by KOCH, 1960).

mm, giving a linear dispersion of 10.3 Å/mm at Hδ. Of these, 13 spectrograms were taken during the primary eclipse, one during the secondary eclipse, and the remaining 5 outside eclipses. In view of the duration of eclipse of about 5 hours, the exposure times during the eclipse were limited to less than 30 min, mostly 20–25 min, with rather small trailing. Figure 2 illustrates spectra of R CMa taken in this manner. When the system was outside eclipse, one spectrogram was taken with the $f/10$ Schmidt camera having a focal length of 100 cm and the same grating, giving a dispersion of 4.1 Å/mm at Hδ. The exposure time for this was about 2 hours. All the spectrograms were carefully calibrated photometrically by means of a tube sensitometer at the Okayama Astrophysical Observatory. Table I gives the data for all the spectrograms studied.

From the spectrograms taken within the primary eclipse, no line could be detected for the secondary component. A preliminary estimation indicates that the components should be separated by more than 3 Å at maximum elongation, and therefore any metallic line due to the secondary component, if present to an observable extent, should be found. However, this trial was not successful at all. Thus, the continuum of the secondary component must be very faint as compared to that of the primary in the measured spectral region 3900–4450 Å.

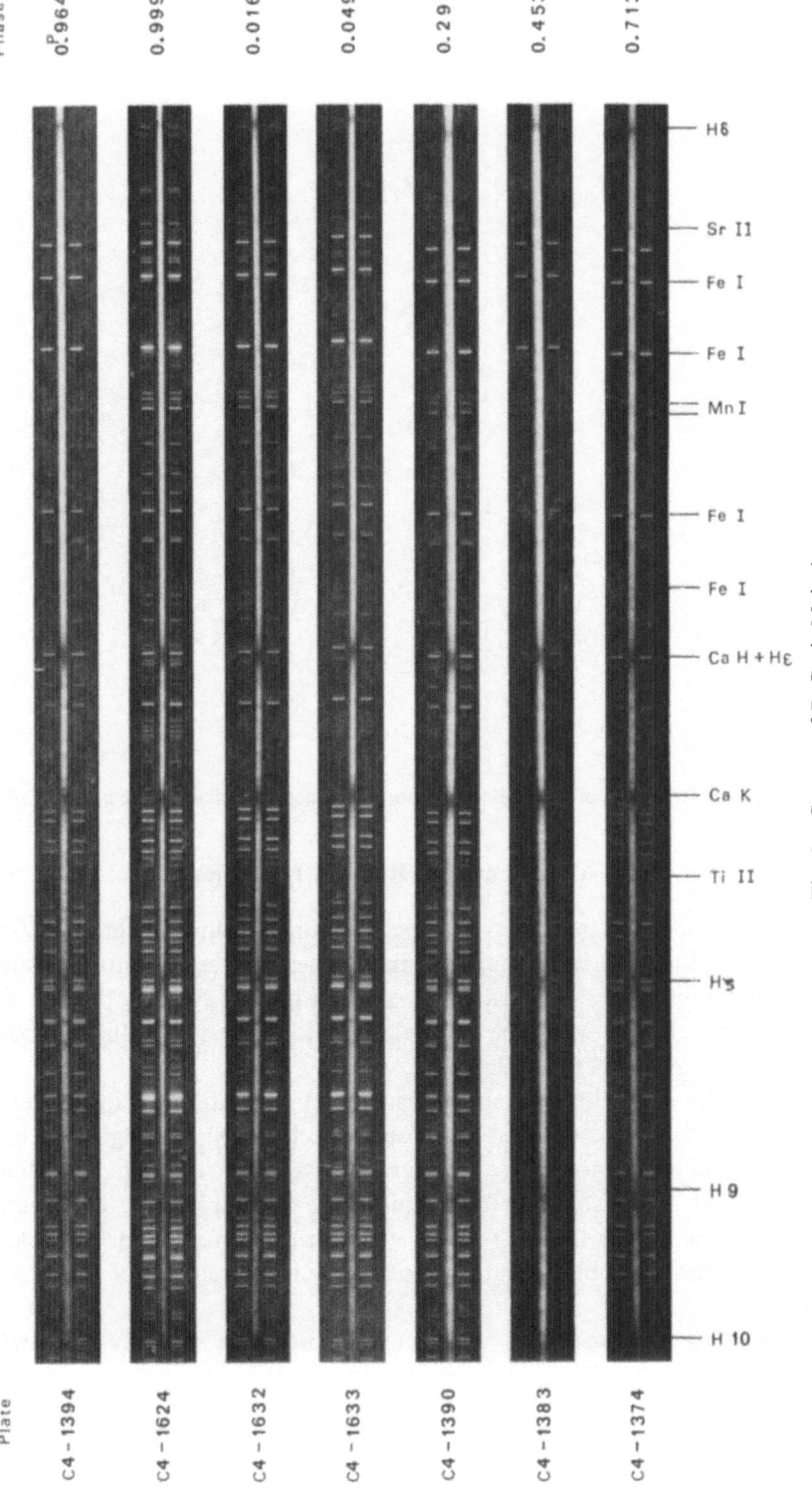

Fig. 2. Spectra of R Canis Majoris.

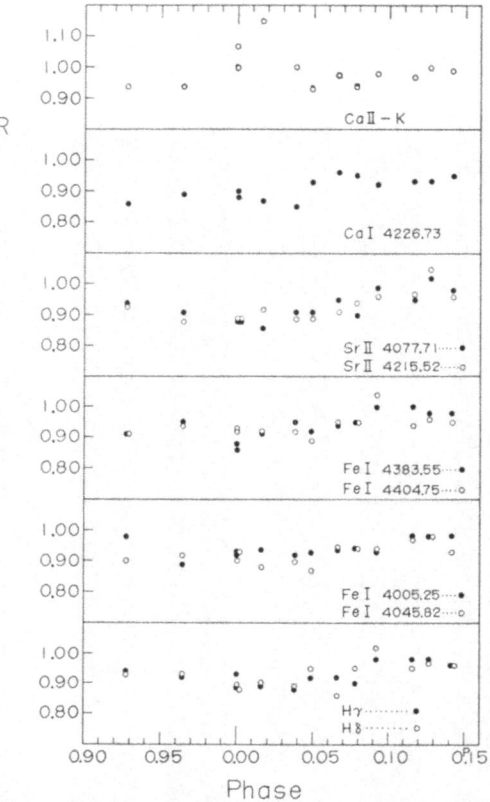

Fig. 3.   Variation of R-values of some selected absorption lines with phase in the primary minimum.

3. Measurement of Residual Intensities

After microdensitometer tracing of the spectrograms, 22 lines without any risk of appreciable blending with neighbouring metallic lines were selected for measurement of their residual intensities. The list of the selected lines is given in the first column of Table II. The selection was done with reference to the results of the analysis for σ Bootis (F2V) by WRIGHT et al. (1964).

In Table II we show the results of measurement of the residual intensities of the relevant lines on the 6 spectrograms taken outside eclipse. Numbers given in brackets beneath plate numbers denote phases of the respective plates. No systematic difference in the residual intensities could be found among the 6 spectrograms. The mean value of the residual intensities for each line is given in the last column of Table II. In calculating the mean, double weight was put on the measurements of plate C10/1565 (4.1 Å/mm).

The remaining 14 plates taken around the eclipses were similarly measured, and the ratio of the measured intensity for each line to the corresponding one outside eclipse

$$R = r_{ecl}/\bar{r}_{out} \tag{1}$$

TABLE I

Data of spectrograms of R CMa

Plate No.	Date	JD	Phase	Range
C4-1374	1966, Feb. 2	2439159.123	0.^{p}713	A1
C4-1375	Feb. 2	.149	0.736	A1
C4-1383	Feb. 4	161.100	0.453	A1
C4-1388	Feb. 5	.995	0.241	A1
C4-1389	Feb. 5	162.025	0.268	A1
C4-1390	Feb. 5	.051	0.291	A1
C4-1393	Feb. 7	163.911	0.928	A1
C4-1394	Feb. 7	.951	0.964	A1
C4-1395	Feb. 7	.992	0.999	A1
C4-1400	Feb. 7	164.082	0.078	A1
C4-1401	Feb. 7	.097	0.092	A1
C4-1403	Feb. 7	.125	0.116	A1
C4-1404	Feb. 7	.138	0.127	A1
C4-1405	Feb. 7	.153	0.141	A1
C4-1624	1967, Jan. 24	514.997	0.999	A2
C4-1625	Jan. 24	515.040	0.038	A2
C4-1626	Jan. 24	.073	0.066	A2
C4-1632	Jan. 25	516.151	0.016	A2
C4-1633	Jan. 25	.190	0.049	A2
C10-1565	Jan. 26	517.06	0.82	B

Range of Wavelength: A1: 3650–4445 Å, A2: 3730–4560 Å, B: 3700–4130 Å.
Dispersion: 10.3 Å/mm for A1 and A2, 4.1 Å/mm for B.

TABLE II

Residual intensities of spectral lines of R CMa outside eclipse

Plate no. λ	C4/1388 (0.^{p}241)	C4/1389 (0.^{p}268)	C4/1390 (0.^{p}291)	C4/1374 (0.^{p}713)	C4/1375 (0.^{p}736)	C10/1565 (0.^{p}819)	Mean[a]
Ti II 3913.46	0.82	0.83	0.78	0.77	0.81	0.79	0.80
Ca II 3933.66	.16	.16	.14	.14	.16	.15	.15
Al I 3944.01	.81	.80	.78	.82	.81	.81	.81
Fe I 4005.25	.74	.73	.68	.68	.71	.72	.71
Mn I 4030.76	.73	.69	.67	.70	.70	.70	.70
Fe I 4045.82	.64	.63	.63	.66	.65	.64	.64
Fe I 4063.60	.76	.74	.75	.73	.74	.76	.75
Fe I 4071.74	.74	.73	.73	.74	.71	.71	.72
Sr II 4077.71	.70	.70	.71	.68	.71	.68	.69
Hδ 4101.74	.25	.25	.21	.24	.27	.23	.24
Fe I 4132.06	.70	.72	.71	.72	.73	.69	.70
Mg I 4167.27	.75	.76	.73	.75	.78		.75
Fe I 4191.44	.78	.77	.79	.80	.80		.79
Fe I 4202.03	.76	.76	.75	.74	.77		.76
Sr II 4215.52	.75	.72	.71	.74	.76		.74
Ca I 4226.73	.64	.62	.60	.60	.63		.62
Fe I 4235.94	.76	.78	.75	.75	.76		.76
Cr I 4254.35	.80	.77	.77	.79	.80		.79
Fe I 4260.48	.69	.71	.68	.71	.71		.70
Fe I 4325.76	.71	.72	.69	.68	.67		.69
Hγ 4340.48	.31	.30	.29	.29	.28		.29
Fe I 4383.55	.68	.69	.68	.69	.71		.69
Fe I 4404.75	.80	.80	.81	.79	.82		.80

[a] Probable errors of the mean values for the respective lines are less than 0.01.

TABLE

Ratio of residual intensities to

Plate No. λ	C4/1383 0.ᴾ453	C4/1393 0.ᴾ928	C4/1394 0.ᴾ964	C4/1395 0.ᴾ999	C4/1400 0.ᴾ078	C4/1401 0.ᴾ092
Ti II 3913.46	0.98	0.94	0.93	0.93	0.96	0.94
Ca-K 3933.66	0.98	0.94	0.94	1.07	0.94	0.98
Al I 3944.01	0.97	0.97	0.88	0.89	0.96	0.95
Fe I 4005.25	0.95	0.98	0.89	0.93	0.94	0.93
Mn I 4030.76	0.99	1.00	0.94	0.93	0.95	0.97
Fe I 4045.82	1.00	0.90	0.92	0.93	0.94	0.94
Fe I 4063.60	0.95	0.92	0.94	0.89	0.92	0.88
Fe I 4071.74	1.05	0.99	0.95	0.94	0.94	1.00
Sr II 4077.71	1.03	0.94	0.91	0.88	0.90	0.99
Hδ 4101.74	1.02	0.93	0.93	0.89	0.95	1.02
Mg I 4167.27	1.00	0.98	0.95	0.91	1.00	1.00
Fe I 4191.44	1.00	0.98	0.93	0.91	0.99	1.01
Fe I 4202.03	0.99	0.94	0.89	0.88	0.97	1.00
Sr II 4215.52	1.01	0.93	0.88	0.89	0.94	0.96
Ca I 4226.73	0.98	0.86	0.89	0.88	0.95	0.92
Fe I 4235.94	0.98	0.94	0.94	0.90	0.97	0.96
Cr I 4254.35	0.97	0.92	0.97	0.91	0.98	1.00
Fe I 4260.48	0.98	0.91	0.93	0.91	0.98	1.02
Fe I 4325.76	0.98	0.92	0.88	0.88	0.94	0.98
Hγ 4340.48	1.05	0.94	0.92	0.93	0.90	0.98
Fe I 4383.55	0.96	0.91	0.95	0.86	0.95	1.00
Fe I 4404.75	1.00	0.91	0.94	0.92	0.95	1.04
Mean (p.e.)	0.993 ± 4	0.938 ± 5	0.922 ± 4	0.904 ± 4	0.947 ± 4	0.975 ± 5

was taken, where $\bar{r}_{out}$ is the one given in the last column of Table II. The R-values reduced thus are given in Table III. In Figure 3, the values of R for some selected lines are plotted with phase for illustration. It is found from Figure 3 that only the Ca II K line behaves in an opposite sense compared with the other lines, although the latter show more or less similar change with phase. This seems to be evidence of emission at the center of the K line.

At the bottom of Table III, the statistical means of the residual intensities of all the lines used, excepting the K line, are given for each plate. These mean values are also plotted in Figure 4, where the arrows represent the respective statistical (probable) errors. It is found that the R-values change distinctly in accordance with the eclipsing phases. The value of R at mid-minimum of the primary eclipse is found to be 0.906 ± 3. It is also noticeable that the eclipse appears to continue over the phase, around $0.ᴾ08 \sim 0.ᴾ09$, representing the end of eclipse as expected from the photometric elements r_g, r_s, i.

4. Interpretation of Variation of R-Values within Eclipse

As mentioned in Section 2, the continuum of the secondary component is practically negligible compared with that of the primary in the spectral region under consideration.

rresponding means for outside the eclipse

| C4/1403 | C4/1404 | C4/1405 | C4/1624 | C4/1625 | C4/1626 | C4/1632 | C4/1633 |
0.$^{\rm P}$1160	0.$^{\rm P}$127	0.$^{\rm P}$141	0.$^{\rm P}$999	0.$^{\rm P}$038	0.$^{\rm P}$066	0.$^{\rm P}$016	0.$^{\rm P}$049
0.97	0.97	0.97	0.90	0.93	0.90	0.90	0.93
0.97	1.00	0.99	1.00	1.00	0.98	1.15	0.93
0.97	0.95	0.97	0.90	0.88	0.95	0.90	0.91
0.98	0.98	0.98	0.92	0.92	0.94	0.94	0.93
0.99	1.03	0.96	0.91	0.90	0.97	0.93	0.94
0.97	0.98	0.93	0.90	0.90	0.94	0.88	0.87
0.95	0.96	0.94	0.93	0.89	0.92	0.87	0.88
1.06	1.08	1.04	0.93	0.95	0.95	0.93	0.95
0.95	1.02	0.98	0.88	0.91	0.95	0.86	0.91
0.95	0.97	0.96	0.88	0.89	0.86	0.90	0.95
1.01	1.01	1.04	0.92	0.94	0.98	0.94	0.94
0.96	1.05	1.01	0.94	0.94	0.98	0.94	0.97
0.96	0.96	1.01	0.93	0.93	0.94	0.92	0.95
0.97	1.05	0.96	0.89	0.89	0.89	0.92	0.91
0.93	0.93	0.95	0.90	0.85	0.96	0.87	0.93
1.03	1.04	1.02	0.91	0.96	0.98	0.94	0.96
1.03	0.96	1.10	0.91	0.94	0.95	0.93	0.94
1.00	1.04	1.04	0.93	0.95	0.95	0.93	0.94
0.98	0.97	0.98	0.91	0.88	0.92	0.92	0.86
0.98	0.98	0.96	0.88	0.88	0.92	0.89	0.92
1.00	0.98	0.98	0.88	0.95	0.94	0.91	0.92
0.94	0.96	0.95	0.93	0.92	0.95	0.92	0.89
0.980 ± 5	0.994 ± 6	0.987 ± 5	0.908 ± 3	0.914 ± 4	0.940 ± 4	0.912 ± 4	0.924 ± 4

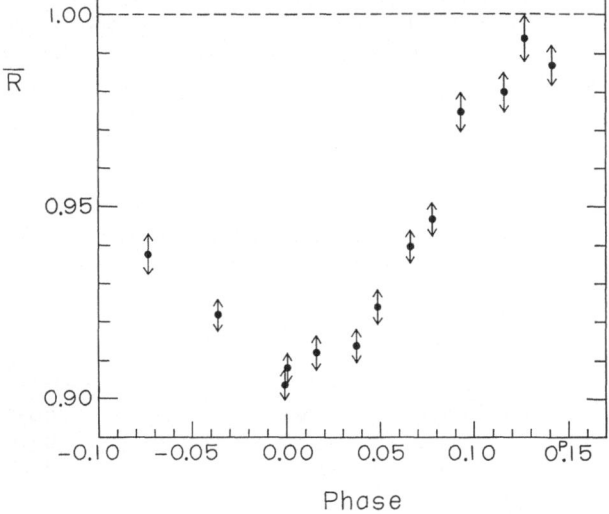

Fig. 4. Variation of the mean R-values with phase in the primary minimum.

This has also been confirmed from the light-curve analysis. For example, the fractional luminosities of the components in B-color were derived by KOCH (1960) to be $L_g = 0.969$ and $L_s = 0.031$ for the greater (brighter) and the smaller (fainter) component, respectively. The residual intensity of a line due to the secondary component is a further small fraction of its continuum L_s. Thus, neglecting the contributions of the secondary component to the observed residual intensities of the spectral lines, we can write quite safely for the residual intensity of a normal absorption line

$$r_{out} = \frac{L_g}{L_s + L_g} I_g \tag{2}$$

while the system is outside eclipse, and

$$r_{ecl} = \frac{(1 - f_g) L_g}{L_s + (1 - f_g) L_g} I_g \tag{3}$$

while the system is within eclipse. Here I_g denotes an average contribution of the unit area of the disk of the primary component to the residual intensity, and f_g the fractional loss of light of the eclipsed component expressed in terms of its total light. Dividing Equation (3) by Equation (2), we obtain

$$R \equiv \frac{r_{ecl}}{r_{out}} = \frac{1 - f_g}{1 - L_g f_g} < 1.0. \tag{4}$$

Formula (4) can explain the variation of R with phase within the primary eclipse of R CMa. If L_g is further given, Equation (4) should give the value of f_g directly. Thus, from spectroscopic measurement of intensities of normal absorption lines within an eclipse we can determine directly the fractional loss of light of the eclipsed component. With the value of R at mid-eclipse, 0.906, obtained in Section 3, we have deduced values of f_g for various values of L_g in blue color (cited in the second column of Table IV) derived from the light-curve analysis, as given in the last column of Table IV. In the fourth and fifth columns of Table IV, we show the values of f_g calculated with the respective geometrical elements r_g, r_s, i, for two extreme cases of

TABLE IV

Comparison of values of f_g deduced spectroscopically and those calculated from the geometrical elements (r_g, r_s, i)

Worker	L_g	$\frac{1}{2}D$	$(f_g)_{cal}$		$(f_g)_{sp}$
			$u = 0$	$u = 1$	
WOOD (1946)	0.928	0.ᴾ081	0.431	0.457	0.591
PLAUT (1950)	0.965	0.088	0.402	0.460	0.746
KOPAL and SHAPLEY (1956)	0.938	0.082	0.422	0.446	0.627
KOCH (1960)	0.969	0.086	0.404	0.431	0.770
KITAMURA and TAKAHASHI (1962)	0.919	0.095	0.431	0.458	0.569

limb-darkening. In the third column are also given values for a half duration of the eclipse as calculated from the respective geometrical elements.

It is quite apparent from Table IV that the fractional loss deduced spectroscopically is considerably larger than that calculated from the geometrical elements. Thus, we have at present

$$(f_g)_{sp} > (f_g)_{cal}. \tag{5}$$

It should be noted here that the geometrical elements used for calculation of $(f_g)_{cal}$ were deduced mostly by avoiding humps or plateaus on the respective light curves. This is easily recognized from comparison of the apparent duration of the eclipse on those light curves with the durations calculated from the respective geometrical elements.

From the spectroscopic evidence represented by (5) and the apparently greater duration of eclipse from the spectroscopic measurements, it seems to be reasonable to assume that the humps and plateaus observed so far are caused by some additional obscuration of the light of the brighter F1V component.

5. Discussion

As mentioned in Sections 1 and 4, some photometric humps or plateaus observed at phases just preceding or following the primary minimum of R CMa seem to be related to such spectroscopic features as the evidence given by (5) and the greater duration for the spectroscopic eclipse. Therefore if we may assume that the body of the secondary component is surrounded by some non-stationary, relatively denser, gaseous material, the photometric asymmetry as well as the present two spectroscopic features could be

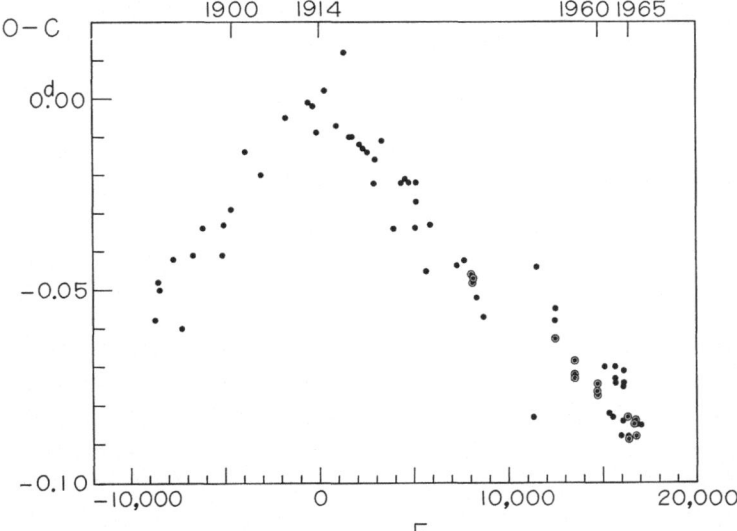

Fig. 5.   Residuals from linear elements for R CMa. ○ = photoelectric determination; • = visual estimate.

explained as the additional obscuration of light of the brighter component. We may recall here that this picture seems to be consistent with the result of the study by SAHADE (1963) who discussed this system by assigning the primary component a normal main-sequence mass and reached the conclusion that the secondary component overflows its Roche limit.

It is well known that R CMa suddenly changed its period around 1914. Figure 5 is a plot of the residuals of all observed heliocentric times of minimum light from the elements of DUGAN and WRIGHT (1939): JD 2420213.1381 + 1.1359435 E. All data

TABLE V

Times of minimum light of R CMa

JD 2430000. +	Method	E	O-C	Observer
6958.0038	pe	14741	− 0.^{p}0774	⎫
6959.1426	pe	14742	− .0746	⎬ KITAMURA and TAKAHASHI
6982.9953	pe	14763	− .0767	⎭
7378.310	vis. est.	15111	− .070	⎫
7696.362	vis. est.	15391	− .082	⎬ DUEBALL and LEHMANN
7746.342	vis. est.	15435	− .083	⎭
8089.411	vis. est.	15737	− .070	FERNANDES and HOFFMANN
8105.310	vis. est.	15751	− .074	⎱ OBURKA
8114.399	vis. est.	15759	− .073	⎰
8384.738	vis. est.	15997	− .088	SHY and TELES
8399.519	vis. est.	16010	− .075	OBURKA
8400.645	vis. est.	16011	− .084	SHY and TELES
8406.335	vis. est.	16016	− .074	OBURKA
8440.417	vis. est.	16046	− .071	ORLOVIUS
8817.533	vis. est.	16378	− .088	MARRACO
8818.6683	pe	16379	− .0884	RUIZ and TABOADA
8832.305	vis. est.	16391	− .083	ORLOVIUS
9140.1442	pe	16662	− .0845	⎱ SATO (unpublished)
9163.9998	pe	16683	− .0837	⎰
9169.678	vis. est.	16688	− .085	BALDWIN
9518.410	vis. est.	16995	− .088	KNIPE
9528.636	vis. est.	17004	− .085	BALDWIN

published before 1956 were from the papers by WOOD (1946) and KOCH (1960). Twenty-two additional times of minimum light listed in Table V are added here. If at the time of the sudden period change a catastrophic phenomenon accompanied by ejection of a considerable amount of mass did occur, some of the ejected material may still be in the neighbourhood of the components. The emission-like feature of the CaII K line observed at mid-minimum supports the idea that even the primary component may be surrounded by gaseous material.

Anyhow, not only the photometric asymmetry but also the spectroscopic features observed here seem to be quite favourable to the assumption that both components are surrounded by gaseous matter.

If we adopt a normal main-sequence mass of 1.7 $M_\odot$ for the primary component of spectral type F1V and use the mass-function $f(m) = 0.0037$ by STRUVE and SMITH (1950), the mass ratio of the components turns out to be about 0.15. For such a small mass ratio, the dimensions of the inner and outer Lagrangian lobes around both components would be considerably different, so it may be considered that the gaseous matter moving inside the outermost contact surface should be much denser around the secondary component than that around the primary. According to almost all discussions published so far (e.g. KOPAL, 1959), the body of the secondary component probably fills its Roche limit, and so the gaseous matter distributed rather densely in the neighbourhood of the secondary's body would obscure more effectively the

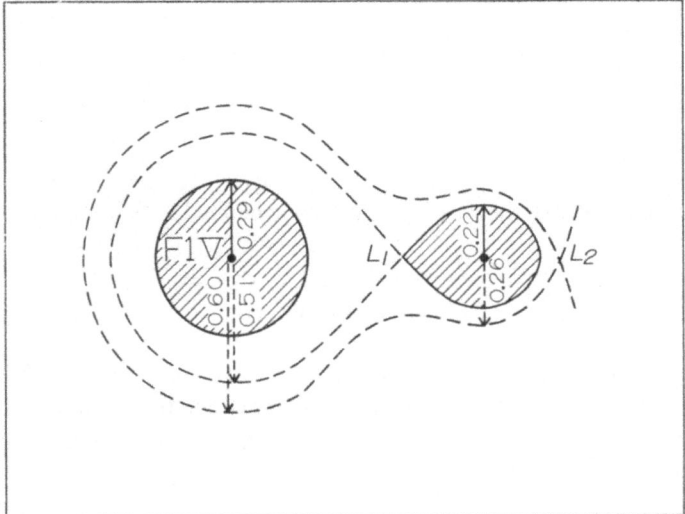

Fig. 6.   A schematic view of the innermost (Roche limit) and outermost contact lobes for R CMa. Numbers written are the respective fractional dimensions.

light of the brighter component at the primary minimum. Such gaseous material must be moving inside the innermost contact surface, and therefore a possible mass loss should still occur through the Lagrangian point L_2. Figure 6 is an illustration of the dimensions of the innermost and outermost Lagrangian lobes, taken from the work by PLAVEC (1964) for the mass ratio 0.15.

Acknowledgements

I wish to thank Mr. M. Kondo for his useful suggestion and discussion on the possibility of determination of fractional loss of light within eclipse from measurements of spectral line intensities. I am also grateful to Mr. K. Sato for his collaboration in spectrographic observations and to Dr. J. Jugaku, Messrs M. Kiyokawa and S. Kikuchi for their stimulating discussions. Thanks are also due to Miss K. Mori for drawing the figures and to Miss. M. Yoneda for typing the manuscript.

References

DUGAN, R. S. and WRIGHT, F. W.: 1939, *Princeton Contr.*, No. 19, 34.
FRINGANT, A. M.: 1956, *Contr. Inst. Astron. Paris*, A, No. 216.
JORDAN, F. C.: 1916, *Allegheny Publ.* **3**, 49.
KITAMURA, M.: 1960, *Atti Convegno Celebr. Morte Schiaparelli, Milano*.
KITAMURA, M.: 1967, *Publ. Astron. Soc. Japan* **19**, 615.
KITAMURA, M. and TAKAHASHI, C.: 1962, *Publ. Astron. Soc. Japan* **14**, 44.
KOCH, R.: 1960, *Astron. J.* **65**, 326.
KOPAL, Z.: 1959, *Close Binary Systems*, Chapman-Hall and John Wiley, London and New York.
KOPAL, Z. and SHAPLEY, M. B.: 1956, *Jodrell Bank Ann.* **1**, 141.
PICKERING, E. C.: 1904, *Harv. Ann.* **48**, 184.
PLAUT, L.: 1950, *Kapteyn Astr. Lab. Publ.*, No. 54.
PLAVEC, M.: 1964, *Bull. Astr. Inst. Czech.* **15**, 165.
SAHADE, J.: 1963, *Ann. Astrophys.* **26**, 80.
SATO, K.: unpublished.
SITTERLY, B. W.: 1940, *Astron. J.* **48**, 190.
SMAK, J.: 1961, *Acta Astron.* **11**, 171.
STRUVE, O. and SMITH, B.: 1950, *Astrophys. J.* **111**, 27.
WENDELL, O. C.: 1909, *Harv. Ann.* **69**, 66.
WOOD, F. B.: 1946, *Princeton Contr.*, No. **21**, 31.
WOOD, F. B.: 1957, in *I.A.U. Symposium No. 3: Non-Stable Stars* (ed. by G. H. Herbig), p. 144.
WRIGHT, K. O., LEE, E. K., JACOBSON, T. V., and GREENSTEIN, J. L.: 1964, *Publ. Dominion Astron. Obs.* **12**, 173.

Discussion

Nariai: You said that the spectroscopic eclipse is larger than the photometric eclipse. But I presume that it is within the plateau of the photometric eclipse. Is that correct?

Kitamura: Yes. What I mentioned is that the phases associated with photometric humps and plateaus are still within the spectroscopic eclipse.

A GROSS SECULAR EXPANSION OF THE PRIMARY
IN RW PERSEI*

DOUGLAS S. HALL

Dyer Observatory, Nashville, Tenn., and Kitt Peak National Observatory**, Tucson, Ariz., U.S.A.

Abstract. On the basis of several old light curves and two new photoelectric light curves, it appears that the A5 star in RW Per has more than doubled in size since 1900. The most striking manifestation of this is the steady decrease in the duration of totality and the fact that the eclipse was partial in 1967. The U-B and B-V indices imply an ultraviolet excess in both stars, which can be accounted for by assuming an L_3 of about 5% in U. For the A5 star M_v is shown to be $0^m \pm 1^m$. On the basis of this the approximate absolute dimensions are 2.1 $M_\odot$ and 3.6 $R_\odot$ for the hot star, 0.4 $M_\odot$ and 5.6 $R_\odot$ for the cool star. The period changed abruptly around 1925. Consideration is given to the well-known gas stream as a possible cause for the secular expansion and the ultraviolet excess.

RW Persei ($P = 13^d.2$) is a binary in which a G or K subgiant eclipses a smaller, brighter A star. In this paper will be discussed observations, many of them new, which suggest that the primary component appears to have been expanding since the turn of the century. The variability of RW Per was discovered by Enebo (SCHROETER, 1906) in 1905. Shortly thereafter PICKERING (1906) published a light curve based on estimates of plates taken at Harvard between 1890 and 1906. The eclipse seemed to be total, with totality lasting 9 or 10 hours or more. A few visual observations were made by WENDELL (1913) between 1906 and 1908. GAPOSHKIN (1953) determined another light curve by estimating plates in the Milton Bureau collection, but it is difficult to determine the duration of totality from his mean light curve due to an apparent asymmetry at the bottom. KORDYLEWSKA (1961) published a mean light curve based on visual estimates made by Kordylewski between 1924 and 1955; this curve showed totality lasting 6 or 7 hours, certainly not longer. STRUVE (1945), while studying RW Per spectroscopically in 1945, noticed that totality lasted approximately 3 or 4 hours and remarked about the short duration.

Recent UBV photoelectric observations, obtained at Kitt Peak National Observatory in 1965 with the number 4 16-inch telescope, showed that totality could not have lasted longer than 2 hours. Soon after that, observations made at Dyer Observatory in 1967 with the 24-inch telescope suggested that there was no total phase at all, and they showed that the eclipse was $0^m.1$ shallower than in 1965; this implied that the eclipse had become partial with $\alpha_0 = 0.98$. In both 1965 and 1967 BD + 42°938 was used as a comparison star and BD + 41°844 as a check star; the faint companion approximately $\frac{1}{4}'$ of arc from the variable was consistently excluded from the diaphragm. These observations are listed in Table I, where the first column contains the phase based on the observed time of primary minimum, and the last three contain the differential V, B, and U magnitudes in the sense RW Per minus BD + 42°938. The

* Dyer Observatory Reprint No. 47 and Kitt Peak National Observatory Contribution No. 375.
** Operated by the Association of Universities for Research in Astronomy, Inc., under contract with the National Science Foundation.

DOUGLAS S. HALL

TABLE I
Differential magnitudes of RW Persei

hel. JD 2439000+	phase (true)	RW Per − BD + 42°938 ΔV	ΔB	ΔU	hel. JD 2439000+	phase (true)	RW Per − BD + 42°938 ΔV	ΔB	ΔU
39.9725	+0.20346	2.732	2.817	3.057	802.6458	−0.01333		4.200	
62.9280	−.05729	2.782	2.861	3.169	802.6468	−.01326	3.849		
63.7404	+.00426	4.438			802.6528	−.01280	3.906		
63.7453	+.00463	4.419			802.6538	−.01273		4.243	
63.7613	+.00584	4.345			802.6548	−.01265			4.462
63.7648	+.00611	4.366			802.6558	−.01258		4.268	
63.7807	+.00731	4.321			802.6568	−.01250	3.939		
63.7828	+.00747	4.303			802.6618	−.01212	3.941		
63.7988	+.00868	4.228			802.6638	−.01197			
63.8036	+.00905	4.199	4.692	4.905	802.6648	−.01189		4.315	
63.8057	+.00921	4.230			802.6668	−.01174			4.523
63.8321	+.01121	4.098	4.478	4.738	802.6678	−.01167		4.343	
63.8363	+.01152	4.069			802.7180	−.00841	3.962		
63.8398	+.01177	4.060			802.7128	−.00826	4.153		
63.8418	+.01194	4.002			802.7138	−.00818		4.677	
63.8536	+.01283	3.974			802.7158	−.00803			4.925
63.8571	+.01310	3.955			802.7168	−.00796		4.685	
63.9023	+.01652	3.727			802.7508	−.00538	4.158		
64.0099	+.02468	3.320			802.7528	−.00523	4.293		
64.0113	+.02468	3.301			802.7538	−.00515		4.884	
76.6751	−.01573	3.759			802.7558	−.00500			5.191
76.6785	−.01547	3.778			802.7568	−.00492	4.295	4.910	
76.6938	−.01431	3.879			802.7618	−.00455	4.311		
76.6966	−.01410	3.874			802.7628	−.00447		4.930	
76.7237	−.01205	4.032			802.7638	−.00439			5.220
76.7258	−.01189	4.025			802.7658	−.00424		4.938	
76.7640	−.00900	4.234			802.7668	−.00417	4.311		
76.7667	−.00879	4.243			802.7688	−.00402		4.951	
76.7938	−.00674	4.349			802.7698	−.00394			5.248
76.7959	−.00658	4.321			802.7718	−.00379		4.970	

Table 1 (continued)

RW Per – BD +42°938

hel. JD 2439000+	phase (true)	ΔV	ΔB	ΔU
76.8181	−0.00489	4.425		
76.8202	−.00474	4.469		
76.8230	−.00452	4.454		
76.8320	−.00384	4.470		
76.8341	−.00368	4.474		
76.8674	−.00116	4.484	5.157	5.528
76.8709	−.00089	4.504		
76.8778	−.00037	4.469		
76.8792	−.00026	4.504	5.186	5.512
76.8841	+.00011	4.489		
76.8855	+.00021	4.469	5.168	5.506
76.9341	+.00389	4.490		
76.9355	+.00400	4.461		
76.9480	+.00495	4.405		
76.9591	+.00579	4.369		
76.9598	+.00584	4.390		
76.9605	+.00589	4.380		
76.9730	+.00684	4.364		
76.9751	+.00700	4.370		
77.0084	+.00952	4.221		
77.0098	+.00963	4.222		
782.9218	+.49225	2.826	2.847	2.992
783.9428	−.43039	2.776	2.831	2.983
784.9618	−.35319	2.753	2.804	2.996
785.8444	−.28631	2.718	2.802	2.988
785.9430	−.27877	2.725	2.816	2.949
791.8778	+.17081	2.733	2.814	2.983
797.6268	−.39360	2.726	2.819	2.974
799.6018	−.24397	2.722	2.791	2.980
799.9408	−.21828	2.731	2.805	2.959
799.9548	−.21722	2.724	2.795	2.978

RW Per – BD +42°938

hel. JD 2439000+	phase (true)	ΔV	ΔB	ΔU
802.7728	−0.00371	4.318		
802.8068	−.00114	4.370		
802.8078	−.00106		5.027	5.297
802.8098	−.00091			
802.8108	−.00083		5.052	
802.8118	−.00075	4.364		
802.8128	−.00068		5.042	5.350
802.8148	−.00053			
802.8168	−.00038		5.052	
802.8178	−.00030	4.360		
802.8498	+.00212	4.342	5.027	
802.8518	+.00227			5.284
802.8528	+.00235		5.010	
802.8548	+.00250			
802.8568	+.00265	4.342		
802.8588	+.00280		5.006	
802.8598	+.00288			5.289
802.8618	+.00303		4.995	
802.8628	+.00311	4.339		
802.8928	+.00538	4.271		
802.8938	+.00546		4.871	5.157
802.8958	+.00561			
802.8978	+.00576	4.267	4.856	
802.8998	+.00591			
802.9018	+.00606		4.821	
802.9028	+.00614			5.082
802.9048	+.00629		4.803	
802.9068	+.00644	4.240		
802.9118	+.00682	4.229		
802.9128	+.00689		4.753	
802.9148	+.00705			4.981

Table I (continued)

hel. JD 2439000+	phase (true)	RW Per − BD + 42°938 ΔV	ΔB	ΔU
800.8388	−0.15024	2.738	2.799	2.960
800.9168	−.14433	2.752	2.807	2.963
802.5698	−.01909	3.555		
802.5718	−.01894		3.720	3.903
802.5738	−.01879			
802.5758	−.01864		3.748	
802.5768	−.01856	3.564		
802.5808	−.01826	3.594		
802.5828	−.01811		3.830	
802.5848	−.01796			4.001
802.5868	−.01780		3.800	
802.5878	−.01773	3.598		
802.5928	−.01735	3.623		
802.5948	−.01720		3.851	
802.5958	−.01712			4.038
802.5978	−.01697		3.884	
802.5998	−.01682	3.667		
802.6308	−.01447	3.797		
802.6318	−.01440		4.090	
802.6328	−.01432			4.295
802.6348	−.01417		4.110	
802.6358	−.01409	3.832		
802.6418	−.01364	3.849		
802.6428	−.01356		4.184	
802.6438	−.01349			4.386

hel. JD 2439000+	phase (true)	RW Per − BD + 42°938 ΔV	ΔB	ΔU
802.9168	+0.00720		4.750	
802.9178	+.00727	4.209		
802.9198	+.00742		4.731	
802.9218	+.00758			4.961
802.9238	+.00773		4.723	
802.9248	+.00780	4.209		
802.9438	+.00924	4.115		
802.9458	+.00940		4.550	
802.9478	+.00955			4.741
802.9498	+.00970		4.517	
802.9508	+.00977	4.062		
802.9518	+.00985		4.509	
802.9538	+.01000			4.709
802.9558	+.01015		4.466	
802.9568	+.01023	4.054		
802.9598	+.01046		4.437	
802.9608	+.01053			4.632
802.9618	+.01061		4.423	
802.9638	+.01076	4.000		
803.8618	+.07880	2.760	2.814	2.968
804.8578	+.15426	2.750	2.818	2.988
804.9538	+.16153	2.737	2.802	2.964
807.9668	+.38224	2.730	2.803	2.965
807.9728	+.39027	2.743	2.807	2.953

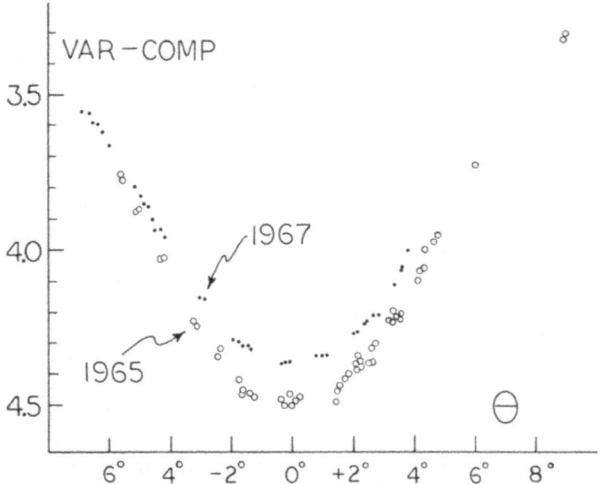

Fig. 1.   Primary minimum of RW Per in the visual. The eclipse appeared total in 1965; but it appeared partial in 1967 and was more than $0^m.1$ shallower.

change over the 2-year period can be seen in Figure 1, which is a plot of the visual observations. This remarkable systematic decrease in duration of totality suggested that perhaps the smaller star has been expanding or else the larger star has been contracting. For exploratory purposes the light curves of Pickering, Kordylewska, and the 1965 light curve at first were solved independently. The resulting elements, though approximate, served to indicate that r_g had remained constant while r_s had expanded.

An attempt was then made to solve all the light curves collectively, allowing only k to vary. Since the 1967 eclipse was partial, it was apparent that $i = 90°$ exactly would be a poor approximation. So the first two curves, which were complete although relatively insecure, were solved for various assumed values of i. The resulting pairs of r_g and i were then tested on the 1965 and 1967 observations, which were secure but did not cover the shoulders. It turned out that $r_g = 0.175$ and $i = 86°.4$ fit the last

TABLE II

Collective light curve solution

	1900 (Pickering)	1940 (Kordylewska)	1945 (Struve)	1965 (Hall)	1967 (Hall)
x_s (assumed)	.6	.6	.6	.6	.6
i	86°.4	86°.4	86°.4	86°.4	86°.4
r_g	.175	.175	.175	.175	.175
L_g: visual	–	.174	–	.200	.200
blue	.132	–	–	.112	.112
ultraviolet	–	–	–	.096	.096
α_0^{oc}	1.00	1.00	1.00	1.00	.98
k	.30	.55	.57	.64	.69

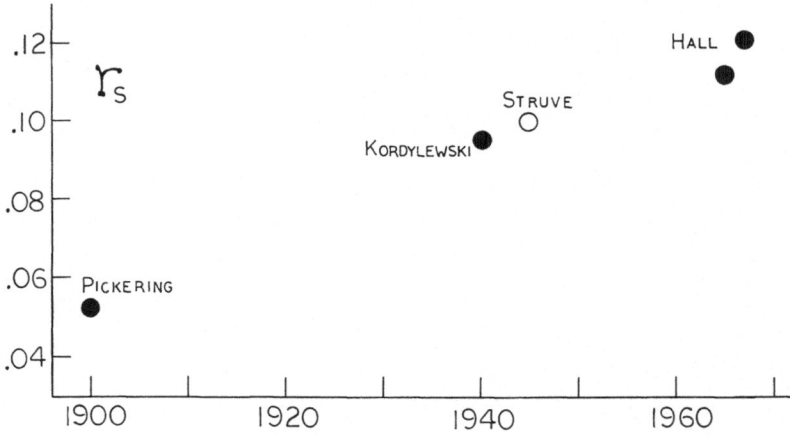

Fig. 2. The relative radius of the A star, r_s, seems to have doubled since 1900. See Table II. The filled circles are based on the four light curves discussed in the text; the open circle is based on the duration of totality observed by Struve.

two curves best; and these values required $k = 0.30$ in 1900, 0.55 in 1940, 0.65 in 1965, and 0.69 in 1967. The duration observed by STRUVE (1945) would be consistent with $k = 0.57$. These elements, summarized in Table II, represent all of the observations reasonably well. If this solution is taken at face value, it seems that the A star has more than doubled in size since 1900. This is illustrated in Figure 2.

This interpretation should be regarded as tentative because it depends quite critically on the older observations, which are of low quality. Specifically the accidental uncertainty is quite large in the old light curves; furthermore, since they are based on estimates of brightness, there is the possibility of serious error in the brightness scale. It is, however, not folly to use these light curves. Logically, the duration in time of the flat portion of a light curve cannot be affected by errors in the brightness scale unless the observers were by chance trying to exceed their limiting magnitude. The use of the branches of the old light curves was admittedly risky but, since r_g was held fixed, it was the duration of totality which was predominantly responsible for determining the value of k in each case.

The B-V and U-B color indices of the A star were derived from the photoelectric observations in both 1965 and 1967. Table III shows that the colors were similar for the two years but there was a noticeable difference in the U-B. With the 1965 observations used as an example, it can be seen in Figure 3 that if the colors are interpreted in the usual way, they imply that the primary is either a moderately reddened A8 or A9 star, or else a considerably more reddened B9 star. Neither interpretation, however, is consistent with a recent spectrum taken in 1967 by BARNES (1968), classified as A5 III by TERRILL (1968). Furthermore, both interpretations result in a very large ultraviolet excess for the secondary star: $\delta_{u-b} = 0^m\!.6$ in the first case and $0^m\!.2$ in the second case. If the 1965 eclipse had been assumed partial, the derived ultraviolet excesses for the secondary star would have been even larger.

TABLE III

Summary of photometric data

	V	B-V	U-B
RW Per in 1965 ($\alpha_0{}^{oc} = 1.00$)			
max	9.70:	.54	.25
min = cool star	11.40	1.15	.32
hot star	9.94:	.43	.24:
RW Per in 1967 ($\alpha_0{}^{oc} = 0.98$)			
max	9.70	.530	.144
min	11.30	1.14	.26
cool star	11.40	1.22	.30
hot star	9.94	.40	.13
BD $+42°938$	6.97	.465	$-$.023
BD $+41°844$	6.21	$-$.065	$-$.323
Pickering e	10.96	.820	.391
Pickering h	12.09	.541	.132

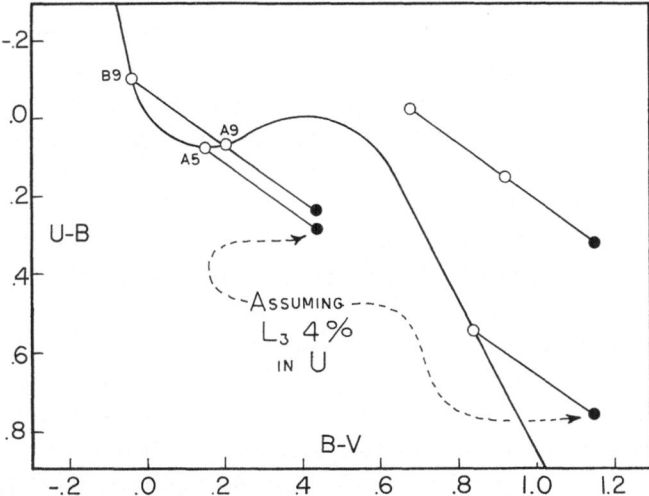

Fig. 3. Filled circles are observed color indices of the hot and cool star; open circles represent unreddened indices. The colors of the hot star are not consistent with its A5 spectrum, and the cool star appears to have a large ultraviolet excess. The assumption of $L_3 = 0.04$ in U renders the colors of both stars normal. See Table III.

Ultraviolet excesses have indeed been found in the subgiant components of other Algol eclipsing binaries. EGGEN (1960) found $\delta_{u-b} = 0.^m12$ in U Cephei; SHAO (1965) found $\delta_{u-b} = 0.^m2$ in SX Cassiopliae; and POPPER (1964) refers to ultraviolet excesses of $0.^m8$ and $0.^m6$ in KU Cygni and RZ Ophiuchi respectively. In the case of RW Per, it appears that not only the subgiant but also the A star has an ultraviolet excess, if its spectrum is assumed to be truly A5. This immediately suggests a fairly attractive explanation. It is possible to render the U-B indices of *both* stars normal simply by assuming a third source of light predominantly in the ultraviolet, visible more or less at all phases. And the amount of light it is necessary to postulate is only about 5% of the total light of the system in the ultraviolet. This, also, is shown in Figure 3. Perhaps related to this explanation is the finding by Popper that the ultraviolet excess in KU Cygni is caused by a strong continuous emission shortward of 4000 Å, which is not coming from regions near the photosphere of the primary star.

Secondary minimum was apparent in the light curve of Gaposhkin and in the 1967 light curve; although rather shallow in both cases, it was presumed real because its depth was consistent with the assumed value of k. In both light curves the secondary eclipse occurred at approximately 180°. This means that orbital eccentricity or a rotating line of apsides is not likely to be important here as an explanation of the photometric peculiarities.

If the A star has truly doubled in radius since 1900, then its luminosity should have increased accordingly and/or its spectral type should have become later. Struve classified RW Per as A5 in 1945 but CANNON and MAYALL (1949) classified it as A2 on a plate taken in 1923. This would imply that RW Per has not changed much in brightness between 1924 and 1945 but may have become brighter since 1945 by approximately 0^m5. Critical comparison of old and recent apparent magnitudes is out of the question until all light curves are transformed to the same standard photometric system. Therefore, as soon as RW Per becomes observable in the fall of 1968, it is planned to obtain photoelectric magnitudes of the same nearby stars which Pickering, Gaposhkin, and Kordylewski used for their magnitude sequences, and in this way transform their old light curves.

Although difficult to do so, it is of considerable interest to estimate the absolute dimensions of this binary system. The distance or absolute magnitude can be estimated in several ways.

(1) Interstellar reddening. The UBV magnitudes of four stars nearby RW Per are given in Table III. If all are assumed on the main sequence, an approximate distance – reddening relation of $E_{b-v} = 0^m3/kpc$ is obtained, in quite good agreement with that given by ALLEN (1963) for stars selected according to their visibility (rather than according to a volume of space, or specifically avoiding obscuration). If the intrinsic B-V of RW Per is assumed to be that of an A5 star, then the appropriate value to use for the reddening is $E_{b-v} = 0^m27$. Using the above relation the absolute magnitude implied is $M_v = 0^m4$.

(2) Spectroscopic parallax. Although the MK luminosity classification was an uncertain one, the implied M_v according to the calibration of BLAAUW (1963) would be approximately 0^m.

(3) Narrow-band photometry. In 1967, uvby photometry of RW Per was obtained by Barnes. He found the $[c_1]$ index of the A star implied $M_v = 1^m6$ if he assumed the spectral type was A5 and the color excess in b-y was 0^m21. If the $[c_1]$ index were

to be corrected for the suspected ultraviolet excess in the A star, the derived M_v would be more luminous, hence more in line with the above two estimates.

It can probably be concluded that $M_v = 0^m \pm 1^m$ for the primary of RW Per. On the basis of the previously mentioned $r_g = 0.175$, the mass function $f(M) = 0.0089 \, M_\odot$ determined by Struve, and the approximate effective temperature of an A5 star, the absolute dimensions of the system – as of 1965 – were calculated. These are presented in Table IV.

TABLE IV

Approximate absolute dimensions in 1965

Star	Mass ($M_\odot$)	Radius ($R_\odot$)
hot	2.1	3.6
cool	0.4	5.6

According to our current understanding of close binary evolution (see e.g. GIANNONE *et al.*, 1968), the subgiant was originally the more massive star, evolved first, exceeded its Roche lobe and transferred mass to the other star. The original mass ratio can be assumed to be $1\frac{1}{4}$, which is the average mass ratio of unevolved close binaries. The usual assumption (or approximation) can be made that the mass exchange was complete with no mass loss from the system. This implies that the present mass of the A star is approximately twice its original mass. Further it implies that the subgiant has lost approximately $\frac{3}{4}$ of its original mass. It could be that the large ultraviolet excess in the subgiant of RW Per is a result of our seeing a star which has lost its hydrogen envelope and is now just the hydrogen-poor core of the original star, but this would leave unexplained the ultraviolet excess of the A star.

It would no doubt be oversimplifying the problem not to consider the well-known ring or stream around the A star. Struve discussed in great detail the spectroscopic manifestations of the stream. He said it is very important to recognize that the hydrogen lines in RW Per are not simply those of a normal reversing layer but are in part produced in an absorbing stream or envelope. Further he noted that the spectroscopically observed rotational disturbance in hydrogen did not coincide in time with the observed time of minimum light; from that fact he concluded that the rotational disturbance of hydrogen is largely a phenomenon of the shell and not so much one of the partial eclipse of the A star. All of this gives us the picture that the stream is not a trivial complication; the entire primary source of light is probably a stream-star combination rather than a simple star. This is consistent with the idea that approximately half of the present mass of the A star is a result of mass exchange from the other star.

Struve described the shell absorption lines visible during totality, which would be difficult to explain unless the stream is supposed to circle around the cool star and be seen projected on its disk during primary eclipse. If, then, the stream encompasses

TABLE V
Times of primary minimum

hel. JD 2 400 000.+	Epoch	O-C (days)	Reference
14 619.73	− 1106	− 0.334	PICKERING (1906)
18 711.41	− 796	− .184	SHAPLEY (1915)
16 032.007	− 999	− .294	KORDYLEWSKI (1934)
29 217.587	0	.000	WOODWARD (1943)
31 500.922	+ 173	− .003	STRUVE (1945)
28 544.515	− 51	+ .041	LAUSE (1949)
28 782.096	− 33	+ .059	,,
28 808.48	− 31	+ .054	,,
28 834.868	− 29	+ .037	,,
28 861.247	− 27	+ .019	,,
28 953.654	− 20	+ .037	,,
28 966.809	− 19	− .007	,,
28 980.009·	− 18	− .005	,,
35 130.469	+ 448	− .039	WHITNEY (1959)
35 420.840	+ 470	− .034	,,
35 698.020	+ 491	− .023	,,
24 136.160	− 385	− .011	KORDYLEWSKA (1961)
25 152.443	− 308	− .011	,,
25 192.055	− 305	+ .006	,,
25 324.070	− 295	+ .036	,,
25 469.220	− 284	+ .002	,,
25 482.422	− 283	+ .006	,,
25 838.758	− 256	− .017	,,
25 865.162	− 254	− .010	,,
26 419.528	− 212	+ .020	,,
27 699.748	− 115	− .013	,,
28 188.114	− 78	− .001	,,
28 359.695	− 65	+ .009	,,
28 755.653	− 35	+ .013	,,
28 834.816	− 29	− .015	,,
30 431.820	+ 92	− .027	,,
33 005.546	+ 287	− .006	,:
33 744.637	+ 343	− .030	,,
35 341.655	+ 464	− .028	,,
38 456.53	+ 700	+ .004	KORDYLEWSKI (1964)
35 750.830	+ 495	+ .008	WHITNEY (1968)
39 063.684	+ 746	+ .028	HALL (this paper)
39 076.883	+ 747	+ .028	,,
39 802.822	+ 802	+ .051	,,

the entire binary system, *it* may be the source of the third light in the ultraviolet.

All available times of primary minimum have been listed in Table V and their residuals calculated from the period and epoch of WOODWARD (1943). Figure 4 shows that the period abruptly decreased in 1925 and may not have remained constant since then. In any case, it is evident that some mass redistribution has been occurring between 1900 and the present.

Photometric solution of the available light curves of RW Per has suggested that

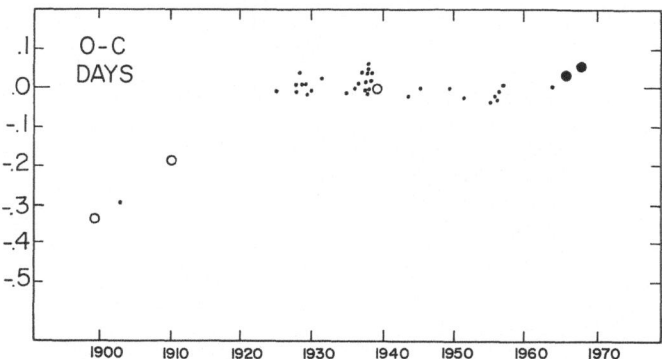

Fig. 4.   Observed times of primary minimum minus times calculated with the elements of Woodward. Open circles are based on mean photographic light curves; small dots have lower weight; filled circles are the photoelectric determinations. See Table V. The period abruptly increased around 1925 and may not have remained constant since then.

the primary source of light, whatever it is in precise detail, may have grown in size. There are two ways this may be interpreted:

(1) The A star has actually expanded. Although a doubling in size in about 100 years is two or three orders of magnitude faster than a normal single star would ever expand through the Hertzsprung gap, it is conceivable that a doubling of mass drastically accelerated its rate of evolution.

(2) The recently acquired mass is becoming luminous, on a 100-year time scale, as it is becoming accommodated by the A star. If this is the true explanation, then it may help fill a major gap in our understanding of close binary evolution·– the actual assimilation of the newly acquired mass.

Detailed observations of secondary minimum will be very useful, and are not to be neglected. The color change during secondary eclipse may reveal more directly the color of the cooler star and confirm the existence of the third light. The depth of secondary eclipse is required for a secure photometric solution, now that the eclipses are partial. A precise determination of the duration of secondary eclipse may serve to discriminate between the first and second interpretation of the apparent expansion.

Acknowledgement

The National Science Foundation generously supported this work through research grant GP-8084 and a travel grant.

References

ALLEN, C. W.: 1963, *Astrophysical Quantities*, 2nd ed., Athlone Press, London, p. 251.
BARNES, R. C.: 1968, private communication.
BLAAUW, A.: 1963, in *Basic Astronomical Data* (ed. by K. Aa. Strand), University of Chicago Press, Chicago, p. 401.

CANNON, A. J. and MAYALL, M. W.: 1949, *Harvard Ann.* **112**, 6.
EGGEN, O. J.: 1960, *Astron. J.* **68**, 483.
GAPOSHKIN, S.: 1953, *Harvard Ann.* **113**, 2.
GIANNONE, P., KOHL, K., and WEIGERT, A.: 1968, *Z. Astrophys.* **68**, 107.
KORDYLEWSKA, J.: 1961, *Acta Astron.* **11**, 43.
KORDYLEWSKI, K.: 1934, *Rocznik Astr. Obsv. Krakowskiego* **12**, 45.
KORDYLEWSKI, K.: 1964, *Inf. Bull. Var. Stars* No. 46.
LAUSE, F.: 1949, *Astron. Nachr.* **227**, 41.
PICKERING, E. C.: 1906, *Harvard Circ.* No. 114.
POPPER, D. M.: 1964, *Astrophys. J.* **139**, 143.
SCHROETER, J. FR.: 1906, *Astron. Nachr.* **170**, 357.
SHAO, C-Y.: 1965, *Astron. J.* **70**, 147.
SHAPLEY, H.: 1915, *Princeton Univ. Observ. Contrib.* **3**.
STRUVE, O.: 1945, *Astrophys. J.* **102**, 74.
TERRILL, C.: 1968, private communication.
WENDELL, O. C.: 1913, *Harvard Ann.* **69**, 149.
WHITNEY, B. S.: 1959, *Astron. J.* **64**, 258.
WHITNEY, B. S.: 1968, private communication.
WOODWARD, E. J.: 1943, *Harvard Bull.* **917**, 7.

Discussion

Kitamura: Did you not find any appreciable proximity effect from observations outside eclipse?

Hall: The 1967 observations, which were the most complete, showed no ellipticity or reflection effects to within $0.^{m}01$. This is not surprising because the period is $13.^{d}2$.

Plavec: I would not trust the old observations at all. Also more recent observations disagree very much. Evidence of secular expansion is doubtful: (1) because of the uncertainty of the observational evidence, and (2) because it is theoretically unacceptable that in 60 years or so the star could double its radius even if it is already in the giant stage (and with a radius like this, it must be just at the beginning of giant evolution) because even there evolution is much slower than your figures indicate. More probably there is exchange of material which can simulate different conditions at the primary minimum.

Sahade: If I remember correctly Struve's velocity curve of RW Persei was a 'wild' one, and he described the spectrum of the star as an A5 object without saying anything about it being a giant; actually I think he must have considered that it was a main sequence object. Would it be possible that the star is now surrounded by a shell and the spectrum gives the impression of being that of a giant?

Hall: Struve emphasized that the spectrum was not that of a normal stellar reversing layer but also of a gas stream. What you suggest may be possible. However, two things support the A5 III classification. First the $[c_1]$ measure of Barnes, which is a measure of the *continuum*, suggests giant luminosity. Second, if you assume $2M_\odot$ for the A star (certainly it could not be much less), the implied $3\frac{1}{2}\,R_\odot$ radius is indicative of a giant. My feeling is that we should start thinking of many Algol primaries as extremely complicated stream-star combinations and not as basically normal main-sequence stars with a relatively trivial ring or shell around them. But this has yet to be demonstrated.

Fracastoro: The apparent secular variation of the radius of one component of the system of RW Persei (as well as of ζ Aur) may be, perhaps more reliably, explained by means of a precession of the orbital plane of the eclipsing system, having the consequence of shifting its geometry from a partial (or grazing) eclipse, to a central one. This appears particularly likely to be true for RW Persei since Hall assumes the presence of a 'third light', i.e., of a third body.

BOSS 5481 DURING THE SHELL EPISODE OF 1965-67

AUGUSTO MAMMANO

Osservatorio Astrofisico, Asiago

and

ALDO MARTINI

Laboratorio di Astrofisica, Frascati

Abstract. Boss 5481, with a spectrum similar to that of VV Cep, has developed an absorption shell spectrum, observed at Asiago from 1965 to the end of 1966. The radial velocities of M and B components have been found to change, while the [Fe II] emissions remain stationary. Equivalent width of selected shell and stellar lines have been measured. Hα appears quite featureless at the end of 1965. Photoelectric observations of Loiano exclude a photometric eclipse. On the other hand, the radial velocity curve of the B component does not agree with the extremely long period of about 75 years announced by COWLEY (1967). Possible explanations are: (1) we are confronted with a grazing eclipse, and (2) the velocity curve of the B component does not reflect the orbital motion.

Riassunto. Lo spettro complesso di Boss 5481, tipo MepIb + B con righe in emissione del [Fe II], si è arricchito, nel 1965, di un insieme di righe di assorbimento dovute ad uno shell alquanto diluito, percettibili fino alla fine del 1966. Si riportano le velocità radiali determinate sugli spettri M, B e nebulare su un intervallo di circa 30 mesi e le larghezze equivalenti variabili delle righe di shell. Hα, alla fine del 1965 era caratterizzata dall'assenza di una chiara evidenza di assorbimento o di emissione. Le variazioni irregolari di piccola entità (minori di $0^m.10$) osservate fotoelettricamente a Loiano in una campagna coordinata con le nostre osservazioni spettroscopiche, non giustificano l'ipotesi di un'eclisse fotometrica. La curva di velocità radiale della componente B, con un massimo di $+30$ km/sec ed un minimo di -70 km/sec non si accorda col lunghissimo periodo (75 anni circa) annunciato brevemente da COWLEY (1967). Le righe dello spettro M presentano piccole variazioni di velocità mentre quelle del [Fe II] risultano praticamente stazionarie. È possibile che le caratteristiche mostrate da Boss 5481 siano dovute ad un'eclisse radente, ma, anche in tal caso, si deve ritenere che le velocità radiali della componente B non rispecchino il moto orbitale.

1. Introduction

Boss 5481A (HR 8164; BD+57°2249; HD 22239) is the bright component ($5^m.8$) of the double star ADS 14864 (separation 5″). Boss 5481 was classified as MepIb + B among a group of 7 binaries of VV Cep type, by BIDELMAN (1954). These stars show a late-type spectrum, often occurring with a superimposed B spectrum, and emission lines of [Fe II] and of hydrogen. Very few of these stars are known to exhibit radial velocity variations.

The ultraviolet absorption shell spectrum of Boss 5481 was first noted by COWLEY and COWLEY (1965) and described later (1966) with some details. Hα, reported by Bidelman as a "sharp absorption line with definite narrow emission edges" in 1949, was found by Cowley (between 1963 and 1965) to have a very strong violet emission component while the redward edge was only marginally present. Also prominent were the [Fe II] lines, together with absorptions of H and He I by the blue component.

Recently COWLEY (1967) reported briefly that the star is a spectroscopic binary with an extremely long period, about 75 years.

M. Hack (ed.), Mass Loss from Stars. All rights reserved

We describe below the spectroscopic behaviour of Boss 5481 observed at Asiago from 1965 to 1968 with the purpose of detecting variations of some of shell lines, in view of possible photometric eclipses like that occurring in the huge systems VV Cep and AZ Cas.

2. The Observations

Boss 5481 has been observed spectroscopically at Asiago from December 1965 to April 1968, mostly with the 'camera III' (2-prisms; dispersion 42 A/mm at Hγ). The projected slit was of 0.02×0.5 mm; details are given in Table I.

In order to ascertain whether an eclipse was taking place, photo-electric observations in V and in B were performed at Loiano by BARTOLINI and BATTISTINI (1968), who kindly permitted us to report the results given in Table II. These values have been plotted on the top side of Figure 3.

TABLE I

Spectr. no.	Date	UT	Camera	Emulsion	Exposure (min.)
6741	1965, Dec. 14	21^{h}05^m	III	IIa-O bak.	60
6742	Dec. 14	21 50	III	IIa-O bak.	20
6747	Dec. 15	17 05	III	X 103a-O	7
6748	Dec. 15	18 35	III	X 103a-O	170
6749	Dec. 15	21 10	III	103a-F	40
6758	Dec. 31	18 37	III	103a-O	180
6760	1966, Jan. 2	19 10	III	103a-O	220
6831	Jan. 6	18 50	III	103a-F	200
6956	Mar. 9	3 48	III	IIa-O bak.	104
7046	May 5	1 55	III	IIa-O bak.	140
7098	Aug. 11	0 52	I	IIa-O bak.	100
7110	Sep. 22	19 15	III	X 103a-O	90
7174b	Nov. 22	18 34	I	IIa-O bak.	8
7175	Nov. 22	21 53	I	IIa-O bak.	80
7209	Dec. 13	19 46	I	IIa-O bak.	150
7214	Dec. 15	19 37	III	103a-F	10
7215	Dec. 15	20 03	III	103a-F	40
7218	Dec. 17	18 15	III	IIa-O bak.	60
7395	1967, Mar. 4	4 10	III	IIa-O bak.	80
7396	Mar. 4	4 59	III	IIa-O bak.	14
7581	Jul. 15	2 09	III	Ia-O	68
7614	Jul. 30	0 30	III	103a-O	130
7615	Jul. 30	1 38	III	103a-O	15
7954	Nov. 13	20 05	III	IIa-O bak.	150
7955	Nov. 13	21 27	III	IIa-O bak.	8
8034	Dec. 9	20 59	III	IN sens.	12
8035	Dec. 9	21 12	III	IN sens.	4
8036	Dec. 9	21 15	III	IN sens.	1
8037	Dec. 9	21 24	III	IN sens.	12
8038	Dec. 9	21 34	III	IN sens.	4
8047	Dec. 11	21 03	III	IN sens.	12
8048	Dec. 11	21 10	III	IN sens.	4
8487	1968, Apr. 20	26 32	III	IIa-O bak.	100

TABLE II

Photoelectric observations of Boss 5481 (Loiano)
Comparison star: BD + 57°2312

Date			JD	ΔV	n	ΔB	n
1965, Nov.	21		2439086.3	−1.335	6	−1.371	6
Nov.	28		093.3	−1.310	5	−1.363	4
Nov.	30		095.2	−1.314	6	−1.377	13
Dec.	12		106.2	−1.263	3		
Dec.	14		109.3	−1.304	6	−1.353	4
1966, May	16		262.6	−1.316	5		
May	24		270.5	−1.339	5		
June	19		296.5	−1.288	3	−1.371	4
July	8		315.3	−1.347	4	−1.402	7
August	12		350.4	−1.341	5	−1.388	4
August	13		351.4	−1.330	5	−1.381	6
August	15		353.5	−1.338	3		
Sept.	9		378.4	−1.332	6	−1.381	5
Sept.	26		760.5	−1.285	7	−1.357	9

3. Identifications

Table III lists the proposed identification and an estimate of the intensity (in an arbitrary scale) of all observed lines in the range 3685–4861 Å. A few emission lines of [Fe II] are prominent, together with relatively strong and broadened He I absorption lines, which are absent in VV Cep. No other emission lines are seen in the range 3685–8800 Å. We could not see any clear emission even at Hα (December 1965); with the dispersion of most of the spectra, the feature, however, is so diffuse that Cowley's statement cannot be disproved. The spectral type of the blue component appears, at the end of the shell phase, quite different from B2V as reported by COWLEY (1967). We would rather favour a B5IV spectrum.

4. The Shell Spectrum

No certain changes occurred in the spectrum of Boss 5481 between 1933 and 1952 (BIDELMAN, 1954). The ultraviolet shell absorption spectrum was first recorded in October 1965; it was lacking in March 1964 and gradually increased from then on. COWLEY and COWLEY (1966) noted that all of the shell lines reported in their Table III arise from very low metastable levels. Sharp, redward displaced absorption cores were superimposed on the broad hydrogen lines.

The observations reported in this paper cover the region H13–Hβ. Shell lines have been detected in the Balmer series from H12 to Hγ and their equivalent widths are reported in Table VI, together with those of the normal stellar lines. The decreasing of intensity of some hydrogen shell lines is also clearly shown by Figures 1 and 2. Significant weakening have been also found for the Ti II (13) doublet at

TABLE III

Identification and estimated intensities of the lines in Boss 5481

λ_0	Ident.	1965, Dec. 14	1965, Dec. 15	1965, Dec. 31	1966, Jan. 2	1966, Mar. 9	1966, May 5	1966, Nov. 22	1966, Dec. 17	1967, Mar. 4	1967, Nov. 13
		i	i	i	i	i	i	i	i	i	i
4861.33	Hβ	5D	3D	4s		3d	4D		5d		5d
4646.17	Cr I (21)	1d									2.5D
4629.34	Fe II (37)	1d									–
4580.05	Cr I (10)	1d									2d
4571.07	Ti II (82)	1d									2.5d
4549.47	Fe II (38)	1s									–
.62	Ti II (82)										
4535.7	Ti I (42)	1D		2d		2D			2D		–
4522.80	Ti I (42)	1s		–		–					–
.64	Fe II (38)										
4518.02	Ti I (42)	1s				1.5s					2d
4501.27	Ti II (31)	1s				1.5d					
4496.86	Cr I (10)	–									
4489.74	Fe I (2)	1d		–		–	2D		2d		2.5s
.18	Fe II (37)										
4481.23	Mg II (4)	1D	2D	2D	1.5D	1.5d	2D	2d	2D	1.5d	2s
4471.48	He I (14)	2d	2d	2D	2.5d	2D	2.5D	2d	2D	1.5d	2d
4461.76	Fe I (2)	1d	1.5d	2D	2D	2D	1.5d	2d	2D	1.5d	2D
4459.12	Fe I (68)	1d	–		–	–	–	–			–
4447.72	Fe I (68)	1s	2s	2s	–						–
4443.80	Ti II (19)	1D	2.5D				2D	–		–	2.5D
4435.68	Ca I (4)	1D	2.5d				2D			2D	2D
4427.32	Fe I (2)	2s	2.5d		2s	1.5d	2d	2s		1.5s	1.5d
4417.42	Ti II (40)	1s	1.5s	1.5s							
4415.14	Fe I (41)	2s	2s	2d	2s	1.5d	2s	1.5s	2.5s	2d	2s
4413.78	Fe II (7F)	2ed								2en	
4412.25	Cr I (22)	1s									
4408.20	V I (22)	2s	1.5s	2d		1.5s			2s		–
.42	Fe I (68)										

Table III (continued)

λ_0	Ident.	1965, Dec. 14	1965, Dec. 15	1965, Dec. 31	1966, Jan. 2	1966, Mar. 9	1966, May 5	1966, Nov. 22	1966, Dec. 17	1967, Mar. 4	1967, Nov. 13
		i	i	i	i	i	i	i	i	i	i
4404.74	Fe I (41)	2d	1.5d	2D	1.5D	2d	1.5d	2D	2d	2d	2d
4400.57	V I (22)	-	2.5d	-	-	1d	2s	-	-	-	2d
.35	Sc II (14)										
4395.22	V I (22)	2s	-	-	-	1d	2s	-	1.5d	-	2.5d
.03	Ti III (19)										
4389.97	V I (22)	1s	-	1s	1s	-	-	-	-	-	1s
4384.72	V I (5, 22)										
83.84	Fe I (41)	2D	2.5D	-	-	2D	2.5D	-	2.5D	2.5d	2.5D
84.98	Cr I (22)										
4379.24	V I (22)	1s	-	-	-	-	-	-	-	-	-
4374.45	Sc II (14)	2D	1.5D	2d	-	2D	2.5D	-	2D	2d	2.5D
4371.28	Cr I (22)	1s	-	1.5s	-	-	-	-	-	-	-
4367.91	Fe I (41)	1s	2.5D	-	-	-	2.5D	-	2.5d	1.5d	2d
4351.77	Cr I (22)	3s	-	-	-	2s	-	-	2s	-	1.5d
4347.24	Fe I (2)	1s	-	-	-	-	-	-	-	-	-
4344.51	Cr I (22)	-	-	2n	-	-	-	-	-	-	2d
.29	Ti III (20)										
4340.47	Hγ	10n	9n	9b	4b	8b	7n	6s	10n	8n	6b
4337.57	Cr I (22)										
.92	Ti III (20)	3s	2.5d	3n	1s	3d	3d	2s	3d	2d	3.5d
.05	Fe I (41)										
4325.76	Fe I (42)	2D	2.5D	-	-	2D	3d	2d	-	2D	2.5D
.01	Sc II (15)										
4320.74	Sc II (15)	1s	-	-	-	-	-	-	1.5s	-	2s
.96	Ti I (41)										
4314.66	Sc II (15)	3D	2d	2d	2d	2s	2.5s	2d	2s	2d	2.5d
4307.91	Fe I (42)	1D	1.5d	1d	1s	1.5s	1.5D	1.5d	1d	1s	-
.90	Ti III (41)										
4300.05	Ti III (41)	-	-	-	-	2D	-	-	-	-	-

Table III (continued)

λ_0	Ident.	1965, Dec. 14	1965, Dec. 15	1965, Dec. 31	1966, Jan. 2	1966, Mar. 9	1966, May 5	1966, Nov. 22	1966, Dec. 17	1967, Mar. 4	1967, Nov. 13
		i	i	i	i	i	i	i	i	i	i
4294.13 / .10	Fe I (41) / Ti III (20)	2s	1.5s	–	–	–	–	–	–	–	1.5s
4291.47 / 90.22	Fe I (41) / Ti III (41)	–	–	3b	–	2D	–	2D	–	–	–
4289.72	Cr I (1)	2d	–	–	–	–	1.5s	–	–	–	–
4287.40	Fe II (7F)	3eD	2eD	2eD	–	3eD	3eD	–	2.5eD	–	2.5ed
4276.83	Fe II (21F)	–	–	–	–	–	2eD	–	–	–	2.5s
4274.81	Cr I (1)	2s	–	–	–	2.5s	–	–	–	–	–
4271.76	Fe I (42)	2d	–	1d	–	2.5D	2.5s	–	–	–	–
4260.43	Fe I (152)	1s	1s	1s	–	1.5s	1d	–	1s	1s	–
4254.33	Cr I (1)	1.5d	–	–	2s	2.5s	2d	1d	1.5s	2.5d	2.5d
4250.79	Fe I (42)	–	–	–	–	1.5s	–	–	2d	–	–
4246.83	Sc II (7)	3eD	2ed	3eb	–	3ed	–	–	3eD	3.5eD	1.5d
4243.98	Fe II (21F)	–	–	–	–	–	–	–	1d	–	–
4233.17	Fe II (27)	1D	2s	1.5d	–	1.5s	2d	–	–	–	–
4231.56	Fe II (21F)	1eD	2eD	–	–	1.5eD	2D	2.5D	1.5D	2eD	–
4226.73	Ca I (2)	1D	–	–	–	2s	2D	–	–	–	–
4215.76	Sr II (1)	2s	2s	2s	1.5d	–	2.5n	–	2s	2d	1d
4190.71	Co I (1)	1d	1D	–	–	–	2s	–	–	–	–
4161.52 / .20	Ti III (21) / Zr II (42)	–	–	–	–	1.5s	–	–	–	–	–
4147.67	Fe I (42)	–	–	–	–	1s	–	–	–	–	–
4143.87 / 76	Fe I (43) / He I (53)	5D	5d	5d	2s	5d	5D	3d	4D	4D	3d
4134.47	V I (27)	1.5s	1.5s	1.5d	1s	1.5s	1d	–	1.5s	1s	1s
4127.85	V I (27)	–	1s	–	2s	2s	1s	–	2s	2s	1s
4123.57	V I (27)	–	1s	–	–	2s	1.5s	–	–	–	–
4120.84	He I (16)	2d	1d	1.5s	1d	1.5s	1d	2D	1.5s	2d	1d
4105.17	V I (27)	–	1.5s	–	1s	2.5d	2d	–	–	2s	–
4101.74	Hδ	12D	12dn	11dn	6dn	11dn	10dn	9dn	8dn	8dn	–

Table III (continued)

λ₀	Ident.	1965, Dec. 14	1965, Dec. 15	1965, Dec. 31	1966, Jan. 2	1966, Mar. 9	1966, May 5	1966, Nov. 22	1966, Dec. 17	1967, Mar. 4	1967, Nov. 13
		i	i	i	i	i	i	i	i	i	i
4095.98	Fe I (217)	–	–	–	–	2s	–	–	1.5d	–	–
4092.51	V I (27)	2D	1d	1.5d	1.5s	2d	1.5d	–	1.5s	2D	1d
4082.40	Sc I (6)	–	2d	–	–	2D	–	–	–	–	–
4071.75	Fe I (43)	2d	1.5s	1.5d	2d	2D	1.5D	1.5D	2d	2d	1d
4067.27	Fe I (217)	–	–	–	–	2D	1.5D	–	–	–	–
4063.89	Fe I (43)	–	–	–	–	1.5s	1.5D	–	2D	–	–
4054.55	Sc I (6)	–	1s	–	–	–	–	–	–	–	–
4045.83	Fe I (43)	2d	1.5d	2d	1.5D	2D	–	2s	1.5D	–	2D
4030.67	{Mn I (2) / Fe I (560)}	1s	1.5d	2d	2s	2D	2D	1.5d	1D	1.5D	–
4026.14	He I (18)	2d	3D	2.5D	2d	2D	2.5d	–	2D	2d	2.5D
4024.78	{Fe II (127) / Ti II (11)}	2s	2d	–	–	1.5s	1.5d	–	–	–	–
4020.90	Co I (16)	–	–	–	–	1.5s	–	–	–	–	–
4012.37	Ti II (11)	–	2d	–	–	–	–	–	–	–	–
4005.20	Fe I (43)	–	1.5d	1s	2d	2D	2d	1d	–	1D	1d
3970.07	Hε	6dn	9dn	–	6dn	12dn	8dn	blend	blend	6dn	3.5D
3968.47	Ca II (1)	6d	12n	–	6s	13n	9n	blend	blend	7n	4n
3964.73	He I (5)	1.5s	1.5d	1.5s	1s	1.5d	1s	1d	–	1.5s	1d
3933.66	Ca II (1)	6s	14n	–	6s	13n	10n	4n	9n	10n	6n
3889.05	H8	5dn	7dn	–	5dn	10dn	7dn	7dn	–	5dn	–
3835.39	H9	–	6dn	–	5dn	7dn	5dn	–	–	3.5dn	–
3819.68	He I (22)	–	5D	–	4D	6D	5D	–	–	–	4D
3797.90	H10	–	3.5D	–	5dn	6dn	5dn	–	–	–	–
3770.63	H11	–	5dn	–	5dn	4dn	4dn	–	–	–	–
3761.32	Ti II (13)	–	3s	–	3.5s	2D	1.5d	–	–	–	–
3759.29	Ti II (13)	–	3.5s	–	3.5s	2.5D	1.5s	–	–	–	–
3750.15	H12	–	6dn	–	5dn	5dn	4dn	–	–	–	–
3734.37	H13	–	–	–	3D	–	–	–	–	–	–
3685.19	Ti II (14)	–	–	–	3.5s	–	–	–	–	–	–

i = intensity; s = sharp; d = diffuse; D = very diffuse; n = narrow; b = broad; e = emission.

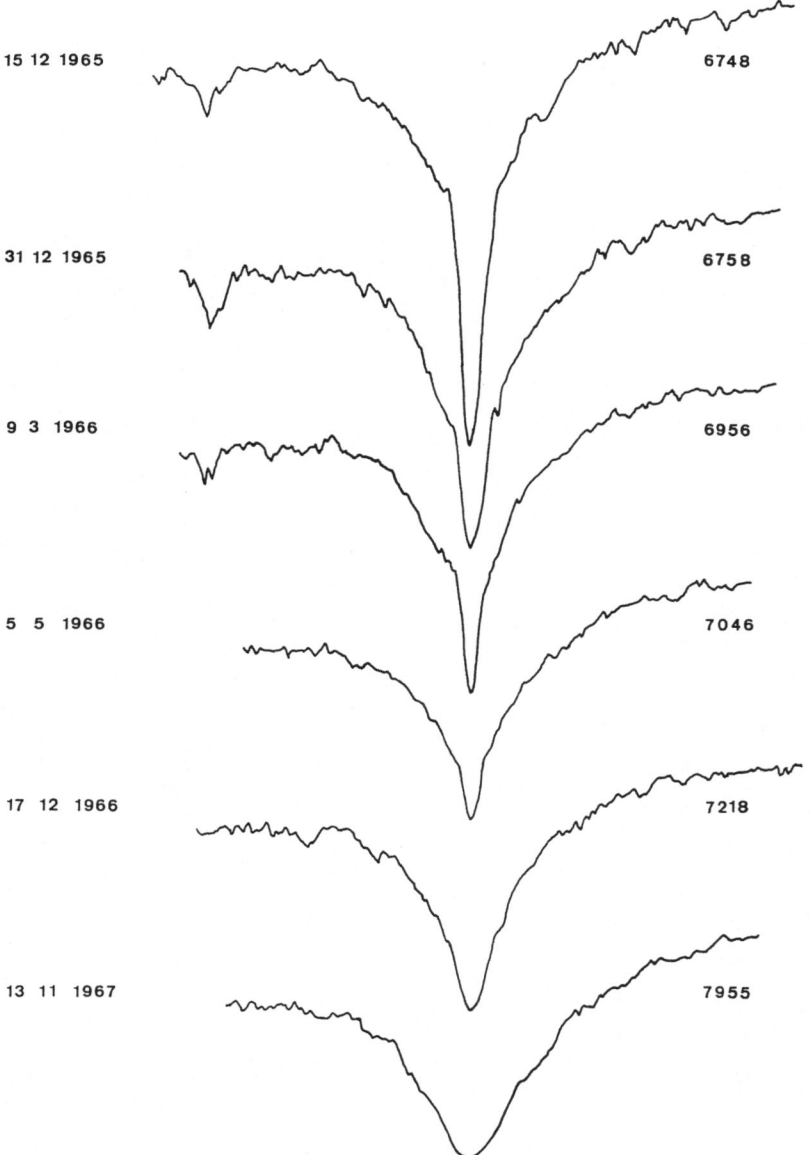

Fig. 1.   Microphotometric tracings (transmission) near H8.

$\lambda\lambda$ 3759–61, whose lower level is metastable. Figure 2 reproduces some tracings including such lines. The variation of the sharp K line of Ca II is less apparent; it was still present in April 1968, when all the shell lines have disappeared. The only line of helium for which a consistent set of measurement was possible is that at 3819 Å and its intensity did not change.

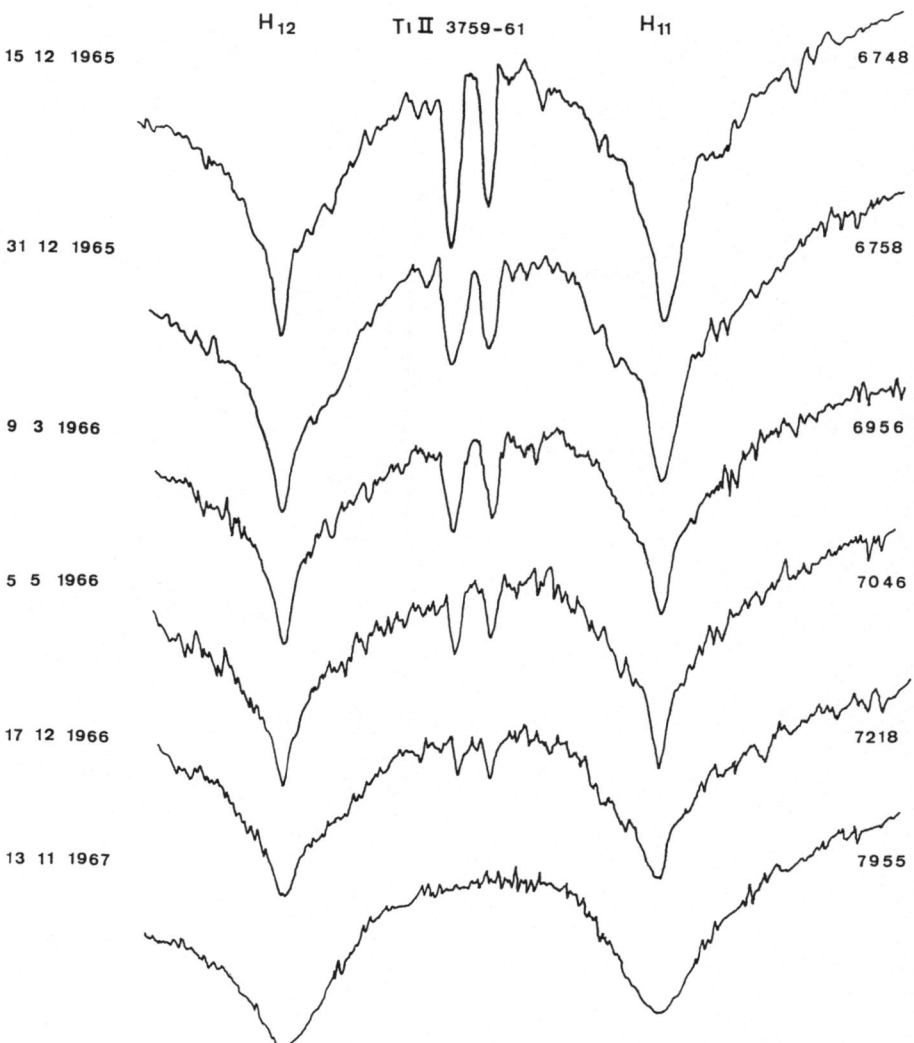

Fig. 2.   Microphotometric tracings of the H11–H12 region including the TiII (13) doublet.

5. Radial Velocities

Table IV lists radial velocities derived by lines of the M and B spectra. The adopted wavelengths for the M component have been taken from WRIGHT (1955). For the B component we were forced to include also lines of the region crowded by features of the cold star. The resulting mean square errors are obviously larger than for the M component, as can easily be seen from Table IV. The weighted mean value of both components together with their mean square errors are plotted in Figure 3.

Although at the employed dispersion, individual radial velocities are of low value because of blending, we have attempted to plot in Figure 4 velocities of the shell

TABLE IV

Radial velocities of components M and B in Boss 5481 corrected for earth motion (km/sec)

| JD−2439000 | 110.274 | 126.276 | 128.297 | 193.858 | 250.580 | 452.273 | 452.412 | 477.260 | 553.673 | 808.337 | 967.606 |
Spectr. no.	6748	6758	6760	6956	7046	7174	7175	7218	7395	7955	8487
4461 Fe I	−3	−12		+59	−14	−42		−15	+7	−34	+3
4454	+3	−25		−7	−27	−75		+9	−34	−25	+6
4427 Fe I	−59	−17		−45	−46					−42	−16
4415 Fe I	−15	−36	−13	−35	−19	−36		−30	−31	−35	−19
4404 Fe I	−44	−24	−13	−16	−27	−45	−23	−36	−32	−34	−26
4314 Ti I	−74	−18		−3	−43	−47	−29	−18	+37	−28	−44
4307 Fe I, Ti II		−26			−50		−25				
4282		−47			−14				−29	−26	+40
4260 Fe I	−22	−4		+17	−21			−23	+15		+7
4254 Cr I	−34	−15		+5	−14	−49	−37	−47	−12	−33	−10
4215 Sr II	−40	−32		+1	−25			−33	−51	−31	−32
4134 V I	−39	−82	−42	−16	−24				−30	−28	+9
4128 V I	−24			−25	−43			−10	−11	−22	
4092 V I	−23			−10	−38			−27	−16	−26	−23
4071 Fe I	−19		−21	+7	+5		−33	−14	−17	−20	
4045 Fe I	−19	−71	−26	+5			−67	−44			
4035	−17	−51	−6	−9	+2		−84		−8		+3
4030 Mn I, Fe I	−17	+3	−14	+12	−17		−31	−57			
4005 Fe I		−26			−44		+12	−26			
V_M	−27	−26	−17	−4	−25	−48	−38	−26	−18	−20	−4
m.e.	±3	±3	±3	±3	±2	±3	±5	±3	±3	±2	±3
4481 Mg II	−27	0	−8	+75	+2					+2	+8
4471 He I	+27	0	+35	+32	−9	−86	−54	−12	+9	+9	+8
4121 He I	+25	+46	+35	+24	−21	−74	−70	−56	+5	+1	+27
4026 He I	+35	+56	−4	+35	−13			−45	+22	+43	+38
3964 He I	+1	−39		−18	+9		−71		−11	+34	+35
3819 He I	+14		+16	−2	−33					+9	+49
V_B	+20	+17	+24	+33	−15	−79	−64	−45	+1	+20	+32
m.e.	±6	±14	±10	±9	±5	±7	±7	±11	±4	±10	±6

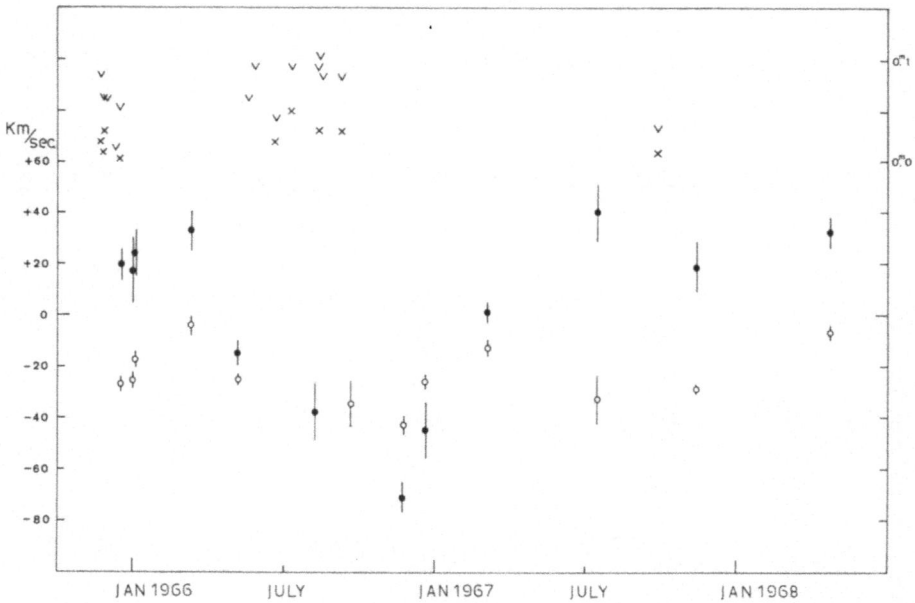

Fig. 3.   Radial velocities of the M and B components (bottom) and photoelectric observations (top).
○: M velocity;   ●: B velocity;   v: V photoelectric observations;   x: B photoelectric observations.

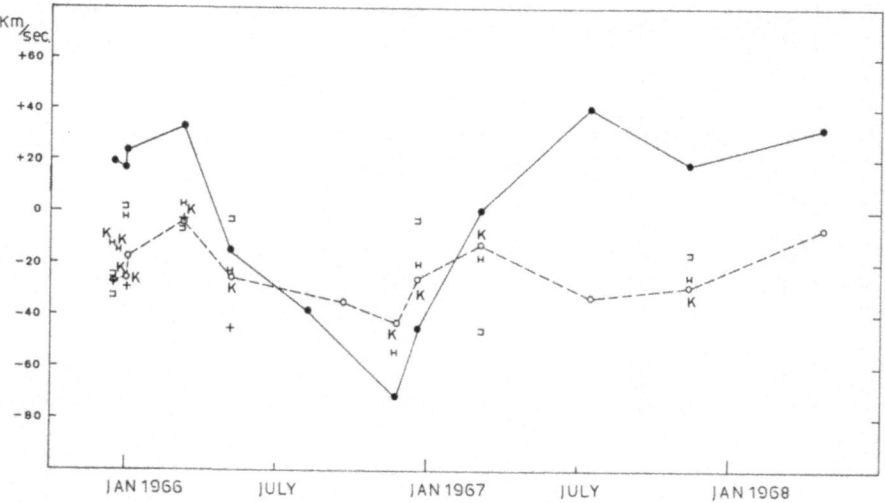

Fig. 4.   Radial velocities of stellar and shell lines.   ○: M velocity;   ●: B velocity;   +: Ti II 3759–61
(shell); H: mean of Balmer shell lines (from Hγ to H13); k: Ca II (H and K shell lines); ∃: [Fe II].

lines (from Table V) for comparison with the stellar features. The same has been
done for a few emission lines of [Fe II].

TABLE V

Radial velocities of selected lines (km/sec)

Spectr. no.	6741	6748	6758	6760	6956	7046	7175	7218	7395	7955
Ca II (H)	−14	−20		−31	+ 3	−35			−13	−26
Ca II (K)	−11	−27		−22	− 4	−27	−43	−23	−18	−23
Ti II 3761		−30		−28	− 1	−54				
Ti II 3759		−24		−34	−10	−36				
Hγ ≈ H13	−13	−14	− 2	−18	+ 3	−24	−54	−20	−18	−19
[Fe II]	−26	−33	+ 2	− 5		− 3		−46		−17

TABLE VI

Equivalent widths of stellar and shell lines in Boss 5481, in Å units

Date	1965, Dec. 14	1965, Dec. 15	1965, Dec. 31	1966, March 9	1966, May 5	1966, Dec. 17	1967, March 4
Spectr.	6741	6748	6758	6956	7046	7218	7395
H14	1.19	1.22		1.01	1.18		
H13	1.44	1.61	1.16	1.44	1.80	1.67	1.56
H12	0.21	0.20		1.42	0.10		
H12 shell				0.12	0.06		
3759 (Ti II)		0.21	0.20	0.15	0.09		
3761 (Ti II)		0.14	0.15	0.11			
H11	1.98	2.24	2.01	2.10	2.11	1.96	1.99
H11 shell		0.50	0.41		0.22		
H10	2.47	2.69	2.45	2.84	2.32	2.47	2.32
H10 shell	0.41	0.44	0.55	0.29	0.22		0.18
3819 (He I)	0.66	0.64	0.66	0.76	0.72	0.65	0.69
H9	2.75	2.74	2.65	2.96	2.80	2.64	2.60
H9 shell	0.51	0.49	0.48	0.45	0.28		
H8	2.21	2.29	2.30	2.57		2.62	2.34
H8 shell	0.47	0.56	0.49	0.40		0.30	0.26
3933 (Ca II)	0.52	0.48	0.44	0.43	0.36	0.36	0.31
3968 (Ca II)	0.93	1.00		0.78	0.73		
Hε shell	0.59	0.66		0.46	0.38		
4024	0.39	0.27			0.30		
4026 (He I)	0.44	0.56			0.34		
Hδ	2.07	2.48		1.98	1.72		
Hδ shell	0.68	0.70		0.51	0.51		

6. Discussion

From Tables III and VI it is clear that only the lines of Ca II (1) and Ti II (13, 14), besides the shell lines of hydrogen, have shown variations from 1965 to 1967.

Apart from hydrogen, the lower level of the shell lines is the ground state (Ca II) or a metastable level (Ti II), which is confirmed by the observations of COWLEY and COWLEY (1965) below 3600 Å. The dilution effect in the shell is further confirmed by the [Fe II] emission lines.

No helium lines have been detected in the shell, although by analogy with other hot chromospheres in cold giants, He I 10830 might be present (VAUGHAN and ZIRIN, 1966).

From Table III and from Figures 1 and 2 it can be inferred that the shell lines lost most of their strength at the end of 1966. The brightness of the star, during and after the shell episode, did not vary except for a few hundredths of magnitude, erratically, so that we can exclude a photometric eclipse. While the radial velocity of the M component change very little, that of the B star vary by about 80 km/sec, which greatly exceeds the observational errors. The kind of variation plotted in Figure 3 is apparently in conflict with the length of the orbital period (75 years) announced by COWLEY (1967), probably on the basis of lines of the M star alone. On the other hand, we know that large variations of radial velocity of unknown origin, are common among shell stars (UNDERHILL, 1966) and also occurred for the B component of Boss 1985 (COWLEY and COWLEY, 1966). The velocity distribution of the shell lines, plotted in Figure 4, favours the idea that they belong to the M component, while the [Fe II] lines are likely to have their origin in a stationary envelope. It is worthy to mention the absence of permitted emission lines of Fe II in the blue-visual region, while forbidden lines of the same ion are present, although quite faint.

As a conclusion, we can conjecture that the long shell episode had origin from the slow passage of the B star behind the atmosphere of the M star. The dilution effect of some shell lines may indicate that the line connecting us with the B star passed very far from the photosphere of the M star, which is confirmed by the absence of eclipse. The strange behaviour of the radial velocity of the B star, on the other hand, cannot be reconciled with Cowley's orbit.

The cooperation of the Electronic Computing Center, Padua University, is gratefully acknowledged.

References

BARTOLINI, C. and BATTISTINI, P.: 1968, private communication.
BIDELMAN, W. P.: 1954, *Astrophys. J. Suppl.* **1**, 175.
COWLEY, A.: 1967, *Astron. J.* **72**, 1161.
COWLEY, A. and COWLEY, C.: 1965, *I.A.U. Circ.* 1935.
COWLEY, A. and COWLEY, C.: 1966, *Astrophys. J.* **144**, 824.
UNDERHILL, A. B.: 1966, *The Early Type Stars*, Reidel Publ. Co., Dordrecht, Holland.
VAUGHAN, A. H. and ZIRIN, H.: 1966, *Annual Report, Palomar Observatory* **272**, 1966–67.
WRIGHT, K. O.: 1955, *Victoria Publ.* **9**, 167.

MASS MOTIONS IN THE SYSTEM OF VV CEPHEI

K. O. WRIGHT and SYLVIA J. LARSON

Dominion Astrophysical Observatory, Victoria, B.C., Canada

Abstract. A series of 65 spectra of the red region of VV Cephei have been obtained at Victoria between 1956 and 1968. Intensity tracings of the region near Hα have been made to study the emission and absorption profiles of this line. It is found that the observed velocity of the central absorption line agrees well with that given by neutral atomic lines in the M-type star, although it may be up to 10 km/sec. more negative for a few years after total eclipse. The radial velocities of the broad emission line (assumed to be symmetrical) show large departures from apparent orbital motion, but it would seem that the mass ratio of the system is more nearly unity than 2. Additional absorption lines can be detected over a large part of the cycle. Their positions can be interpreted as gas moving from the M-type star towards the secondary star.

VV Cephei is one of the least understood of the well-observed stellar systems. It has a period of 7430 days and primary eclipses have been observed in 1936 and 1957. The principal star has been classified as M2 I ab, though FREDERICK (1960) has suggested that it may not be a supergiant and that the strong, broadened lines in the spectrum may be the result of strong magnetic fields which BABCOCK (1958) found to be about 800 gauss. The secondary spectrum is considered to be produced by a B-type star, although the only real indication of this is the strong continuous spectrum in the ultraviolet. The hydrogen lines are strong with sharp absorption cores and, as usually described, emission on the short-wavelength side of the absorption; they have been described as 'reversed P-Cygni type' lines. Measurements of Victoria spectra by Wright (unpublished) show that the intensities of the continua of the M- and B-type stars are equal at 4000 Å. No trace of the helium lines at 3820, 4009 and 4026 Å has been found although the same method of subtracting the M-type spectrum from the composite spectrum was used as was successful in detecting these lines in the B-type spectra of the other, somewhat similar systems ζ Aurigae and 31 and 32 Cygni.

Discussions of the orbit have been given by GAPOSCHKIN (1937), GOEDICKE (1939), FREDERICK (1960) and PEERY (1966); Goedicke and Peery have also studied the structure of the M-type atmosphere from observations of the chromospheric lines made a few months before and after primary eclipse. Peery's orbit, based mainly on 268 Michigan spectrograms, shows an eccentricity of 0.34, which is larger than the value, 0.26, given by Goedicke. The scatter of the velocities about the computed curve is at least partially the result of irregular motions in the atmosphere which seem to be characteristic of M-type stars.

High-dispersion spectra (3–10 Å/mm) have been obtained at Victoria since April, 1956, just before the last eclipse, when according to LARSSON-LEANDER (1957, 1958) total eclipse lasted 498 days from 1956, July 28 to 1957, December 7 (J.D. 2435682–2436180). Brief descriptions of some of these spectra have been given by WRIGHT and MCKELLAR (1956), MCKELLAR *et al.* (1957) and by WRIGHT (1967). The present communication gives results obtained from measurements of the Hα line on 65 plates

M. Hack (ed.), Mass Loss from Stars. All rights reserved

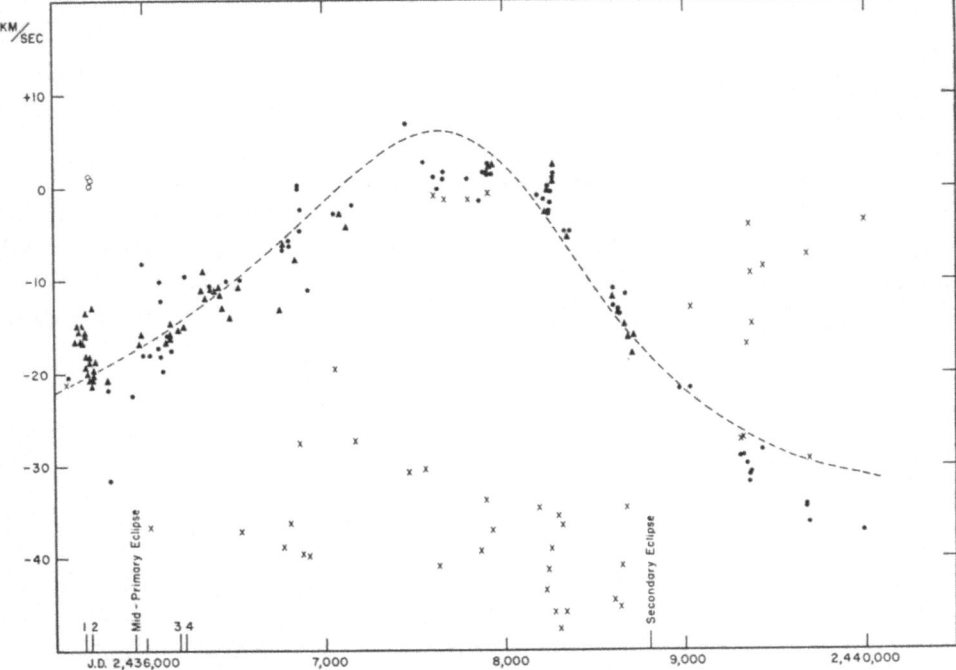

Fig. 1. Radial-velocity observations of VV Cephei: The data for the M-type star are represented by dots (●) for the red region and triangles (▲) for the violet region; the crosses (×) represent velocities given by the Hα emission line. The dotted curve (---) represents Peery's computed velocity curve for the M-type star.

taken between 1956 and 1968. Radial velocities measured on an equal number of spectrograms in the ultraviolet and violet regions are also discussed briefly.

Radial velocities obtained from Victoria spectrograms of VV Cephei since 1956 are shown in Figure 1. Data obtained from the M-type spectrum are shown as dots (●) for the red plates and as triangles (▲) for the violet plates. Results obtained from the emission Hα line, discussed below, are shown as crosses (×). The computed velocity curve adopted by Peery is shown as a dotted line (---). The Victoria observations follow Peery's curve fairly well although the semi-amplitude (K) may be somewhat less; the Victoria values of maximum positive velocity are below Peery's curve and they are also more negative at maximum negative velocity. The Victoria velocities are not yet considered definitive since there seem to be real differences between velocities given by different atoms even on the same spectrogram. The values shown here are based almost entirely on lines of neutral atoms. Since the red region of the spectrum, with the exception of the hydrogen lines, is produced by the M-type star, and since the velocities obtained from the red and violet plates agree well, it is probable that these velocities are a fair representation of the orbital motion of the star. However, there do seem to be systematic differences from the computed velocity curve which may be ascribed in part to atmospheric motions. The scatter of Perry's observations

about his velocity curve is much greater than that of the Victoria observations; the
higher dispersion available at Victoria may explain part of this result but part of the
scatter in Peery's observations may be because the Michigan data cover a much longer
period of time and atmospheric motions probably do not occur at the same phase in
each cycle.

Peery attempted to derive the mass-ratio of the system by obtaining a velocity
curve for the secondary star from measurements of the centre of the Hβ emission
line which, on his plates, could frequently be seen on both short- and long-wavelength
sides of the absorption core, but nearby absorption lines undoubtedly affect the Hβ
emission profile. The presence of the broad emission line can be seen on the Victoria

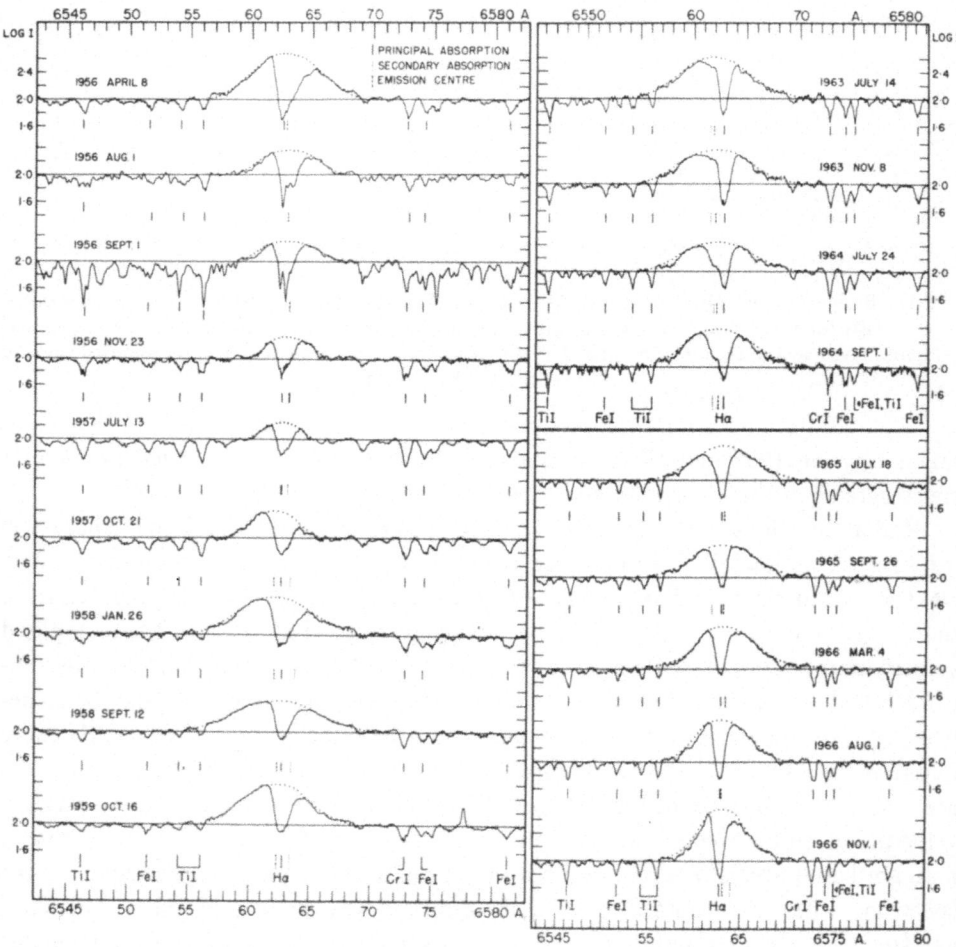

Fig. 2.   Logarithmic intensity tracings of the Hα profile obtained from Victoria spectra of VV Cephei.
The vertical solid lines (|) give the position of the principal absorption line, the dashes (¦), that of the
centre of the emission line, and the dots (:), the estimated position of the additional
absorption component.

spectra of Hβ, but it is much more prominent at Hα and therefore, in this paper, we have concentrated our efforts on a study of the profile of Hα. Representative intensity tracings of Hα derived from Victoria spectra are shown in Figure 2; the tracings have been made on a logarithmic intensity scale in order to show both the strong emission and absorption features. The original dispersions were 7.3 Å/mm for the spectra obtained with the 72-inch telescope (shown on the left side of the diagram), and 8.5 and 9.7 Å/mm for the spectra obtained with gratings have 600 and 1200 grooves per mm at the coudé focus of the 48-inch telescope (shown at the top and bottom of the right side of the diagram). In order to study the Hα profile and the velocities of the several components, a mean continuum was drawn through regions within fifty angstroms of the hydrogen line. A symmetrical emission profile relative to this continuum was then drawn on each tracing and the position of its centre was measured relative to nearby lines in the M-type spectrum; the centre of the emission was found by fitting the two halves of the line, and also by taking the mid-point between the positions where the wings joined the continuum; the two methods agreed quite well, but greater weight was given to the former. The positions of the absorption components of the Hα line were also measured. Two components were measured when they were clearly shown on the tracings and also when the asymmetries were quite pronounced. As noted above, the radial velocities of the M-type lines were measured over most of the plate and the velocities agreed well with those used as standards for the hydrogen line.

The emission Hα line can easily be drawn as a symmetrical profile on most of the tracings and the broad feature can be explained as an expanding envelope surrounding the secondary star. The velocities ($+20$ to -40 km/sec) are quite different from those obtained by Peery for Hβ, which ranged from -30 to -110 km/sec; for the Michigan observations of this line his systemic velocity, V_0, was 37 km/sec more negative than the value obtained for the M star. If the mean position of the emission line of Hα can be considered to represent the velocity of the secondary star, the best curve that can be fitted to the secondary velocities (shown as ×'s in Figure 1) should represent the orbital motion; and the relative values of the semi-amplitudes, K, should give the mass-ratio of the system. Peery found $M_M/M_B = 2.04$. The observations shown in Figure 1 show a large scatter and do not well represent such motion but they do indicate that the range in velocities of the two stars is nearly the same and therefore that the mass-ratio is more nearly unity than two. The present observations should be given greater weight than those obtained by Peery since the spectrograms were obtained with higher dispersion and also since the stronger emission at Hα permits the profiles to be drawn much more readily, especially on the long wavelength side of the line, at all phases of the cycle. It will, however, be desirable to measure the profiles of Hβ and possibly those of some of the later members of the Balmer series before the velocities of the emission lines in the secondary spectrum can be considered definitive. The emission velocities do not follow a curve representing orbital motion very well. This may be the result of the difficulty of measuring lines that are cut up by absorption, even on intensity tracings, or it may be related to motions of gas

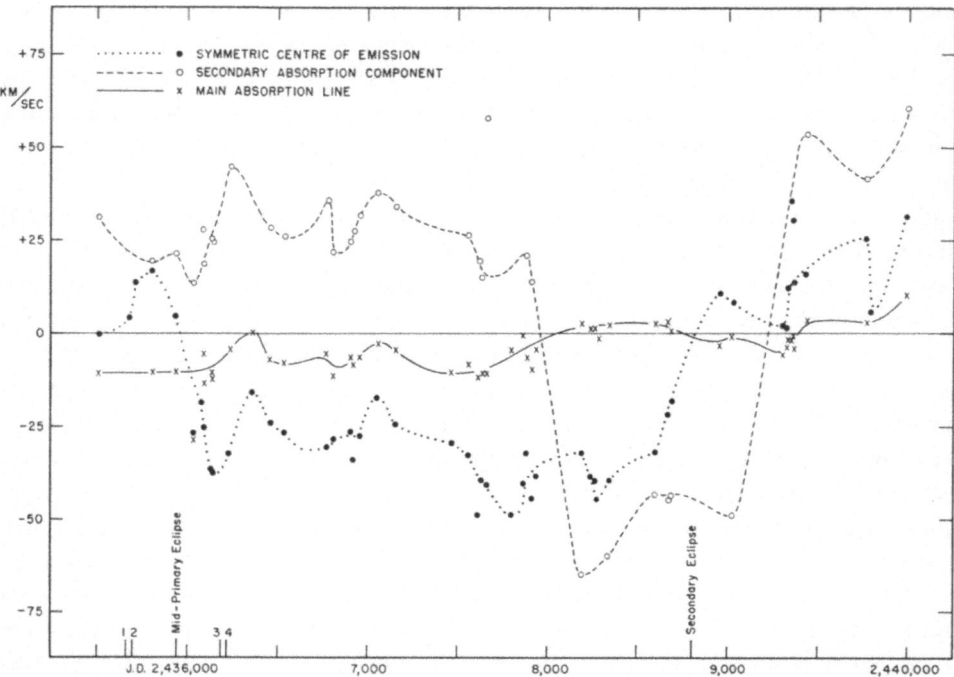

Fig. 3. Velocity displacements of the components of Hα relative to atomic lines in the M-type spectrum of VV Cephei. Crosses (×) represent data for the principal absorption line, dots (●), the centre of the emission line, and open circles (○), the velocities of the additional absorption line.

around the B star or in other parts of the system such as have been found, for example, for H.D. 47 129 (Plaskett's star) by STRUVE *et al.* (1958) and by ABHYANKAR (1959).

In order to study the velocities of the various components of absorption and emission in the Hα line, the data have been plotted against time in Figure 3. The zero velocity represents the velocity of the M star and is the position on the tracings found by interpolation from the positions of nearby lines of FeI and TiI. In this diagram, the crosses (×) represent the velocities of the principal Hα-absorption component. There seem to be real displacements from the zero velocity, averaging nearly −10 km/sec during the first half of the period covered by these observations from primary eclipse till about 3 years before secondary eclipse (J.D. 2435 500–2438 000); after that time the Hα velocities and those of the M-type lines seem to agree reasonably well. The dots (●) represent the relative velocities of the emission line which is assumed to be symmetrical. The velocities are nearly the same as those of the lines produced in the M-type star near primary eclipse (J.D. 2435 930) and about J.D. 2438 800 which, presumably, is near the time of secondary eclipse, although no observations of this eclipse seem to have been made. As noted earlier, the average negative displacement is about −40 km/sec and the observed positive value up to 1968 is about +25 km/sec.

The principal indication of mass motions in the system is given by the additional absorption component of the Hα line which can be estimated from the profiles drawn in Figure 2. These components are often difficult to measure accurately when the displacement is small and the line appears only asymmetrical. However, they can often be separated quite readily from the principal absorption component as shown in the tracings of Figure 2, where the displacement is to the longer wavelengths on the earlier series of plates taken with the Littrow spectrograph of the 72-inch telescope, on the left-hand side of the diagram, during and after primary eclipse, and to the shorter wavelengths in the later series, closer to secondary eclipse, shown at the right-hand side of the diagram. The velocity of the additional absorption line, shown as a series of open circles ($\bigcirc$) in Figure 3, seems to range from about $+15$ to $+35$ km/sec from primary eclipse until J.D. 2437900 when there is quite a rapid shift from the long- to the short-wavelength side of the principal absorption line. The velocity stays between -65 and -45 km/sec from J.D. 2438200 to 2439000. It then becomes positive again in the range $+40$ to $+55$ km/sec from J.D. 2439400 until the present time, J.D. 2440100. Thus the gas producing the absorption represented by this component is moving away from the observer during the first part of the period under discussion, when the B star is beyond the M star, and is coming towards the observer when the B star is between the M star and the observer. The simplest explanation of this phenomenon is that there is a stream of gas moving from the M star towards the B star which is of sufficient density to produce absorption which affects the emission profile of the envelope surrounding the B star whether it is flowing towards the observer or away from him. Although the general trend of these motions seems to be well marked, there seem to be small variations which need to be confirmed by measurements on the other hydrogen lines of the Balmer series.

The observations of VV Cephei could not have been obtained at Victoria without the cooperation of the observing staff. Numerous discussions concerning the system have been held with Dr. J. B. Hutchings and Dr. A. H. Batten, and Messrs. J. M. Fletcher, T. V. Jacobson and J. D. Francis have measured many of the plates for radial velocity.

References

ABHYANKAR, K. D.: 1959, *Astrophys. J. Sup.* **4**, 157.
BABCOCK, H. W.: 1958, *Astrophys. J. Sup.* **3**, 201.
FREDERICK, L. W.: 1960, *Astron. J.* **65**, 628.
GAPOSCHKIN, S.: 1937, *Harvard Circ.*, No. 421.
GOEDICKE, V.: 1939, *Publ. Obs. Univ. Mich.* **8**, 1.
LARSSON-LEANDER, G.: 1957, *Arkiv Astron.* **2**, 135.
LARSSON-LEANDER, G.: 1958, *Arkiv Astron.* **2**, 301.
McKELLAR, A., WRIGHT, K. O., and FRANCIS, J. D.: 1957, *Proc. Astron. Soc. Pacific* **69**, 442.
PEERY, B. F.: 1966, *Astrophys. J.* **144**, 672.
STRUVE, O., SAHADE, J., and HUANG, S. S.: 1958, *Astrophys. J.* **127**, 148.
WRIGHT, K. O.: 1967, in *Colloquium on Late-Type Stars* (ed. by M. Hack), Trieste, p. 153; *Contr. Dom. Ap. O.*, No. 109.
WRIGHT, K. O. and McKELLAR, A.: 1956, *Proc. Astron. Soc. Pacific* **68**, 405.

PERIODS OF ECLIPSING NOVALIKE VARIABLES*

GEORGE S. MUMFORD

Tufts University, Medford, Mass., U.S.A., and
*Kitt Peak National Observatory**, Tucson, Ariz., U.S.A.*

Abstract. Recent observations of minima of the novalike variables T Aurigae, EM Cygni, U Geminorum, DQ Herculis, EX Hydrae, V Sagittae, and WZ Sagittae are presented. A new period for EM Cygni has been derived; possible period changes for U Geminorum and DQ Herculis are discussed.

1. Introduction

If, as many authors suggest (see, e.g., KRUSZEWSKI, 1966), orbital periods of close binaries change because of mass loss, mass transfer, or some other cause, it is of especial interest to keep nova-like variables that are known to eclipse under close surveillance. These objects give evidence for both the transfer and loss of matter; at least one has been suggested as a possible radiator of gravitational waves (KRAFT *et al.*, 1962); and periods of revolution are generally less than 7 hours (MUMFORD, 1967a). Thus, when orbital periods are initially reliable, not only should changes be expected to occur after a few years, but they should be measurable since in that span of time so many cycles elapse.

During the past 4 years, a part of the author's effort at Kitt Peak National Observatory has been directed toward this problem. Evidence is beginning to accumulate that the periods of some nova-like stars may be changing. On the other hand, it may be even more significant that among the stars with the shortest orbital periods and which have been observed over the greatest number of cycles, no changes in the initial, or improved values of the orbital periods appear to be indicated.

2. Data

Save for the exceptions noted elsewhere, all minima presented here were derived from unpublished blue light curves based on photoelectric observations made with either one or the other 91-cm reflectors at Kitt Peak.

Pertinent data are collected in Tables I–III. The column headed 'Julian Date' gives the observed moment of minimum; that labeled 'Deviation' gives the residual in the sense observed time of minimum minus the computed.

3. Discussion of Individual Stars

T Aurigae: This old nova was found to be an eclipsing binary by WALKER (1963). Subsequent observations by MUMFORD (1967a) suggested a slight revision in the light

---

* Contribution from the Kitt Peak National Observatory, No. 405.
** Operated by the Association of Universities for Research in Astronomy, Inc., under contract with the National Science Foundation.

M. Hack (ed.), Mass Loss from Stars. All rights reserved

TABLE I

Minima of certain nova-like variables

Star	Julian date (2400000 +)	Cycle	Deviation (day)
T Aur	39768.980	10544	+ 0.001
	39912.657	11247	0.000
DQ Her	40006.9007	26092	+0.0046
	40007.8686	26097	+0.0044
EX Hya	39530.9283	26834	− 0.0001
	39625.7053	28223	+0.0001
	40003.6526	33762	+ 0.0001
	40004.6758	33777	− 0.0002
	40006.6554	33806	+0.0006
	40009.6570	33850	− 0.0001
V Sge	39767.7640	3652	+ 0.0085
	40006.8600	4117	+ 0.0038
WZ Sge	39769.6658	39196	0.0000
	39769.7221	39197	− 0.0003
	40005.8839	43363	− 0.0002
	40008.8885	43416	0.0000
	40008.9451	43417	− 0.0001

elements. While low weight should be assigned to the last value in Table I because of the poor quality of the night, these revised elements still give adequate representation of the observations.

EM Cygni: Krzeminski and Mumford (to be published) have found that this nova-like variable likely undergoes eclipses. Previously unpublished times of minima are given in Table II. In the third column are residuals based on an earlier determination of the light elements (MUMFORD, 1967a). It is clear that the more recent observations are not satisfied so another solution has been made. The fourth column gives residuals based on the elements

$$\text{Min. light} = \text{HJD } 2437882.8596 + 0\overset{d}{.}29090942 \, E.$$

For comparison purposes, the sums of the squares of the residuals are 3.847×10^{-5} and 1.872×10^{-5} in each case, respectively. Large individual residuals are attributable to the variable shape and depth of the eclipse leading to difficulty in determining the moment of minimum light.

U Geminorum: Eclipses of this dwarf nova were first detected by KRZEMINSKI (1965). The reader is referred to this paper for the definition of the moment of minimum that attempts to take into account the variable duration of eclipse, as well as the peaks in brightness immediately prior to and following eclipse. Krzeminski's value for the period has been used in calculating the residuals given in the third column

TABLE II
Observed minima of EM Cygni

Julian date (2400000 +)	Cycle	Deviation (day)	Deviation (day)	Observer
37882.8600	0	+ 0.0009	+0.0004	K
37883.7318	3	0.0000	− .0005	K
37906.7127	82	− .0010	− .0015	K
37911.6600	99	+ .0008	+ .0004	K
37936.6775	185	+ .0001	− .0003	M
37966.6410	288	− .0001	− .0005	M
37968.6775	295	0.0000	− .0004	M
37996.6045	391	− .0003	− .0007	M
38174.9332	1004	+ .0008	+ .0005	M
38345.6981	1591	+ .0017	+ .0016	M
38348.6055	1601	0.0000	− .0001	M
38496.9698	2111	+ .0003	+ .0004	M
38561.5520	2333	+ .0009	+ .0007	K
38562.4239	2336	− .0003	− .0001	K
38624.3882	2549	+ .0002	+ .0005	K
38674.7153	2722	0.0000	+ .0003	M
38675.5880	2725	− .0001	+ .0002	M
38676.7510	2729	− .0007	− .0004	M
38878.9340	3424	+ .0001	+ .0005	M
38883.8792	3441	− .0002	+ .0003	M
39052.6040	4021	− .0030	− .0024	M
39054.6425	4028	− .0009	− .0002	M
39230.9332	4634	− .0015	− .0007	M
39232.9:	4641	:	:	K
39293.7701	4850	− .0002	− .0002	K
39767.6621	6479	− .0010	+ .0004	M
39769.6968	6486	− .0027	− .0013	M
40006.7883	7301	− .0026	− .0010	M
40007.9531	7305	− .0014	+ .0002	M
40008.8260	7308	− .0013	+0.0004	M

K = W. Krzeminski, M = G. S. Mumford.

of Table III. Minima since Julian Day 2438031 are due to Mumford, those before and including this date were observed by Krzeminski.

A systematic trend among the more recent observations is evident. No simple linear solution gives a satisfactory representation of the residuals. The best agreement is obtained by using a slightly revised period and including a second-order term. Residuals in the last column were computed from the elements

$$\text{Min. light} = \text{HJD } 2\,437\,638.8270 + 0^d.176\,905\,89 \, E$$
$$+ 0^d.000\,000\,000\,018 \, E^2 .$$

TABLE III
Observed minima of U Geminorum

Julian date (2430000+)	Cycle	Deviation (day)	Deviation (day)
7638.8269	0	− 0.0001	−0.0001
7639.0038	1	− .0001	− .0001
7639.8883	6	− .0002	− .0002
7691.7221	299	+ .0002	+ .0002
7696.6754	327	+ .0001	+ .0001
7725.8649	492	+ .0002	+ .0002
7732.7640	531	− .0001	− .0001
7748.6858	621	+ .0002	+ .0002
7783.7129	819	− .0001	− .0001
7937.9750	1691	+ .0001	0.0000
8025.0124	2183	− .0002	− .0003
8030.8506	2216	+ .0001	0.0000
8031.0275	2217	+ .0001	0.0000
8345.9198	3997	− .0002	− .0003
8409.9601	4359	+ .0002	0.0000
8410.8443	4364	− .0001	− .0003
8411.9059	4370	0.0000	− .0001
8493.6367	4832	+ .0003	0.0000
8496.6440	4849	+ .0002	+ .0001
8674.9658	5857	+ .0008	+ .0004
8879.6464	7014	+ .0013	+ .0006
9053.8987	7999	+ .0013	+ .0003
9054.9596	8005	+ .0008	− .0002
9231.6897	9004	+ .0018	+ .0006
9527.6533	10677	+ .0019	0.0000
9527.8300	10678	+ .0017	− .0002
9913.8392	12860	+ .0022	− .0005
9919.8548	12894	+0.0030	+0.0003

While the validity of the last term is yet to be tested through additional observations, a simple calculation indicates that it is of the appropriate order of magnitude if the figures given for mass ejection in a dwarf-nova outburst (see, e.g., PAYNE-GAPOSCHKIN, 1957) are to be believed.

DQ Herculis: WALKER (1961) also established this old nova to be an eclipsing binary. At that time he commented that there seemed to be some preliminary evidence for an increase in the period of the system. Previously AHNERT (1960) from observations prior to the 1934 outburst had found the period to increase by +0.0004122 day.

Residuals given in the final column of Table I have been calculated from Walker's elements. An increase in the period seems verified, though more observations are necessary to establish this conclusion beyond any doubt. The present residuals are diminished to very small values by using a second-order term with a coefficient somewhat less than half of that given above for U Germinorum.

EX Hydrae: Eclipses of this star were first detected by Krzeminski (unpublished). Residuals in Table I have been computed using the elements given by MUMFORD

(1967b). While there appears to be some difference in appearance between former minima and these (MUMFORD, 1969), there is nothing to suggest a period change.

V Sagittae: This nova-like variable was found to be a spectroscopic and eclipsing binary by HERBIG *et al.* (1965). Later SMAK (1967) suggested from an analysis of earlier photographic observations that the period of the system was changing. The sense of this change seems borne out by the residuals given in Table I. Low weight must be given the final value as the eclipse on this night was very shallow and the minimum was ill-defined.

WZ Sagittae: Discovered to be an eclipsing binary by KRZEMINSKI (1962), this recurrent nova has the shortest orbital period of any nova-like variable, about 1 hour and 22 min (KRZEMINSKI and KRAFT, 1964). With a nearby red, optical companion, the system presents a difficult observational problem. To negate, so far as possible, the effects of the red star when using a small telescope, observations are generally made in the ultraviolet with both stars in the diaphragm. In part because of this problem, no significance is attached to the slight trend suggested by the residuals in Table I. The period has remained essentially constant. In this conjunction it should be noted that the period of WZ Sagittae is known to one less decimal place than that of EX Hydrae.

4. Summary

The present data suggest changes in the periods of three nova-like variables: U Geminorum, DQ Herculis, and V Sagittae. The latter has the longest orbital period of any of the stars considered here. The other two stars have fairly comparable periods, somewhat longer than 4 hours, and well-defined eclipse curves.

For two other stars, T Aurigae and EM Cygni, no period changes seem indicated at present. Both of these have longer periods than U Geminorum or DQ Herculis, but the eclipses are variable in depth and shape. Moreover, the cycle count is relatively small.

On the other hand, there may be some significance in the fact that stars with the shortest periods, EX Hydrae and WZ Sagittae, show no evidence for a period change over a relatively large number of cycles. Likewise WALKER (1965) has concluded that the period of VV Puppis, about 1.67 hours, appears to be unchanging. Clearly additional observations relating to these very interesting problems are highly desirable.

Acknowledgments

This work has been supported by a grant from the National Science Foundation.

References

AHNERT, P.: 1960, *Astron. Nachr.* **285**, 191.
HERBIG, G. H., PRESTON, G. W., SMAK, J., and PACZYŃSKI, B.: 1965, *Astrophys. J.* **141**, 617.
KRAFT, R. P., MATHEWS, J., and GREENSTEIN, J. L.: 1962, *Astrophys. J.* **136**, 312.

KRUSZEWSKI, A.: 1966, in *Advances in Astronomy and Astrophysics*, Vol. IV (ed. by Z. Kopal), Academic Press, New York, p. 233.

KRZEMINSKI, W.: 1962, *Publ. Astron. Soc. Pacific* **74**, 66.

KRZEMINSKI, W.: 1965, *Astrophys. J.* **142**, 1051.

KRZEMINSKI, W. and KRAFT, R. P.: 1964, *Astrophys. J.* **140**, 921.

MUMFORD, G. S.: 1967a, *Publ. Astron. Soc. Pacific* **79**, 283.

MUMFORD, G. S.: 1967b, *Astrophys. J. Suppl.* **15**, 1.

MUMFORD, G. S.: 1969, *Astrophys. J.* (in press).

PAYNE-GAPOSCHKIN, C.: 1957, *The Galactic Novae*, North-Holland Publishing Co., Amsterdam.

SMAK, J.: 1967, *Acta Astron.* **17**, 55.

WALKER, M. F.: 1961, *Astrophys. J.* **134**, 171.

WALKER, M. F.: 1963, *Astrophys. J.* **138**, 313.

WALKER, M. F.: 1965, *Mitt. Sternw. Budapest*, No. 57.

ABOUT THE INTERPRETATION OF GASEOUS STREAMS IN CLOSE BINARY SYSTEMS

C. J. VAN HOUTEN

Abstract. It is rather generally accepted that gaseous streams in close binary systems are caused by transfer of matter from the less massive component, which fills its lobe of the inner contact surface, to the more massive one, via the first Lagrangian point. This model is satisfactory theoretically but leaves many observational peculiarities unexplained. It is generally believed that the process is amply supported by observational evidence, but a closer inspection shows this not to be the case and a more careful discussion of the available data leads to a different picture. As an example a rediscussion is given of Struve's observations of SX Cas, resulting in the conclusion that not one (as Struve believed) but *two* streams of gas are approaching the surface of the primary component, at roughly opposite sides. The origin of these streams is ascribed to the tidal bulges, which conclusion can be confirmed by observations of many other close binary systems.

This would mean that, if the stream through L_1 really exists, it cannot be observed. It is even possible that it does not exist at all, if supposedly a flat disk around the primary is extending toward the inner contact surface, sealing off the first Lagrangian point. There are observational indications of the existence of such a disk.

A qualitative explanation of these phenomena will be given in a forthcoming publication.

Discussion

Williams: Is there any observational evidence showing that the primary is a fast rotating star?

Van Houten: No, you cannot see any lines of the star itself, only the shell lines.

Williams: Would you not expect, on theoretical grounds, that tidal friction would be fairly efficient in slowing down the fast rotation?

Van Houten: This is not a very close system, with a period of 30^d. There is a lot of room in it.

Kitamura: Could you give me any idea on the density of the secondary stream, which you have mentioned, as compared with the primary stream?

Van Houten: Struve did not mention any difference between the intensity of the lines belonging to the two streams, so I suppose they are about the same. He only mentions that the lines are exceedingly sharp, which is typical of a shell spectrum.

M. Hack (ed.), Mass Loss from Stars. All rights reserved

MASS EXCHANGE IN THE BINARY SYSTEM AD HERCULIS
OBTAINED FROM PHOTOMETRIC OBSERVATIONS*

K. WALTER

Abstract. AD Herculis is a typical Algol system the bright A4 component of which, in the primary minimum, is nearly centrally eclipsed by the large gK2 component. The period of revolution is rather long ($P = 9^d.77$), and the components are well separated ($r_A = 0.095$, $r_K = 0.24$). The large parts of the light curve where the components are out of eclipse favour the determination of the constants of rectification for reflection and oblateness. However, a satisfying rectification of light curves observed in B and V could not be reached for the whole light curve but only for the part between primary and secondary minimum. This extraordinary fact was the starting-point for an interpretation of the observed light curves between the minima, which is given here in a preliminary form.

Our hypothesis is that a gas stream originates in the Lagrange point L_1 near the subgiant component and approaches the A4 star along curves typical for low initial velocities at L_1. The orbits and the velocities reached by the particles depending only weakly on the initial velocities, the way of the gas stream is situated along the 'gravitational pipeline' (as it has been named by Kopal). Since as yet no spectroscopic orbit has been measured, a mass ratio has been assumed by which – as in some other typical Algol systems of similar length of period – the subgiant component will be somewhat 'undersized'. Most probably, in our case a good deal of the particles of the gas stream will fall on the A star which means that the gas stream is lying on the one side in relation to the line joining the centres of the components (the side containing the preceding part of the K star), while the other side is nearly unoccupied by absorbing particles. The smooth, undisturbed lapse of the one half of the light curve will be understood in this way.

The reflection coefficients have been found large, the reflected light amounting 0.05 of the system in B and 0.03 in V. Probably the reflection emerges chiefly by masses congested in the surroundings of the Lagrange point L_1 and irradiated by the A star, giving re-emission also at short wavelengths.

When the gas stream is seen in its longitudinal direction which takes place in the phases following the secondary minimum, it has its strongest absorbing effect. The light curves show losses of intensity of 0.04 (B) and 0.012 (V) at these phases. The light which is absorbed or scattered then will, by a good deal, come from the masses near L_1. As an interesting by-effect it has been found that, in these phases, the observations possess larger deviations (rms) than at the other phases. It is suggested that this is due to a cloudy structure of the gas stream.

When the gas stream approaches the A star, its density will diminish about as the velocity is growing and it seems possible that the absorption of the light of the A star will be overcompensated by the enlargement of radiation which must occur by the impact of particles on that side of this star which is visible before primary minimum.

For such a model of AD Herculis it was attempted to estimate the masses of the gas stream needed to explain the observed photometric effects. If we assume that the gas stream consists of hydrogenium and postulate that in its longitudinal direction the gas stream absorbs a good deal of the light behind it, but will not take off much light from the A star, a number of 2×10^{12} H-atoms per cm^3 is needed at a position within the gas stream, which is situated halfway between the components.

Thomson scattering will contribute most of the absorption. Regarding that the velocity at the considered point of the gas stream is about 150 km/sec and using an estimated value of about 5×10^{22} cm^2 for the cross-section of the stream, we get 2.4×10^{18} gr/sec or 4×10^{-8} solar masses per year which are passing over from the K- to the A-star.

Acknowledgements

The observations were made in Sicily 1964–65 with support of the DFG. I have to thank the Osservatorio di Catania for hospitality. The discussion of the observations was prosecuted together with Dipl. Phys. D. Korsch (Tübingen).

* The full text of this paper will appear in *Astron. Nachr.*

Discussion

Underhill: Are there any spectroscopic observations of this system? The density and path length suggested for the stream are sufficiently great that one would expect to see a strong shell spectrum; which lines seen would depend on the excitation conditions, but one could not escape finding very strong shell line of Na D and Ca II K. In fact I think the suggested density of the gas stream is about the same or greater than in the atmosphere of a K type giant.

Walter: The only spectrographic observations known to me are published by Wyse, about 30 years ago; he determined the spectral types of the components. I agree that spectral observations of AD Her – a star of $9^{m}4$ (v) – are very important.

Kitamura: I think that the photometric elements deduced here would depend upon the rectification. Which type of rectification did you use in your light curve analysis?

Walter: I used the Russell-Merrill method. As the primary minimum is deep and the eclipse is a total one, the photometric solution gives trusty results. The influence of the rectification on the relative dimensions is very small in any case.

Sahade: I would like to congratulate Dr. Walter for his observations because they provide us with a very nice independent confirmation of the model that the spectrographic observations suggest for the Algol-type systems and is currently accepted as describing the state of affairs in them.

PERIOD CHANGES OF W URSAE MAIORIS SYSTEMS*

T. HERCZEG

Hamburger Sternwarte, Hamburg, Germany

Abstract. Period variations observed in the case of numerous eclipsing binaries are among the most sensitive means of detecting evolutionary changes; they are especially suitable for indicating mass exchange between the components or mass loss of the whole system. As to the W Ursae Maioris systems, because of their short periods and a possibly rather strong interaction between the components, period changes can become critical in the course of a comparatively short time span. Assuming photoelectric accuracy of the measurements, any linear variation of the period with time capable of producing a substantial effect ($\Delta P \approx P/2$) during 10^8 years, could be detected with certainty after about 30 or 40 years of observation. In particular, if W Ursae Maioris systems are steadily evolving towards the U Geminorum stage, a decrease of the period, measurable in perhaps 70 or 80 years, should be expected.

A study of the published minima shows, however, that early observations are of relatively little avail because of their modest accuracy. Only the beginning of the 'photoelectric era', roughly 30 years ago, brought along reliable O-C diagrams of the times of minima. The number of well-observed variables is limited to about 15; some ten more pairs can afford useful additional evidence.

A survey of these binaries seems to reveal a characteristic of possibly considerable importance: sudden changes ('jumps') of the period are not only frequently occurring, but they are perhaps the dominating feature of the O-C graphs. (As it is definitely established now, several typical semi-detached systems – like Algol itself – also show this type of period variation.)

Among the W Ursae Maioris stars, a few objects are representing this behaviour singularly clearly; three of them we are going to consider in some details:

AH Virginis exhibited a sudden change of its period by $+0.3$ sec in 1955. Residuals of photoelectric minima against suitable linear ephemeris formulae before and after this 'jump' are of the order of $\pm 0^d.0003$ only. It is, however, somewhat disturbing that the secondary minima followed a slightly different pattern.

ER Orionis seems to show not less than three jumps in 30 years, the sequence of the signs being $+ + -$, with the net effect of a lengthening of the period.

W Ursae Maioris revealed three such abrupt changes until now. The latest of them was, as indicated by Kristenson's and Cester's recent observations, a very sharp shift of the phase without sensible change of the period. In both cases (ER Ori, W UMa) an alternative interpretation of the O-C values, namely as a possible light time effect, cannot be entirely excluded though it is clearly less convincing.

In general, a review of the best observed cases allow us to formulate, with due caution, the following conclusions:

(1) The great majority of W UMa-systems is undergoing period changes of the order of a few tenths of a second.

(2) In almost all cases, representation of the period variations by sudden jumps (and by strictly constant periods between these jumps) is at least as good, indeed, in most cases better, than descriptions of the O-C diagrams by periodic or quadratic terms.

(3) Among the sudden changes, both positive and negative ones occur but the variables considered hitherto suggest, by and large, a tendency towards lengthening of the period. Any definitive establishment of secular changes, e.g. a secular increase of the periods is, of course, greatly complicated by the very existence of these abrupt changes.

It seems that, as a fair estimate, three more decades of accurate and regular observations are necessary for substantiating the above 'rules'.

---

* A full account of this research will be sent to *Astronomy and Astrophysics*.

M. Hack (ed.), Mass Loss from Stars. All rights reserved

Discussion

Wood: In reference to your suggestion that spectral peculiarities might be expected during the intervals when period changes were taking place, Batten at Victoria found peculiarities in the spectrum of one of the stars he was observing (RS CVn, I think) and later observations showed a period change had indeed occurred at about this time. Many years earlier, about 1915, Jordan at Allegheny obtained radial velocity curves of R CMa which were highly distorted and varied from season to season; later photometry showed a period change had occurred at about this time. Thus there is at least some observational evidence in support of your suggestion.

Herczeg: This is just the type of additional information we have to look for and I am very much pleased indeed to learn about this remark.

Sahade: May I know whether the times of mid-eclipse suggested by the light curves of W UMa stars in different colours are the same or whether there are differences.

Herczeg: I don't know about any really clearly established difference between the lines of minima as determined in different colours. The W UMa stars exhibite, on the other hand, slightly changing light curves, even during eclipses. These changes can affect the accuracy of the timing of minima by, roughly speaking, $\pm 0\overset{d}{.}001$ to $\pm 0\overset{d}{.}003$. This certainly leads to a higher scatter in the (O-C) diagrams but I don't believe this can change their basic character. This question is, however, still not settled.

MASS LOSS FROM CLOSE BINARIES.
THEORIES

MASS EXCHANGE IN CLOSE BINARY SYSTEMS
WITH PRIMARY COMPONENTS OF 30 $M_\odot$

G. BARBARO, P. GIANNONE, M. A. GIANNUZZI and C. SUMMA

Istituti di Astronomia e di Fisica dell'Università di Padova, Osservatorio Astronomico di Roma

Abstract. The evolutions of the originally more massive components (primaries) of two close binary systems with very large masses have been computed during the phases of mass transfer to the companions and in the further *detached* stages at constant mass. The two double stars are composed initially of a primary of 30 $M_\odot$ and a secondary of 10 $M_\odot$ with separations of about 119 $R_\odot$ (case I) and 307 $R_\odot$ (case II), respectively. At the contact of the more massive star with its Roche lobe, the central helium reaches the ignition in case I and is at a more advanced burning in case II. At the end of the transfer of matter, the original primaries have masses equal to about 11.7 $M_\odot$ and 12.7 $M_\odot$, respectively, at separations of 97.5 $R_\odot$ and 229 $R_\odot$. The further evolutionary changes lead rapidly to a stage of steady central helium burning. In the H-R diagram the stars settle in a position with effective temperatures strongly depending on the initial separations. The subsequent evolutionary phases, which will be followed in detail, are anticipated.

1. Introduction

In recent years, the evolutions of some close binary systems have been extensively studied by some authors. Exchange of matter between the components of systems of low, intermediate and large total masses and the evolutionary changes of the resulting stars have been computed in detail. At the contact of the more massive (and therefore more evolved) component (*primary*) with its Roche lobe, a rapid phase of mass transfer from the primary to the companion (*secondary*) takes place. In the cases studied up to now, it has been assumed that the contact can occur: (1) during the phase of central hydrogen burning (case A), when the star is evolving away from the main sequence, or (2) after the exhaustion of hydrogen at the center (case B), when the star is rapidly expanding almost horizontally in the H-R diagram.

In case A, at the end of the rapid mass loss, the original primary has become the less massive star and the system then evolves as *semi-detached*. The initially more massive star continues to burn its central hydrogen and to expand, losing matter on a nuclear time-scale. On the contrary, in case B the system normally becomes *detached* and the original primary evolves to a white dwarf, or to a subgiant 'undersize' with central helium burning.

Besides cases A and B, a third case (C), corresponding to the contact of the primary with its Roche lobe after the ignition of helium at the center, can be investigated. As is well known from the evolutions of single stars of intermediate masses (up to values of the order of 10 $M_\odot$), helium is ignited when the star is evolving along the Hayashi track. For more massive models, on the other hand, helium starts to burn while the star is still expanding almost horizontally in the H-R diagram. One can therefore expect that in case C the evolutionary changes will result in a larger set of possibilities in comparison with case A. This is because of a more complicated structure of the actual star model with regard to the hydrogen-burning models.

M. Hack (ed.), Mass Loss from Stars. All rights reserved

Moreover, for more massive stars, semi-convective zones are rather extended in mass and play an important rôle in the evolutions of such models.

Besides these considerations, among *detached* close binary systems with components of early spectral type one can find some double stars with large separations (A) and large total masses $(M = M_1 + M_2)$, and with mass ratios $(m = M_2/M_1)$ of the order of $0.3 \sim 0.4$ (e.g. δ Ori: $A = 40 \ R_\odot$, $M = 41.4 \ M_\odot$, $m = 0.38$).

We have therefore started to study the evolutions of massive close binaries. Two double stars of total masses equal to 40 $M_\odot$, mass ratios 0.33, and with initial separations about 119 $R_\odot$ (case I) and 307 $R_\odot$ (case II) have been chosen. At the contact of the more massive star with its Roche lobe, the system, initially *detached*, becomes *semi-detached*, and the exchange of mass takes place.

In both cases presently studied (see Section 3), the phase of transfer of matter has been followed from the contact up to the end of the mass loss from the original primary. In a very rapid time-scale (of the order of $3 \sim 5 \times 10^3$ years), when the stellar masses have decreased to about 11.7 $M_\odot$ and 12.7 $M_\odot$, the exchange of matter stops. The two systems, again *detached*, are then composed of a less massive star highly overluminous with regard to its mass, and of a more massive companion at separations of 97.5 $R_\odot$ and 229 $R_\odot$. During the further evolutionary stages, central burning grows progressively until a steady phase of nuclear-energy release through the helium reactions takes place. In the H-R diagram the evolutionary tracks turn backwards, and any change in the structure of the star occurs on a nuclear time-scale. The position of the turn-off in the H-R diagram strongly depends, in so far as effective temperature is concerned, on the initial separation.

We intend to continue the study undertaken, both by considering a more advanced stage of central helium burning at the contact of the more massive component and by evolving to further phases the cases already treated. From the preliminary results presented in this paper, we would expect that a second phase of mass loss from the same ex-primary could occur. Additionally, it could also happen that this second transfer of matter takes place either during the depletion of helium near the center or after its exhaustion. This seems to open many possibilities for the resulting configurations of binary systems especially with regard to the structure of the initially more massive component.

2. Numerical Procedure and Assumptions

The well-known system of stellar structure equations has been solved numerically by means of the Henyey method, using for the most part the program of HOFMEISTER *et al.* (1964). The fitting mass between the integrations of the outer layers (atmosphere and subatmospheric zone) and of the interior has been set at $M_F = 0.97 \ M$.

Burning of hydrogen through the proton-proton and the carbon-nitrogen cycles (HOFMEISTER *et al.*, 1964) and helium depletion due to the 3α (DUORAH and KUSHWAHA, 1963) and the subsequent (α, γ) reactions (REEVES, 1964) have been included. Thermodynamical-energy release has also been added.

Opacity has been interpolated from the tables of Cox and STEWART (1965) for the mixtures corresponding to extreme Population I stars ($X=0.602$, $Y=0.354$, $Z=0.044$). Radiation pressure has been taken into account.

During the evolution of the single star of 30 $M_\odot$ attention has been devoted to the treatment of the semi-convective zone (CHIOSI and SUMMA, 1969). For the instability against convection, the Schwarzschild's criterion has been assumed (see KATO, 1966). In the unstable regions with a varying mean molecular weight, a partial mixing leading to the condition of equality of the logarithmic radiative and adiabatic gradients has been performed.

The evolution of the primary away from the main sequence has been assumed to be unaffected by the presence of the companion, until the more massive star has filled up the corresponding Roche lobe. At the contact of the primary, the change of the total stellar mass has been allowed for; redefinition of the independent variable at each mesh point takes into account this decrease in mass.

During the transfer of matter from the primary to the companion the usual hypotheses of conservation of the total mass of the system and of the total angular orbital momentum have been assumed. Moreover, in the Roche scheme for close double stars, one has negligible distortions from spherical symmetry of the star, circular orbits and synchronism of the orbital and the rotational motions. For any given mass ratio and any separation of a binary system, the critical radii for both components can be determined. Correspondingly, the applied rate of mass loss is chosen in order to maintain the difference between the stellar radius of the original primary and the radius of its critical lobe within 4% of the critical radius itself.

3. Results

The two close binary systems under study are composed initially of a primary component of 30 $M_\odot$ and a secondary star of 10 $M_\odot$ with the two separations of:

$$A = 119.2\ R_\odot \text{ (case I)} \quad \text{and} \quad A = 307.2\ R_\odot \text{ (case II)}.$$

The corresponding periods of revolution are $P=23^{\text{d}}88$ and $98^{\text{d}}63$, respectively.

At the contact of the primary with its Roche lobe, the evolutionary stages of the primary correspond to the ignition of helium at the center ($\varepsilon_{3\alpha} \cong 10^3$ erg g^{-1} sec^{-1}) at the age of 4.334×10^6 years in case I, and to an already initiated burning of central helium at 4.342×10^6 years in case II.

In Tables I and II, for the two considered cases, the main quantities corresponding to the contact of the primary (stage a), to the end of the mass transfer (stage b) and to the turn-off in the H-R diagram (stage c) are listed. The indices (e) and (c) refer to the photosphere and to the center, respectively. Moreover, from Figures 1 and 2 for the stages mentioned above, the inner structure of the stellar models can be derived. This can be seen both from the behaviour with mass of the X-profile and from the thick tracks which give, on the abscissae, the extents of the convective regions. The total luminosity and the nuclear-energy flux at each mass shell are also displayed. Their

TABLE I

Characteristic quantities for the binary system of case I

Stage	$t(10^6$ years)	$M/M_\odot$	$A/R_\odot$	$\log L/L_\odot$	$\log T_{eff}$	X_e	$\log T_c$	$\log \varrho_c$	Y_c	X_c^{12}
a	4.3337	30	119.2	5.513	4.257	0.602	8.233	2.631	0.956	–
b	4.3359	11.74	97.5	5.361	4.376	0.415	8.275	2.746	0.952	0.004
c	4.5063	11.74	97.5	5.337	4.642	0.415	8.306	2.792	0.538	0.382

TABLE II

Characteristic quantities for the binary system of case II

Stage	$t(10^6$ years)	$M/M_\odot$	$A/R_\odot$	$\log L/L_\odot$	$\log T_{eff}$	X_e	$\log T_c$	$\log \varrho_c$	Y_c	X_c^{12}
a	4.3416	30	307.2	5.528	4.054	0.602	8.276	2.734	0.939	0.017
b	4.3473	12.72	229.3	5.380	4.200	0.415	8.277	2.734	0.924	0.032
c	4.5396	12.72	229.3	5.392	4.307	0.415	8.315	2.807	0.457	0.441

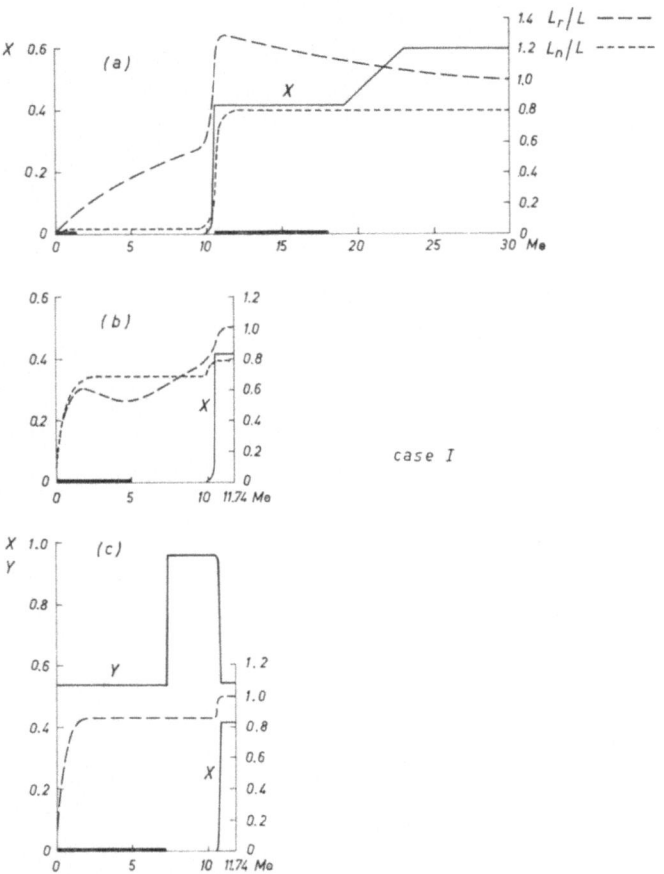

Fig. 1a–c. Case I. The solid lines give the hydrogen content X at stages a, b and c, and the helium content Y at stage c as functions of mass (in units of the solar mass) in the original primary. At the same stages the total luminosity (L_r) and the nuclear-energy flux (L_n) as fractions of the surface luminosity (L) are represented by the dashed and the short-dashed lines, respectively. On the abscissae, the thick tracks show the extents of the central convective regions and of the intermediate convective zone.

comparison allows one to estimate the discrepancy from the thermal equilibrium of the stellar interior.

A. CASE I

At the contact (stage a), the structure of the primary's model consists of:

(1) a convective helium core containing 5.2% of the stellar mass;

(2) an intermediate radiative helium region extending up to a mass fraction equal to 0.317;

(3) an intermediate radiative shell of 3.3% in mass with varying chemical composition;

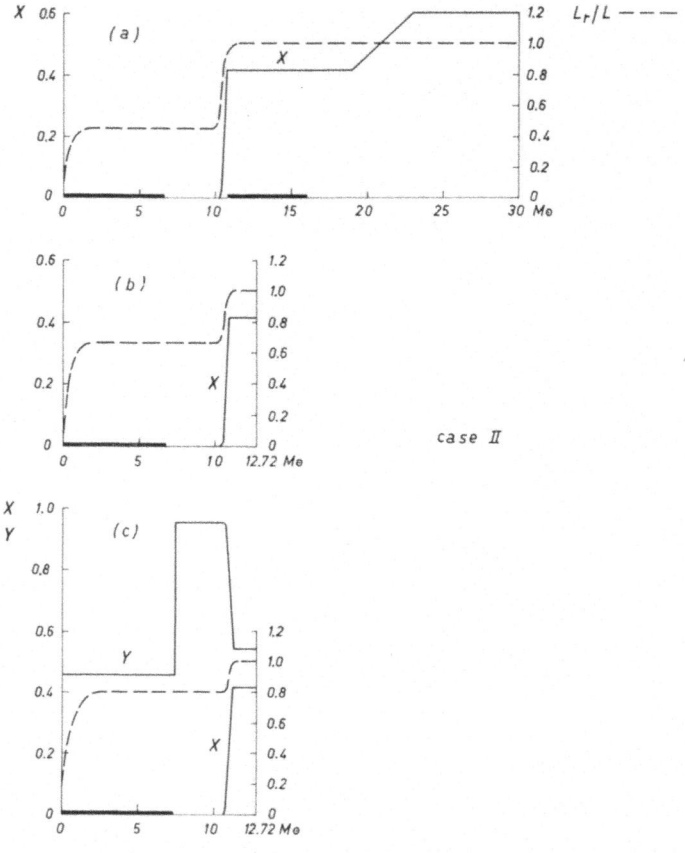

Fig. 2a–c.   Case II. As in Figure 1, but with $L_r \simeq L_n$.

(4) a convective helium-rich zone ($Y=0.526$) including a mass fraction equal to 0.255;

(5) an outer radiative shell of 16% in mass with varying hydrogen content;

(6) an envelope with the initial chemical composition and with a thin convective superficial region.

From Figure 1a one can also see that the helium burning near the center has not yet restored the thermal equilibrium in the stellar interior. From this point of view the actual case approximates a typical case B (see Section 1). The central burning contributes in fact only about 11% to the total energy flux emerging from the innermost helium region. The largest contribution to the surface luminosity is given by the hydrogen burning through the carbon-nitrogen cycle in the intermediate radiative shell. It attains indeed 75% of the superficial luminosity. An appreciable fraction of the surface luminosity (about 0.3) is absorbed by expansion of the outer layers.

When the primary fills up its critical volume a very rapid loss of matter to the secondary takes place. Two following phases of mass exchange, each of them with a transfer of about $8\,M_\odot$, at rates of about $8 \times 10^{-2}\,M_\odot\,\mathrm{yr}^{-1}$ and $9 \times 10^{-3}\,M_\odot\,\mathrm{yr}^{-1}$,

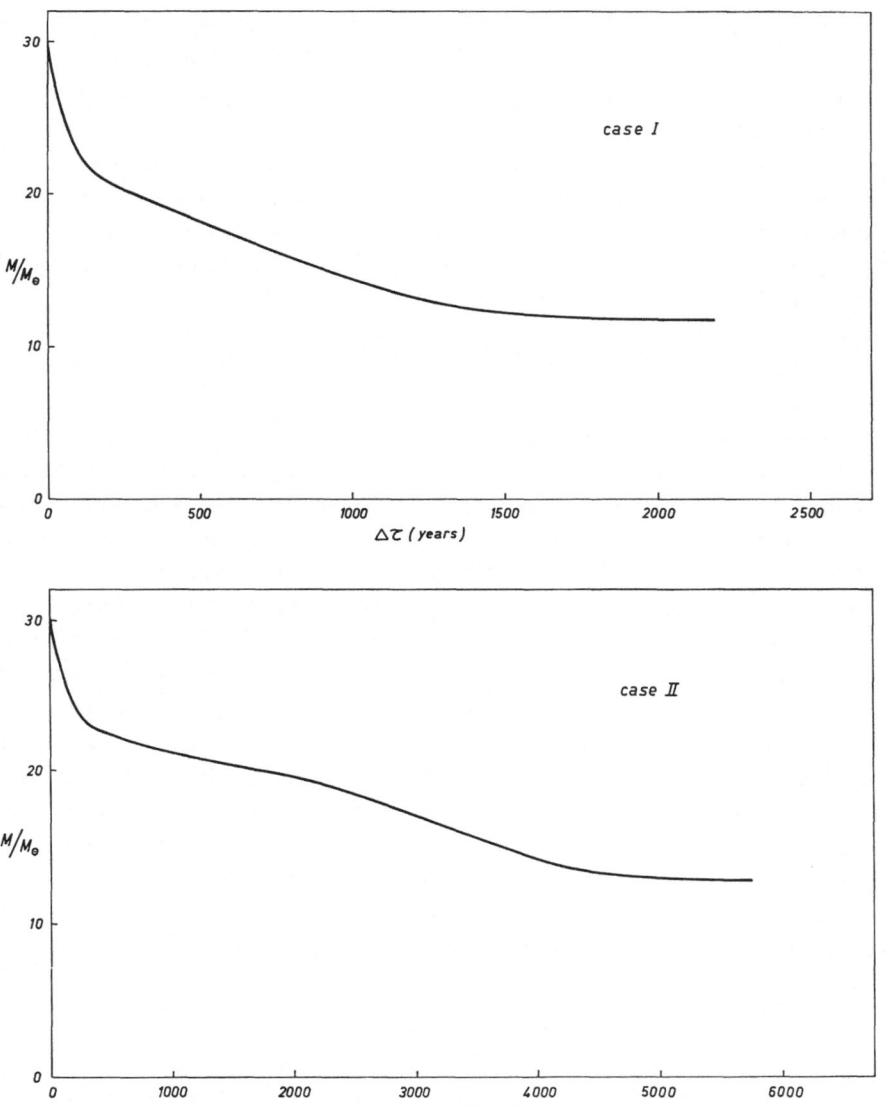

Fig. 3.   Variations with time during the *semi-detached* phase of the masses of the original primary components in the two close binary systems with initial separations 119.2 $R_\odot$ (case I) and 307.2 $R_\odot$ (case II). Times are measured from the contact.

can be easily traced out. In Figure 3 the variation with time of the total stellar mass of the original primary is shown for cases I and II.

The loss of matter then continues at a progressively lengthened rate. During the *semi-detached* phase the surface luminosity of the ex-primary decreases to a minimum and then grows again. Then central burning becomes more efficient, whereas the shell burning diminishes, mainly because of the decrease of temperature. Depending on the loss of matter, the shell source is in fact shifted to relatively more outer regions.

The whole phase of mass loss is finally completed (stage b) after about 2.2×10^3 years from the contact, when the original primary, now the less massive component, is 11.74 $M_\odot$ and the separation has decreased to 97.5 $R_\odot$. The orbital period is now $17\overset{d}{.}66$. The star has lost the whole hydrogen-rich envelope; the outer radiative shell and most of the initial intermediate convective region. Since the phase of mass exchange is very rapid, the remaining envelope is fitted to the helium interior through a shell with a steep X-profile, practically unaltered with regard to the model at stage a (see Figure 1b). In the envelope the hydrogen abundance is now $X_e \cong 0.42$. Moreover, the convective core has enlarged to 42.3% of the stellar mass and because of the 3α reactions near the center a very small amount of helium has been transformed into carbon (see Table I).

The evolution of the original primary has been further followed during the subsequent *detached* phase until the stage of steady burning of helium takes place near the center. At the beginning – compare again Figure 1b – the thermal equilibrium is not yet completely restored in the ex-primary. The whole radiative zone contracts at a rate progressively lengthened with time heating up the stellar interior.

The nuclear sources of energy in the core and at the shell now contribute to the surface luminosity by about 60% and 10%, respectively. With regard to stage a, the burning of helium near the center through the 3α reactions has meanwhile been increasing and has already become the main source of energy. The hydrogen burning in the shell through the carbon-nitrogen cycle has also grown. The increasing gradient of the X-profile compensates in fact the decrease of temperature until shrinking in mass of the shell source causes a new reduction of the energy output from the shell.

In the H-R diagram the evolutionary track moves rapidly to larger effective temperatures at about the same luminosity (see Figure 4). This continues until the energy released by thermodynamical changes becomes negligible with regard to the nuclear-energy flux. As a consequence, the evolutionary model tends to its proper

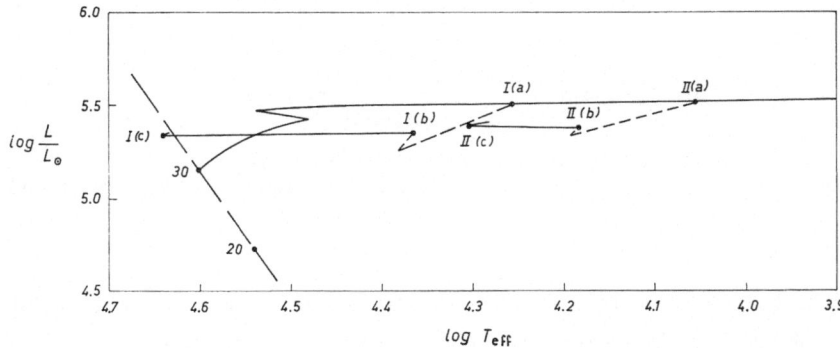

Fig. 4. H-R diagram. The evolutionary paths of the original primaries in the close binary systems of cases I and II are given by the solid lines during the initial and final *detached* phases. The dashed curves refer to the intermediate *semi-detached* phase of exchange of matter between the components. The main sequence of homogeneous hydrogen stars and the evolutionary track of the single 30 $M_\odot$ star are also displayed.

position of steady burning of helium in the core and of hydrogen in the shell (see
KIPPENHAHN and WEIGERT, 1967). This occurs after about 1.7×10^5 years from the
contact at the age of 4.506×10^6 years, when $\log T_{eff} = 4.642$ (stage c).

The location of the model in the H-R diagram is in a fairly good agreement with
the position one can estimate for the corresponding (same total mass and same
q_0-value) equilibrium model with a double energy-source extrapolated from *generalized q_0-sequences* (GIANNONE *et al.*, 1968). It should however be noted that: (1) the
envelope has now a hydrogen abundance in mass equal to $X_e = 0.415$, whereas the
sequences quoted above have been computed with $X_e = 0.602$, and (2) the bulk of the
core is now composed of helium and carbon instead of pure helium. As is well known,
the luminosity value of these equilibrium models is fixed by the total stellar mass,
whereas effective temperature critically depends on q_0. The dependence is such that
at increasing q_0-values, effective temperature increases (for $q_0 = 1$ one has the homogeneous helium-star main sequence).

The evolutionary model is now located in the H-R diagram practically on the
main sequence of homogeneous hydrogen-burning stars, in a position corresponding
to a star model of about 35 $M_\odot$. It is composed of a convective core, an intermediate
radiative helium region, a thin helium-rich shell and an envelope with a mass fraction
equal to about 0.06, which is the remnant of the old intermediate convective zone
(Figure 1c). The convective core extends in mass by about 62% and contains a mixture
of helium, carbon and oxygen.

The stellar interior is in thermal equilibrium, the total luminosity at each mass

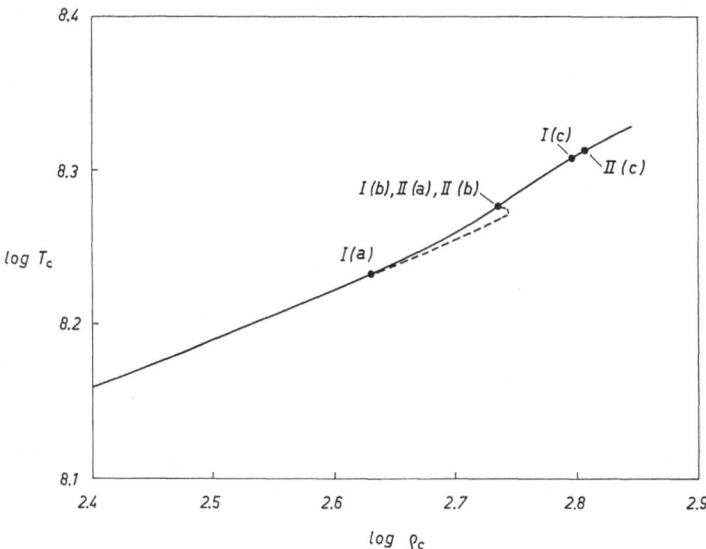

Fig. 5.   Evolution of the central conditions in the $\log T_c$ – $\log \varrho_c$ plane. The solid line refers to the
single 30 $M_\odot$ star. In case I, between stages a and b the center evolves along the dashed curve and
then to stage c on the solid line. In case II the locations of the center at the following stages are
also indicated.

shell coincides with the nuclear-energy flux. In Figure 1c the two curves overlap each other. The convective core contributes almost all the luminosity (85%). The thin hydrogen-burning shell (5% in mass) supplies the other 15% of the total energy flux. Because of the nuclear processes near the center, the conversion of helium into carbon and oxygen leads progressively to an increase of the chemical inhomogeneity between the convective core and the envelope. This fact causes the evolutionary track to turn backwards and to develop to larger radii.

The evolution of the central conditions of the original primary in the two studied cases is shown in Figure 5. The solid line refers to the single star of 30 $M_\odot$. The evolutionary changes during the phase of transfer of matter to the secondary follow the dashed curve in case I. The central region heats up and contracts with a behaviour very nearly homologous. At the end of the *semi-detached* phase (stage b) and in the further evolution, the path develops almost adiabatically. In both cases I and II the evolutionary tracks of the centers then coincide with that of the single 30 $M_\odot$ star, but they are obviously travelled at different times.

B. CASE II

The contact of the primary (stage a) now occurs at a more evolved phase during the expansion of the star along the horizontal track in the H-R diagram (see Figure 4 and Table II). Despite the short time interval between stages a, about 10^4 years, the innermost structure of the model is changed with regard to the corresponding stage of case I because of a more advanced helium depletion near the center ($Y_c = 0.939$). The central burning is in fact more efficient and has grown to values such as to nearly restore the thermal equilibrium in the interior (see Figure 2a). Moreover, due to the shell burning, the X-profile is shifted outwards by 0.5 $M_\odot$ with respect to case I.

As for case I, the rapid phase of mass loss proceeds initially at a rate of about 3×10^{-2} $M_\odot$ yr^{-1} and subsequently at 2.4×10^{-3} $M_\odot$ yr^{-1}. An exchange of 7 $M_\odot$ is associated with each of these rates (see Figure 3). After 5.8×10^3 years from the contact, the system again becomes *detached* (stage b). The ex-primary is now 12.72 $M_\odot$; the separation is 229.3 $R_\odot$ and the period is $63^{d}.68$. The convective core extends in mass by about 54% and contains a little carbon (see Table II). The remnant resembles the corresponding stage of case I, but its position in the H-R diagram is displayed at a lower effective temperature (the magnitude being nearly the same because of the small difference of the total masses). The interior is more adjusted thermally than in case I and the burning of central helium supplies nearly 70% of the surface luminosity (see Figure 2b). The behaviour with time of the shell source is analogous to that of case I. Actually, the hydrogen-burning provides nearly the other 30% of the superficial energy flux. The helium interior enlarges in mass and the star continues to move in the H-R diagram to the left until the central burning grows to larger values.

A turn-off (stage c) at $\log T_{\text{eff}} = 4.307$ is then reached after about 2.0×10^5 years from the contact at the age of about 4.54×10^6 years (see Table II). The convective core extends in mass by about 58% and contains helium, carbon and oxygen (see Figure 2c). With regard to stage c of case I, 20% of the difference of the central helium contents is due to the difference existing at the times of the contacts. A more efficient central burning provides in case II a larger nuclear transformation of helium into heavier elements.

4. Concluding Remarks

For the two considered close binary systems, some comments on the results can be instructive. At first, one can easily see that, in spite of the small interval of time between the contacts in cases I and II, the innermost structure of the primary of 30 $M_{\odot}$ changes deeply with regard to the thermal equilibrium (compare Figures 1a and 2a). According to this, the duration of the further phase of mass exchange is almost triplicated for a mass transfer which is in case II smaller by 1 $M_{\odot}$ in comparison with case I (Figure 3). Moreover, in the H-R diagram (Figure 4), at the end of the *semi-detached* phase, the displacement of the evolutionary model to the position of steady nuclear burning is for case II less shifted to the left than for case I. Therefore, passing from system I to system II, we note that the evolutionary behaviour of such systems tends from an evolution of type B to one of type A (see again Section 1) with a decrease in luminosity during the mass exchange and a further development to lower effective temperatures (i.e. evolution of the ex-primary to the right in the H-R diagram).

In the cases presently studied, stages c, at the turn-off in the H-R diagram, correspond to observable stars, since there the evolutionary track slows drastically. Luminosities of the ex-primaries nearly coincide, whereas effective temperatures strongly depend on the initial separation (i.e. on the structure of the original primary). With regard to the main-sequence models of the same mass, the actual ex-primaries show very large overluminosities, of the order of 3 mag. In both cases, from stages c on, the evolutionary tracks turn backwards to the right in the H-R diagram and the evolution takes place on a nuclear time-scale. This occurs because of the increasing inhomogeneity in the bulk of the core, where the 3α and α-capture reactions transform helium into carbon and oxygen. As is well known, this leads to progressively larger radii.

This fact can open many possibilities for what concerns the further evolutionary changes in binary systems. Because of the increase of radius, we would expect, indeed, that a second phase of mass transfer from the same ex-primary to the original secondary could occur. Moreover, this can take place when: (1) helium is still burning in the core (case C-A), (2) helium has already been exhausted (case C-B), or, finally, (3) carbon has ignited (case C-C). Correspondingly, we may expect that the resulting configurations of the close double stars can differ considerably. In this regard, an open problem is still represented by the ex-secondary, which, in its turn, can fill up its Roche lobe and give back matter to the other component. In our cases, by assuming a quiescent accretion of matter, one would find the original secondaries still on the main sequence at the locations corresponding to their new total masses.

The observable stages of the ex-primaries seem to be consistent with the Wolf-Rayet components of massive close binaries, as already suggested by PACZYŃSKI (1967) for massive primaries evolved according to case B. Such an identification could completely be achieved by fitting atmospheric features, too. However, we point out that the location of these models in the H-R diagram and the mass values appear to be reasonable and that the agreement could be further improved by choosing more suitable values for the free parameters (total mass, mass ratio, and separation) of massive close binary systems.

References

CHIOSI, C. and SUMMA, C.: 1969, in preparation.
COX, A. N. and STEWART, J. N.: 1965, *Astrophys. J. Suppl.* **11**, No. 94.
DUORAH, H. L. and KUSHWAHA, R. S.: 1963, *Astrophys. J.* **137**, 566.
GIANNONE, P., KOHL, K., and WEIGERT, A.: 1968, *Z. Astrophys.* **68**, 107.
HOFMEISTER, E., KIPPENHAHN, R., and WEIGERT, A.: 1964, *Z. Astrophys.* **59**, 215.
KATO, S.: 1966, *Publ. Astron. Soc. Japan* **18**, 374.
KIPPENHAHN, R. and WEIGERT, A.: 1967, *Z. Astrophys.* **65**, 251.
PACZYŃSKI, B.: 1967, *Acta Astron.* **17**, 355.
REEVES, H.: 1964, *Stellar Energy Sources* (Stars and Stellar Systems, vol. VIII), University of Chicago Press, Chicago.

Discussion

Lauterborn: I'd like to show to you a slide [see Figure 6] showing the evolution of a system similar to the ones on which Dr. Giannone has reported. The calculations were carried out by Prof. Kippenhahn at UCLA during the last 4 or 5 months. The system originally consists of a primary of 25 $M_\odot$ accompanied by a secondary of 15 $M_\odot$ at a

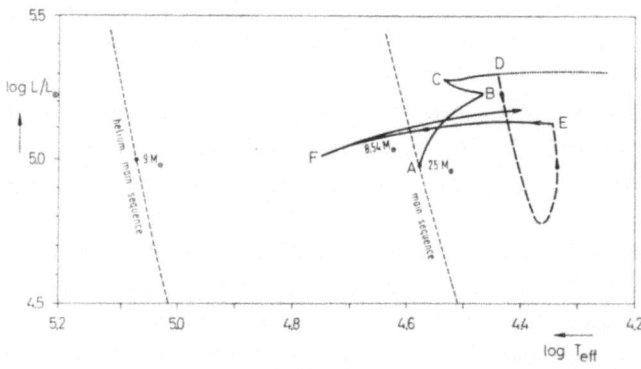

Fig. 6.

separation of about 50 $R_\odot$. After mass loss at point E the 25 $M_\odot$ star has become an 8.54 $M_\odot$ star. The star now begins to contract and tries to reach a state near the 9 $M_\odot$ star on the helium main sequence which is to be seen in the left of the H-R diagram. At F the contraction is stopped because the conditions in the center now become more favorable to carbon burning. The calculations were stopped at the point where a second phase of mass loss begins. I'd like to remark that this is another case of evolution of a close binary system with large mass of the primary which seems indeed to result in a binary system with a WR component. The mass of 8.5 $M_\odot$ fits well into the range of $M = 10 \pm 2$ $M_\odot$ mentioned by Underhill in her paper on WR stars, and the same is for the luminosity which is $\log L/L_\odot = 5.0$ here and was $M_{bol} = -7.5$ in the paper of Underhill.

Giannone: I would like to note that the results you have shown fit fairly well with ours as concerns the general behaviour of very massive binary systems with large separations. By comparison of the evolutionary changes during the rapid phase of mass loss and the further evolutionary paths of the original primary components we can also better understand some features which depend on the different value of the mass of the initially more massive star (e.g. the different decrease of luminosity during the phase of mass loss).

Underhill: The luminosity of Wolf-Rayet stars is found from observations to fall close to $\log L/L_\odot = 4.8$. There is little direct evidence for significantly higher luminosities. These observations look to conflict with the identification suggested by the theoreticians of their evolved mass-exchange binaries, total mass ~ 30–40 $M_\odot$, with Wolf-Rayet systems, for the theoretical results suggest that $\log L/L_\odot$ is greater than 5.0 during stages when the binary may be observed. Total mass in WR-OB systems seems to be $\geqslant 30$ $M_\odot$.

May I again emphasize that a 'Wolf-Rayet spectrum' is a spectroscopic phenomenon. If we are to identify models in a certain part of the H-R diagram with Wolf-Rayet stars, it is necessary to show that the typical type of spectrum will certainly occur and that the postulated theoretical models give not only necessary but also sufficient physical conditions for generating the spectrum.

Dallaporta: From the fact that in the 7 binary systems containing a Wolf-Rayet star its mass value is of the order of 10 $M_\odot$, are you inclined to consider this mass value as a necessary requisite for a star being a Wolf-Rayet? Or would you admit a possible larger range of mass values?

As a second question: in stellar evolution theory we are generally only able to deduce mass, radius, composition and luminosity for any stage of a given star. Should we succeed in getting these parameters to correspond to those of a Wolf-Rayet star, what further determination would you require in order to allow to conclude that the star is a Wolf-Rayet? Would it be necessary in your view to deduce the structure of the diffuse atmosphere and succeed in calculating the line spectrum to be expected from it?

Underhill: The estimate of (10 ± 2) $M_\odot$ for the mass of the Wolf-Rayet star is quite a good answer, I believe, although observational data are available for only 7 systems and 124 WR stars are known. That is 5.6% of the known systems and single stars has been included in the mass estimate. In the case of stars of other spectral types our knowledge of masses is based on actual measurements of a smaller percentage of the known stars. The uncertainty quoted, ± 2, is an estimate of the probable error of the value 10 $M_\odot$.

The question of relating stellar structure models to stars which are recognized by their spectral types, these types having been assigned according to the appearance of the line spectrum, is a very complex one. My conclusions on this subject are summarized in an article in *Vistas in Astronomy*, at present in press. So far as interpreting main-sequence stars of types A0 to O9 by means of classical model atmospheres in which the spectrum is predicted by LTE theory, I believe one can obtain a fair understanding of the continuous spectrum between 3000 and 7000 Å. This enables one to assign an effective temperature. However, attempts to obtain a consistent interpretation of the strong and medium lines, that is the classifications lines, has met defeat. Quite clearly Non-LTE theories of line formation must be used. One illustration of the ambiguities encountered with our definitions of spectral type is that B stars with the same type according to the continuous spectrum, i.e. (B-V) colour, may have quite different line spectra even when it is pretty certain that the stars have about the same value of $\log g$. The 'strong He line' and the 'weak He line' B stars are conspicuous examples of what occurs. Since a Wolf-Rayet type is assigned on the basis of the *line spectrum* I am convinced that in this case, the line spectrum must be predicted correctly before a model is identified with a star.

EVOLUTION OF CLOSE BINARIES AND ORIGIN OF
ALGOL-TYPE SYSTEMS

J. ZIÓŁKOWSKI

Institute of Astronomy, Polish Academy of Sciences, Poland

Abstract. Observational data for Algol-type systems are discussed basing on available evolutionary tracks. It is shown that the observed properties of the massive binaries (with total mass greater than 5 $\mathfrak{M}_\odot$) can be satisfactorily explained assuming that their contact components are burning hydrogen in the core (case A). On the contrary, the majority of contact components of the observed low-mass binaries are burning hydrogen in the shell. The observed distribution of such binaries as a function of different luminosity excesses seems to indicate that their origin is connected with systems in which the primary component reached the Roche lobe before exhaustion of hydrogen in its center. Some mass loss from the system as a whole seems to be necessary in this case.

Under the term 'Algol-type' systems we understand those eclipsing binaries in which the more massive component is a main-sequence star, while the less massive component is a subgiant filling up its Roche lobe. There is a widespread opinion that such systems called also semi-detached systems are products of a large-scale mass transfer between the components. This hypothesis provides an explanation of the evolutionary paradox found in these systems; namely that the less massive component is the more evolved one. There are also some observational indications (cf. PLAVEC, 1967) showing that a large fraction of these systems is still in the course of mass exchange or is losing mass from the system.

The mass transfer is due to existence of the Roche lobe. The more massive star expands during its evolution and at some moment fills up its Roche lobe. Then a rapid mass transfer takes place on the thermal time-scale (MORTON, 1960; KIPPEN-HAHN and WEIGERT, 1967). The rapid mass transfer stops when the thermal equilibrium of the losing mass component is restored. The mass ratio is then usually more than reversed. In further evolution, depending on the initial conditions, the system may remain semi-detached with a slow mass transfer or may become detached.

KIPPENHAHN and WEIGERT (1967) introduced the terms 'case A' and 'case B' for the types of evolution which occur when the primary (more massive) component reaches its Roche lobe during central hydrogen burning (case A) or after exhaustion of hydrogen in the center (case B). The term 'case A' was used in different publications for this phase of evolution in which a subgiant contact component is burning hydrogen in the core. However, after exhaustion of hydrogen in the center of this component we obtain a semi-detached system burning hydrogen in the shell (as in case B) although the Roche lobe was reached for the first time during central hydrogen burning (as in case A). For convenience we shall call this stage of evolution, when hydrogen is burning in the shell, case AB.

A number of evolutionary sequences corresponding to case A were computed for massive binaries (KIPPENHAHN and WEIGERT, 1967; PACZYŃSKI, 1966, 1967a, b; PACZYŃSKI and ZIÓŁKOWSKI, 1967; PLAVEC, 1968; PLAVEC *et al.*, 1968; ZIÓŁKOWSKI,

M. Hack (ed.), Mass Loss from Stars. All rights reserved

1969). Recently, some computations were carried out also for low-mass systems and for few cases AB (ZIÓŁKOWSKI, 1969). In all computations for case A it was found that after rapid mass transfer the system remains semi-detached. Hydrogen is burning in the core and a slow mass transfer takes place on the nuclear time-scale. For case AB we found that the evolution is similar to that in case B, being, however, much slower for the given initial mass of the primary. Also the distribution of time of evolution among different evolutionary phases is in case AB distinct from that in case B.

Evolutionary sequences corresponding to case B were computed by KIPPENHAHN and WEIGERT (1967), KIPPENHAHN et al. (1967), PACZYŃSKI (1967), and ZIÓŁKOWSKI (1969). It was found that for massive binaries (with the initial mass of the primary $\gtrsim 4 \, \mathfrak{M}_\odot$) the mass transfer is very rapid and is terminated by the ignition of helium in the center of losing mass star. After the helium ignition the star contracts and the system becomes detached. For low-mass systems the mass transfer is relatively less rapid, particularly during its final stage, and stops when almost the whole hydrogen-rich envelope is lost.

The aim of this paper is to compare briefly the theoretical results with the observational data. The discussion is based mainly on the results presented in our latest paper (ZIÓŁKOWSKI, 1969). The observational data are taken from PLAVEC (1967) and KOPAL (1959).

In Figure 1 we plotted a histogram showing the distribution of the observed semi-detached systems as a function of the total mass of the system. The observed distribution suggests the existence of massive (with total masses exceeding 5 $\mathfrak{M}_\odot$) and low-mass (with total mass below 4 $\mathfrak{M}_\odot$) binaries as two separate groups. It is convenient to retain this classification because systems belonging to different groups exhibit, as we shall see, different observational characteristics.

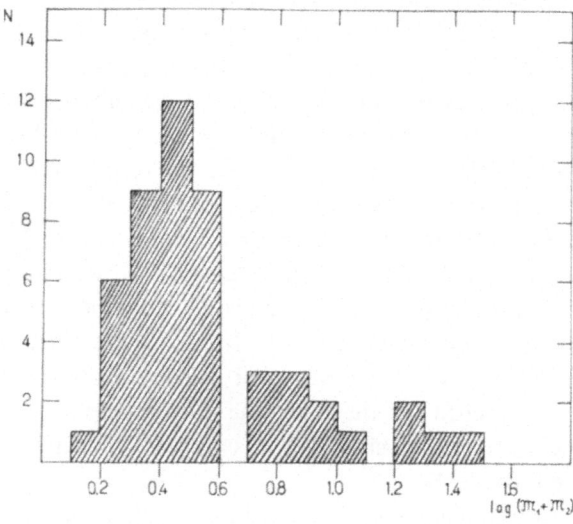

Fig. 1. Distribution of the observed number of the semi-detached systems as a function of the total mass of the system.

1. Massive Binaries

Figure 2 shows the luminosity excess vs. mass ratio diagram for contact components of massive semi-detached binaries. We chose these characteristics because they are less affected by the observational errors than the masses and luminosities. For better

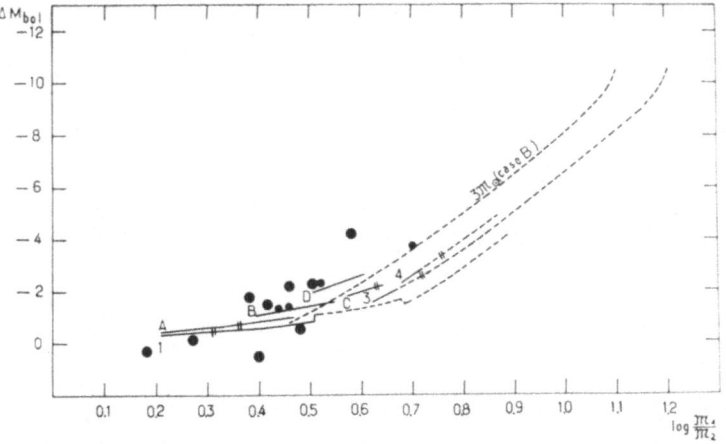

Fig. 2. Luminosity excess of the contact component vs. the mass ratio for the massive semi-detached binaries. $\mathfrak{M}_2$ denotes the mass of the contact component. Solid lines denote theoretical evolutionary paths with contact components in the stage of central hydrogen burning. Broken lines correspond to the contact components, which are burning hydrogen in the shell. Double slashes show the point where the initial secondary component finishes its main-sequence evolution and a contact system forms. Theoretical paths for case B are labelled with the initial mass of the primary, for cases A and AB with the number of the system in Table I. Large circles denote observational points with high weights, small circles – those with low weights.

TABLE I

Initial parameters for the systems evolving in case A

No.	Initial mass of the primary [$\mathfrak{M}_\odot$]	Initial mass ratio	Central hydrogen content of the primary at the beginning of mass transfer
1	4	1.5	0.40
2	4	1.5	0.10
3	4	1.5	0.05
4	4	2.25	0.05
5	2	1.5	0.20
6	2	1.5	0.05
7	1	1.5	0.05
8	1	2.25	0.05
A[a]	8	1.5	0.40
B[a]	8	1.5	0.20
C[a]	8	2.25	0.20
D[a]	16	1.5	0.225

[a] Evolutionary paths computed by PACZYŃSKI (1966, 1967a).

comparison we have added (Table I) some evolutionary paths computed by PACZYŃSKI (1966, 1967a).

As we can see, the contact components of the massive binaries have luminosity excesses rather below 3 mag. The observed properties of such systems can be satisfactorily explained by the assumption that they are burning hydrogen in the center, i.e. that they are evolving in case A, as it was suggested by PACZYŃSKI (1966). The probability to observe such a system in case B is very small because the system remains semi-detached in this case only during a very short time. The advanced evolution in case AB is probably not possible because the contact system forms before the large luminosity excess can be attained. The points where it occurs are marked in Figure 2 by double slashes. However, we should like to emphasize that if there is some mass loss from the system as a whole all corresponding points will be shifted upwards on our diagram. The question of the mass loss from the system cannot be solved at present, this, however, being not decisive for the explanation of the observational data.

2. Low-Mass Binaries

The luminosity excess vs. mass ratio diagram for contact components of these systems is shown in Figure 3. As we can see, the majority of contact components have large (3–10 mag) luminosity excesses. Definitely they cannot burn hydrogen in the core. As it follows from our computations, the luminosity excess during central hydrogen burning (i.e. in case A) never exceeds 3 mag and for low-mass binaries is even smaller (not exceeding 1.5 mag). Larger luminosity excesses can be attained only in cases B or AB. It was suggested (PACZYŃSKI, 1967c, d; GIANNONE et al., 1968) that the observed semi-detached low-mass binaries correspond to case B. We suggest that a

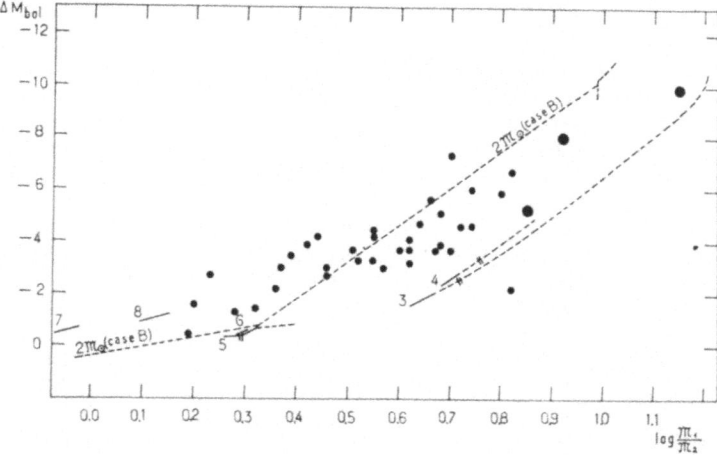

Fig. 3. Luminosity excess of the contact component vs. the mass ratio for the low-mass semi-detached binaries. The meaning of all symbols is the same as in Figure 2. Theoretical path for the system $2 \mathfrak{M}_\odot + 1 \mathfrak{M}_\odot$ is based on the data published by KIPPENHAHN et al. (1967). In addition two evolutionary tracks for more massive systems evolving in case AB (systems number 3 and 4) are shown for comparison (because no corresponding tracks for the low-mass systems were available).

large part of them correspond to case AB. The main argument in support of such a suggestion is the distribution of the observed binaries as a function of the different luminosity excesses. As it may be seen from Figure 3 we observe 24 binaries with moderate (3–6 mag) and 5 binaries with high (above 6 mag) luminosity excesses. The analysis of theoretical data for the system evolving in case B with the initial mass of the primary equal 2 $\mathfrak{M}_\odot$ (KIPPENHAHN *et al.*, 1967) and 3 and 4 $\mathfrak{M}_\odot$ (ZIÓŁKOWSKI, 1968) shows that these systems spent much more time in the region of high than in the region of moderate luminosity excesses. For low mass systems this ratio is near 5:1. The observed distribution is just reversed. It is rather impossible to explain such a difference by the observational selection.

Unfortunately we do not possess complete evolutionary tracks in case AB for low-mass binaries. We intend to carry out such computations in the near future in order to check our present suggestions. However, the results obtained for more massive systems show that in case AB the system spends much more time in the region of moderate than in the region of high luminosity excesses. Thus the agreement with the observations should be more satisfactory.

However, there exists a difficulty connected with our explanation. Namely, the advanced evolutionary phase (necessary for a large luminosity excess) in case AB is possible only either for small central hydrogen content at the beginning of the mass transfer ($X_c \sim 0.05$) or if there is some mass loss from the system. In all other cases the system of W UMa type will form. The question whether the majority of observed systems is really connected with a small range of initial conditions (corresponding to small hydrogen content in the center of primary) or whether there is a large mass loss from the system cannot be decided at the present time.

References

GIANNONE, P., KOHL, K., and WEIGERT, A.: 1968, *Z. Astrophys.* **68**, 107.
KIPPENHAHN, R. and WEIGERT, A.: 1967, *Z. Astrophys.* **65**, 251.
KIPPENHAHN, R., KOHL, K., and WEIGERT, A.: 1967, *Z. Astrophys.* **66**, 58.
KOPAL, Z.: 1959, *Close Binary Systems*, J. Wiley and Sons, New York.
MORTON, D. C.: 1960, *Astrophys. J.* **132**, 146.
PACZYŃSKI, B.: 1966, *Acta Astron.* **16**, 231.
PACZYŃSKI, B.: 1967a, *Acta Astron.* **17**, 1.
PACZYŃSKI, B.: 1967b, *Acta Astron.* **17**, 193.
PACZYŃSKI, B.: 1967c, *Acta Astron.* **17**, 355.
PACZYŃSKI, B.: 1967d, I.A.U. Colloquium: On the Evolution of Double Stars, Observatoire Royal de Belgique, Communications, Série B, No. 17.
PACZYŃSKI, B. and ZIÓŁKOWSKI, J.: 1967, *Acta Astron.* **17**, 7.
PLAVEC, M.: 1967, *Bull. Astron. Inst. Czech.* **18**, 334.
PLAVEC, M.: 1968, *Astrophys. Space Sci.* **1**, 239.
PLAVEC, M., KŘIŽ, S., HARMANEC, P., and HORN, J.: 1968, *Bull. Astron. Inst. Czech.* **19**, 24.
ZIÓŁKOWSKI, J.: 1969, *Astrophys. Space Sci.* **3**, 14.

Discussion

Plavec: In my talk of Friday I mentioned that we did come across a situation which was really case AB, when the stars start with this central hydrogen abundance, then hydrogen becomes completely exhausted and mass loss continues under the conditions of case B, and we get such similar systems. But I want to remark that probably the number of such systems, case AB, must be rather low, because of the narrow limits that occur within these cases.

Grzędzielski: Yes, it is quite true. It may happen that the number of such systems is low. However, it seems at present that the best interpretation of observational data consists in assuming that the AB cases are responsible for the majority of low-mass semi-detached systems.

ON THE TIME-SCALE OF THE MASS TRANSFER
IN CLOSE BINARIES

B. PACZYŃSKI, J. ZIÓŁKOWSKI, and A. ŻYTKOW

Institute of Astronomy, Polish Academy of Sciences, Warsaw, Poland

Abstract. We show that the mass outflow from the more massive component of a close binary that fills up its Roche lobe, may take place on a dynamical time-scale if this component has a deep convective envelope. We suggest that the outbursts of the U Geminorum-type stars might be related to this phenomenon.

It is well known that the majority of semi-detached binaries have the less massive component filling up its Roche lobe. It is generally believed now that this lighter component was originally the more massive one, and that a major mass transfer between the two stars had taken place in the past. MORTON (1960) has demonstrated that the major mass transfer probably takes place on a thermal time-scale, and is therefore very rapid and difficult to observe. Morton's considerations were supported by direct numerical computations by KIPPENHAHN and WEIGERT (1967), and independently by PACZYŃSKI (1967) and PLAVEC *et al.* (1968). This rapid mass exchange is caused by the fact that the Roche lobe around the more massive component shrinks while this component loses its mass.

All published evolutionary tracks of the components of close binaries are such that the initially more massive star has a radiative envelope when it expands and fills up its Roche lobe. PACZYŃSKI (1965) has suggested that the mass exchange may take place on a dynamical time-scale if the more massive component has a deep convective envelope at the beginning of the mass loss. Here we would like to present some additional arguments in favour of that suggestion and analyse in some details a very simple case.

Let us consider a star with the following properties:

(a) spherical symmetry,

(b) convective (adiabatic) equilibrium throughout the interior, and

(c) hydrostatic, but not necessarily thermal equilibrium.

The entropy S per one gram of the stellar interior is determined by a thin region below the surface, where the convection is not adiabatic. In general we have

$$S = S(\mathfrak{M}, R, L),\tag{1}$$

where $\mathfrak{M}$, R, and L are the mass, radius and the luminosity of the star respectively.

The radius of an isentropic depends on its mass and entropy per gram only, if the star is in hydrostatic equilibrium. We have therefore:

$$R = R(\mathfrak{M}, S).\tag{2}$$

M. Hack (ed.), Mass Loss from Stars. All rights reserved

The luminosity of a star is given by the well-known formula

$$L = \int_0^{\mathfrak{M}} \varepsilon \, d\mathfrak{M}_r - \int_0^{\mathfrak{M}} T \, (\partial S/\partial t)_{\mathfrak{M}_r} \, d\mathfrak{M}_r, \qquad (3)$$

where ε is the rate of energy generation in nuclear reactions, and T is the temperature. In our case this formula simplifies to

$$L = \int_0^{\mathfrak{M}} \varepsilon \, d\mathfrak{M}_r - dS/dt \int_0^{\mathfrak{M}} T \, d\mathfrak{M}_r. \qquad (3a)$$

Let us now consider a star that was originally in a thermal equilibrium. Then $L = \int_0^{\mathfrak{M}} \varepsilon \, d\mathfrak{M}_r$. Let us change the mass of the star by $\Delta\mathfrak{M}$ in a time Δt, and let the star remain convective throughout, and in hydrostatic, but not necessarily in thermal equilibrium. Then we may get from Equations (1)–(3)

$$\left. \begin{array}{l} \Delta S \approx \partial S/\partial\mathfrak{M} \, \Delta\mathfrak{M} + \partial S/\partial R \, \Delta R + \partial S/\partial L \, \Delta L \\[4pt] \Delta R \approx \partial R/\partial\mathfrak{M} \, \Delta\mathfrak{M} + \partial R/\partial S \, \Delta S \\[4pt] \Delta L \approx \int_0^{\mathfrak{M}} \Delta\varepsilon \, d\mathfrak{M}_r - \Delta S/\Delta t \int_0^{\mathfrak{M}} T \, d\mathfrak{M}_r. \end{array} \right\} \qquad (4)$$

In general, for an arbitrary ε it is possible to find $\int_0^{\mathfrak{M}} \Delta\varepsilon \, d\mathfrak{M}_r$ as a linear function of $\Delta\mathfrak{M}$, ΔR, and ΔS. Therefore we may eliminate the changes ΔR and ΔL from Equations (4), and get

$$\Delta S + A \, \Delta S/\Delta t + B \, \Delta\mathfrak{M} = 0, \qquad (5)$$

where A and B are constants determined by the stellar model prior to the mass loss. This formula may be changed to

$$\Delta S = \frac{-B \, \Delta\mathfrak{M} \, \Delta t}{A + \Delta t}. \qquad (5')$$

From the last formula we see that for a small $\Delta\mathfrak{M}$ and Δt, the change of entropy ΔS is a quantity of a higher order. This means that at the beginning of the mass loss the process is adiabatic. PACZYŃSKI (1965) has shown that in such a case the decrease of the mass leads to the increase of stellar radius.

Let us now consider a still simpler situation of a star made up of a perfect gas. In that case the star is a polytrope of an index 1.5, and Equation (2) may be written as

$$\tfrac{1}{2} \ln \mathfrak{M} + \tfrac{3}{2} \ln R - (\mu H/k) \, S = \text{constant}, \qquad (6)$$

where the constant depends on the units used, μH is a mean molecular weight, and k is the Boltzmann constant.

The derivatives of Equation (1) may be found from the tables of convective stellar envelopes (BAKER and TEMESVÁRY, 1967). For stars close to the main sequence we have

$$\frac{\partial S}{\partial \ln L} \approx 6 \frac{k}{\mu H}, \qquad \frac{\partial S}{\partial \ln R} \approx -9 \frac{k}{\mu H}, \qquad \frac{\partial S}{\partial \ln \mathfrak{M}} \approx -2 \frac{k}{\mu H}. \tag{7}$$

Let us further assume that $\varepsilon = \varepsilon_0 \varrho^\alpha T^\nu$. After some transformations it is possible to find that Equation (5) may be written as

$$\frac{\mu H}{k} \Delta S \left(1 + C \frac{t_{KH}}{\Delta t}\right) = D \frac{\Delta \mathfrak{M}}{\mathfrak{M}}, \tag{8}$$

where t_{KH} is the Kelvin-Helmholtz time-scale, and C and D are dimensionless constants given by the relations

$$\left. \begin{array}{l} C = \dfrac{\partial S/\partial \ln L}{-\partial S/\partial \ln R + \frac{3}{2}k/\mu H + (\nu + 3\alpha)\, \partial S/\partial \ln L} \\[3mm] D = \dfrac{\frac{3}{2}\partial S/\partial \ln \mathfrak{M} - \frac{1}{2}\partial S/\partial \ln R + (2\nu + 3\alpha)\, \partial S/\partial \ln L}{-\partial S/\partial \ln R + \frac{3}{2}k/\mu H + (\nu + 3\alpha)\, \partial S/\partial \ln L} \end{array} \right\} \tag{9}$$

For $\alpha = 1$, $\nu = 16$ we have $C \approx 0.05$, $D \approx 1.7$. If the nuclear reactions were not affected by the mass loss we would have had $C \approx 0.6$, $D \approx 0.14$. In any case C and D are positive. When the mass is being lost we have $\Delta \mathfrak{M} < 0$, and therefore ΔS is small and negative. From Equation (6) and (8) we have

$$\frac{\Delta \log R}{\Delta \log \mathfrak{M}} = -\frac{1}{3}\left(1 - \frac{2 D \Delta t}{C \cdot t_{KH} + \Delta t}\right). \tag{10}$$

For small Δt we obtain the increase of the stellar radius in the case of mass loss. For large Δt (large compared with the Kelvin-Helmholtz time-scale) we may have a certain decrease of the radius if $D > 0.5$. In any case there is only a narrow range of possible values of $\Delta \log R / \Delta \log \mathfrak{M}$.

Similar results were obtained by means of numerical computations for stellar models of 1 $\mathfrak{M}_\odot$. The model of a single star of that mass was evolved through the exhaustion of hydrogen in the core up to the phase when the convective envelope covered 70% of the mass and 80% of the radius of the star. Then 0.001 $\mathfrak{M}_\odot$ was removed on a time-scale of 10, 33, and 1000 years. For the $\Delta \log R / \Delta \log \mathfrak{M}$ we obtained the values of -0.04 in the first two cases, and $+0.04$ in the last one.

Let us suppose now that the more massive component of a close binary has a deep convective envelope and approaches its Roche lobe. As soon as the surface of the star will exceed the Roche lobe, the mass outflow towards the mate star will begin near the vicinity of the inner Lagrangian point L_1. The radius of the Roche lobe will change according to the approximate formula

$$\frac{d \log \varrho}{d \log \mathfrak{M}} = 2q - 2 + \frac{0.086}{0.38 + 0.2 \log q}(q + 1), \tag{11}$$

where ϱ is the radius of the Roche lobe, and q is the mass ratio. $q > 1$ when the star considered is more massive. This formula was obtained under the assumption that the total mass and the total angular momentum of the binary are conserved. For $q > 1$ we have $\Delta \log \varrho / \Delta \log \mathfrak{M} > 0$. This means that the radius of the Roche lobe may decrease rather rapidly when our star loses its mass. At the same time the stellar radius will not decrease appreciably, if at all, according to Equation (10). The star will overflow its Roche lobe. We shall have therefore a possibility of a very violent mass exchange going on on a dynamical time-scale.

The only binaries known to have the more massive components with low surface temperatures filling up their Roche lobes are the U Geminorum (SS Cygni) stars. These components may have fairly deep convective envelopes. We suggest, following PACZYŃSKI (1965) that the outbursts observed in those stars might be related to the highly violent mass exchange expected in such systems. Perhaps during each outburst the whole convective envelope changes its structure as a result of the rapid mass outflow and becomes radiative. This stabilizes the rate of the mass loss. After each outburst the envelope might slowly return to the convective equilibrium. These are of course only speculations. We intend to study this problem carefully in the future.

We would like to point out the assumptions that we consider to be particularly uncertain. It is not clear, whether one may use the same surface boundary condition (Equation (1)) in a mass-losing star, as in a star with no mass outflow from the surface. Secondly, the star must overgrow a little the Roche lobe in order to lose any mass at all. That means that the very beginning of the mass exchange will always be very gentle. Perhaps this will leave the star enough time to change its structure so much as to avoid a violent mass loss on a dynamical time-scale. Finally, if the mass outflow were very rapid, one should have taken into account the fact that the heat transport by the convection is not infinitely fast. Therefore the envelope would not have been strictly adiabatic. It seems likely that the most rapid mass transfer might eventually take place on a time-scale characteristic for the convective motions in the envelope rather than on a dynamical time-scale of the star as a whole.

References

BAKER, N. H. and TEMESVÁRY, S.: 1967, *Tables of Convective Stellar Envelope Models*, N.A.S.A., Goddard Space Flight Center, New York.
KIPPENHAHN, R. and WEIGERT, A.: 1967, *Z. Astrophys.* **65**, 251.
MORTON, D. C.: 1960, *Astrophys. J.* **132**, 146.
PACZYŃSKI, B.: 1965, *Acta Astron.* **15**, 89.
PACZYŃSKI, B.: 1967, *Acta Astron.* **17**, 193.
PLAVEC, M., KŘIŽ, S., HARMANEC, P., and HORN, J.: 1968, *Bull. Astron. Inst. Czech.* **19**, 24.

Discussion

Bath: I have found confirmations of instabilities to mass loss on a dynamical time-scale considered in this paper, by using non-linear adiabatic perturbation theory of models in hydrostatic equilibrium, similar to that used in pulsation theory. The unstable expansion appears to be associated with the depression of the ratio of specific heats in the ionization regions and continues down to the I or the II region, depending on the mass and the position of the initial model in the $\log L/\log T_e$ plane. The consequent mass loss varies between 10^{-4} and 10^{-7} $M_\odot$ and the variations in bolometric magnitude are now being investigated.

Lauterborn: The increase of radius in the formula $\Delta \ln R/\Delta \ln M = +0.04$ for a time-scale of 10^3 years probably is a result of the fact that you already have thermal equilibrium for such a long time. It is well known then in Göttingen by a work of Endler (Diplomarbeit) that in this case the radius increases with decreasing mass. Concerning the possibility that U Gem stars can be explained by mass loss from stars with convective envelopes on a dynamical time-scale I'd like to remark that this question is also discussed at Göttingen. It seems, however, that to get the final results one should carry out hydrodynamical investigations about the mass flow in binary systems. To do this is a very difficult thing.

MASS EXCHANGE IN CLOSE BINARIES
OF MODERATE MASS AND SHORT PERIODS

M. PLAVEC and J. HORN

Astronomical Institute, Ondřejov, Czechoslovakia

Abstract. The process of mass exchange in binary stars of short period is studied on a family of model binaries with primaries of 5 solar masses. Results obtained by means of stationary models are compared with actual computations of non-stationary model sequences. Changes in mass ratio, luminosity, period, masses, etc., are studied.

1. Introduction

The exchange or loss of mass, occurring in close binaries when the more massive component reaches the Roche limit, is likely to play a dominant role in the evolution of almost all spectroscopic binaries. In investigating this process, our Ondřejov group (Harmanec, Horn, Kříž, and Plavec) have concentrated on binaries of moderate mass and short period.

If the period is between 0.6 and 1.2 days and the primary has a mass of 5 $M_\odot$, then the star is bound to reach its Roche limit when it is still a main-sequence star converting hydrogen into helium in its convective core. Such situation was called case 'A' by KIPPENHAHN and WEIGERT (1967).

At first sight this case might seem rather unimportant. A statistics of 354 spectroscopic binaries contained in the recent catalogue by BATTEN (1968) and brighter than 6.5^m shows that most of them have periods of several days or tens of days, and therefore enter under case 'B' or even 'C'. In only about 10% spectroscopic binaries case 'A' appears relevant.

However, Figure 1 shows also that owing to strong selection effects, the distribution of periods is different for the sample of 162 eclipsing binaries from the Finding List by KOCH *et al.* (1963) brighter than 8.5^m at maximum light and situated North of $\delta = -35°$. Here for about 30% systems case 'A' appears appropriate.

It is known from the work by KIPPENHAHN and WEIGERT (1967), PACZYŃSKI (1966) and our group (PLAVEC, 1968) that mass exchange in case 'A' produces semi-detached binaries similar to Algol. Our main task at the present time is to compare quantitatively our models with the observed systems and to decide whether the assumption about conservation of mass and angular momentum is correct or whether some material must be expected to escape from the system.

The work is nearly but not quite completed. An extensive network of models has been elaborated covering the range of masses of the primary components between 3 $M_\odot$ and 9 $M_\odot$, with all admissible mass ratios and four different values of the parameter representing the degree of chemical inhomogeneity of the primary (or its age). This range of primary masses covers about 15 observed semi-detached eclipsing

M. Hack (ed.), Mass Loss from Stars. All rights reserved

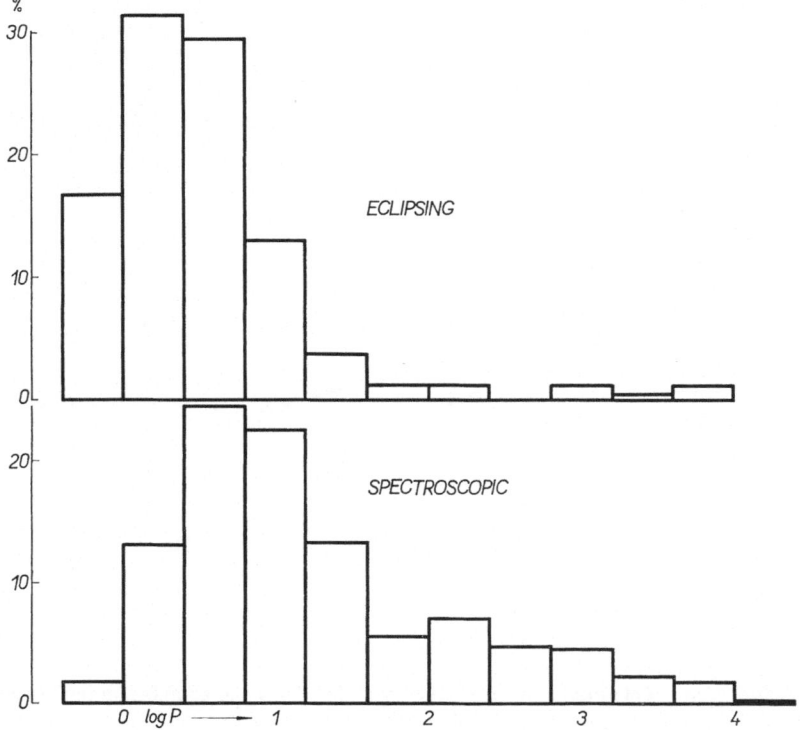

Fig. 1.   Distribution of periods of eclipsing and spectroscopic binaries.

binaries with known elements; these elements are being rediscussed and will be used
for comparison.

It would be impossible to review the whole work here. Instead, one free parameter
will be fixed and the results obtained and problems encountered will be demonstrated
on the family of model binaries having the same mass as the primary component,
namely $5 M_\odot$.

2. The Stationary Models

One method of computing the results of the process of mass exchange can be called
the method of *stationary models* and can be explained by means of Figure 2. The
main phase of mass exchange (the *rapid phase*) is relatively so short that the change
in chemical composition due to nuclear reactions can as a rule be neglected. Thus the
function $X(m)$ describing the variation of hydrogen abundance with absolute mass
inside the primary component is the same as at the beginning of the process of mass
loss. It can be uniquely determined by means of one parameter, for example the
central hydrogen abundance X_c.

Consider now, for example, a star with $X_c = 0.25$ in a binary with mass ratio
$q = 0.6$. Suppose that at b it has reached the Roche limit, and begins to lose material

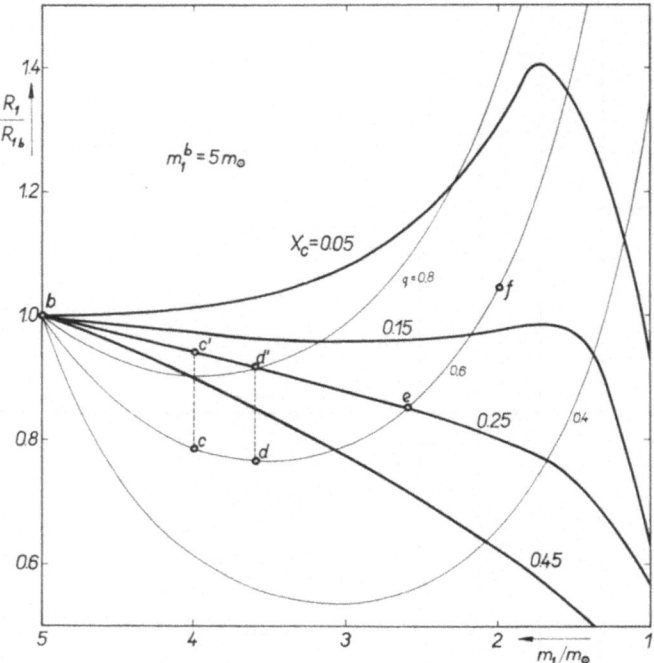

Fig. 2. Actual radii as determined by the Roche limit (thin lines) and equilibrium radii (heavy lines) for stellar models with decreasing mass.

from the outer layers. If it were free to adjust itself to the new condition of smaller mass and restore its thermal equilibrium, its radius would lie on the line $bc'd'e$. However, the star is not free to choose its radius, since its upper limit is determined by the Roche limit. As the mass exchange proceeds, the radius of the Roche limit varies along the curve $bcde$. The star, being forced to adjust its constitution to this radius, deviates strongly from thermal equilibrium and does not return to it until at point e. Here the Roche radius becomes again equal to the equilibrium radius and the rapid mass loss ends.

Thus the final mass of the primary component is determined by the intersection of the two curves in Figure 2, and other quantities characterizing the star at this stage are given by the appropriate model in thermal equilibrium (stationary model). Figure 2 indicates how much all these properties depend on the initial conditions. If we keep the initial mass of the primary m_{1b} fixed, we have two independent initial parameters, for which we can take X_c and q_b (initial mass ratio, $q_b = m_{2b}/m_{1b}$).

The results of computations by means of the stationary models will be shown in a few diagrams. Note that all model sequences have a certain lower limit in q_b, usually 0.4. This is because with lower initial mass ratio the distance between components diminishes so much that the Roche limit shrinks onto the surface of the secondary and this star becomes unstable, too. Our program does not permit to calculate mass exchange in a contact system.

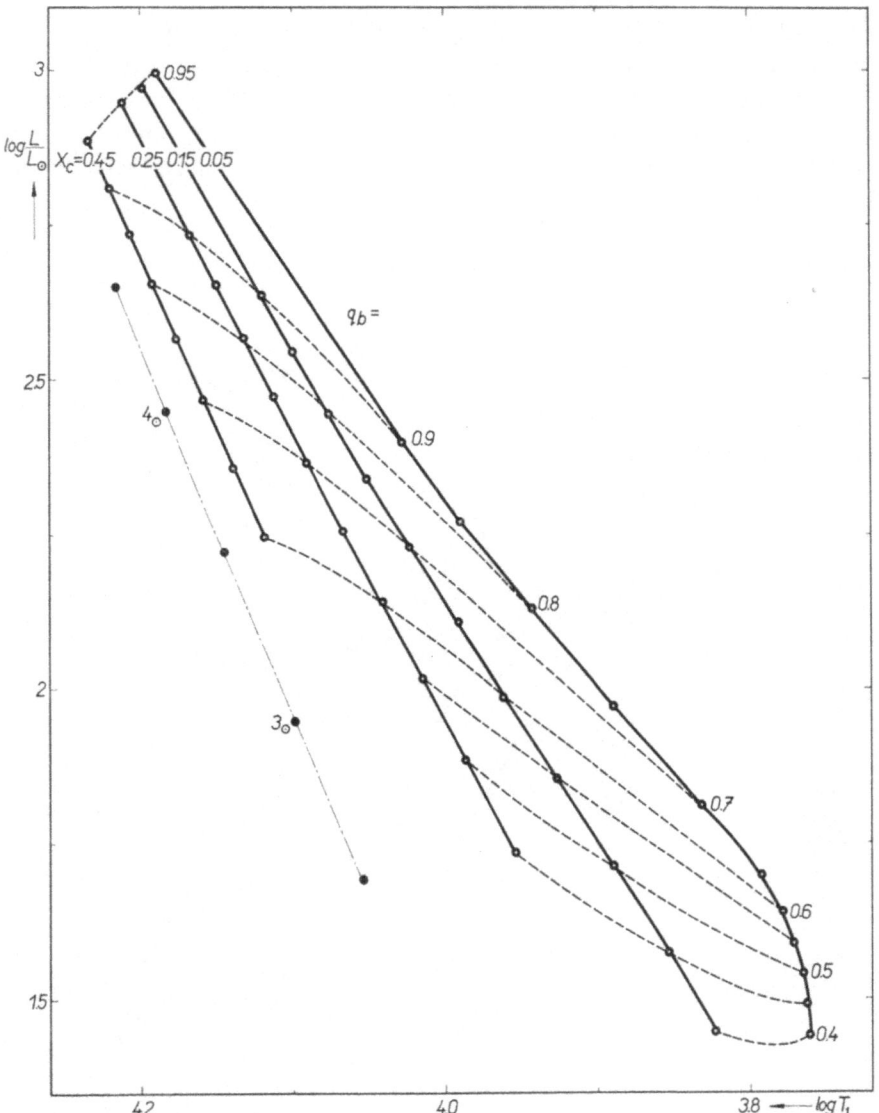

Fig. 3.   Theoretical H-R diagram for stationary models representing the end stage of rapid mass loss.

Figure 3 is a theoretical H-R diagram. It is seen that the mass exchange transforms the original primaries into stars situated above the Main Sequence, or at least above the Zero Age Main Sequence represented here by the homogeneous models. The loci of models with the same X_c run roughly parallel with the Z.A.M.S., so that the transformed primaries preserve the luminosity class of the parental star. This can be seen better in Figure 4, where we have absolute visual magnitudes and the empirical MK luminosity classes. Genuine subgiants are generated with X_c near 0.25, while higher degree of depletion of hydrogen leads rather to giants of luminosity class III.

The smaller was the initial mass ratio q_b, the later is the spectral type of the primary after mass exchange. At the same time, these stars are greatly overluminous for their masses. This is evident from Figure 5, where we plot the difference, in magnitudes, between the calculated bolometric magnitude of the model, M_1, and the 'normal' bolometric magnitude following for the same mass from the standard mass-luminosity relation, $M_s(m_1)$, in the sense

$$\Delta m_1 = M_1 - M_s(m_1). \tag{1}$$

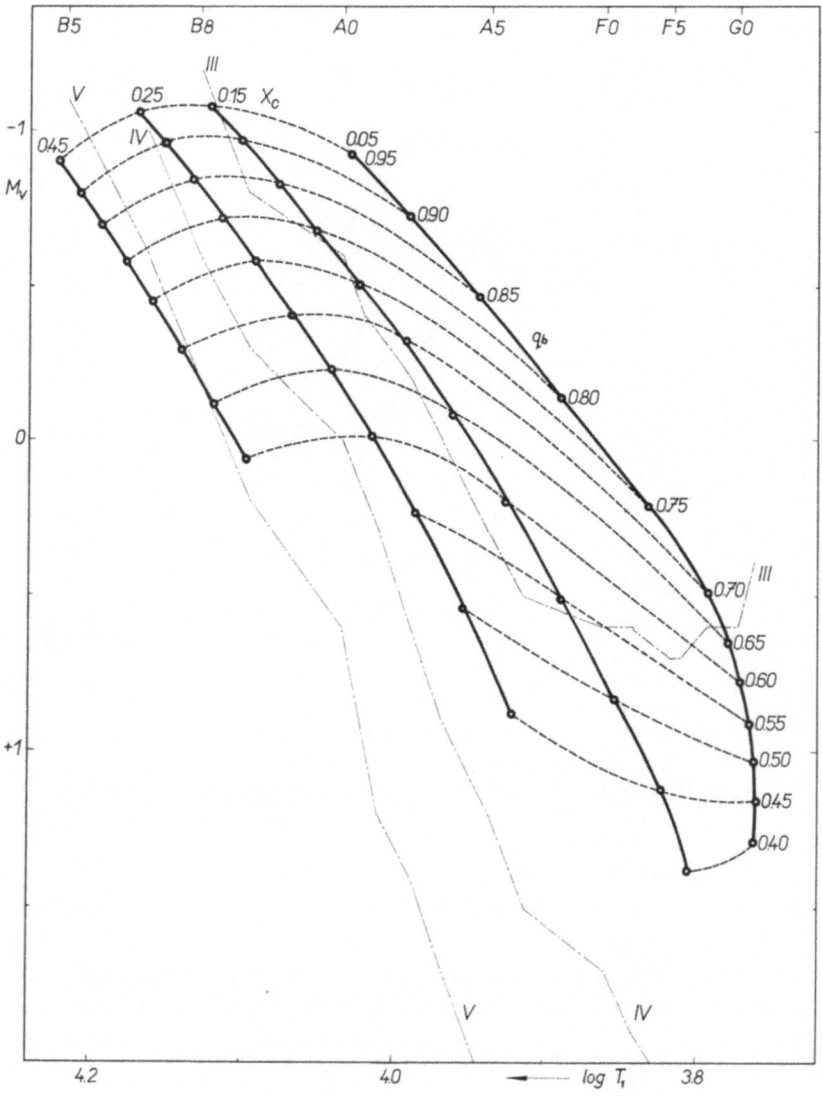

Fig. 4. H-R diagram for same models as in Figure 3.

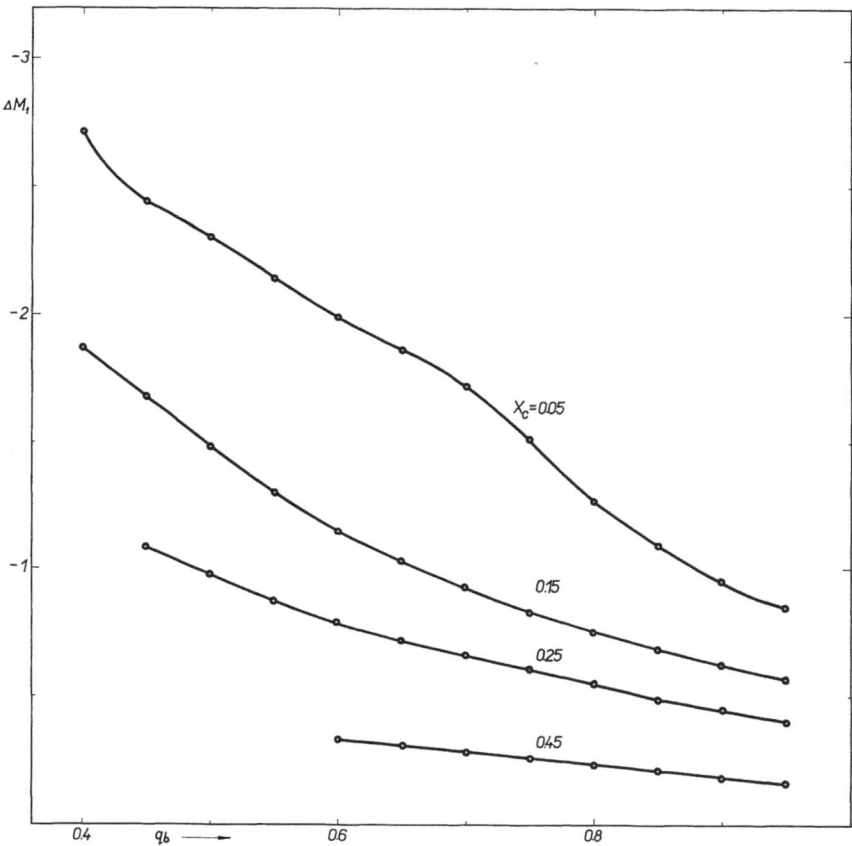

Fig. 5.   Degree of overluminosity for same models as in Figure 3.

The transformed primaries can be overluminous by more than 2^m provided they are generated in systems where the primary is relatively old and the secondary is much less massive than the primary.

It is interesting to note that the amount of material exchanged between the two stars is the larger the larger was the initial disparity of masses. In each single case studied by us, the initially more massive star was found to be converted into the less massive one. For $X_c = 0.45$, the masses are just about interchanged, for lower X_c the mass ratio is more than reversed and the disparity in masses is increased. With low X_c and low q_b, the primary may lose more than 70% of its initial mass.

A curious result follows from this property of the process of mass exchange. Figure 6 shows that the final mass of the secondary, m_{2e}, is almost independent of its initial mass, except for $X_c = 0.05$.

In case of $X_c = 0.45$, the binary returns very nearly to its initial period, while in other cases the period is lengthened by the mass exchange. In the extreme case of $X_c = 0.05$, the period may increase from 1.3 to 4.6 days.

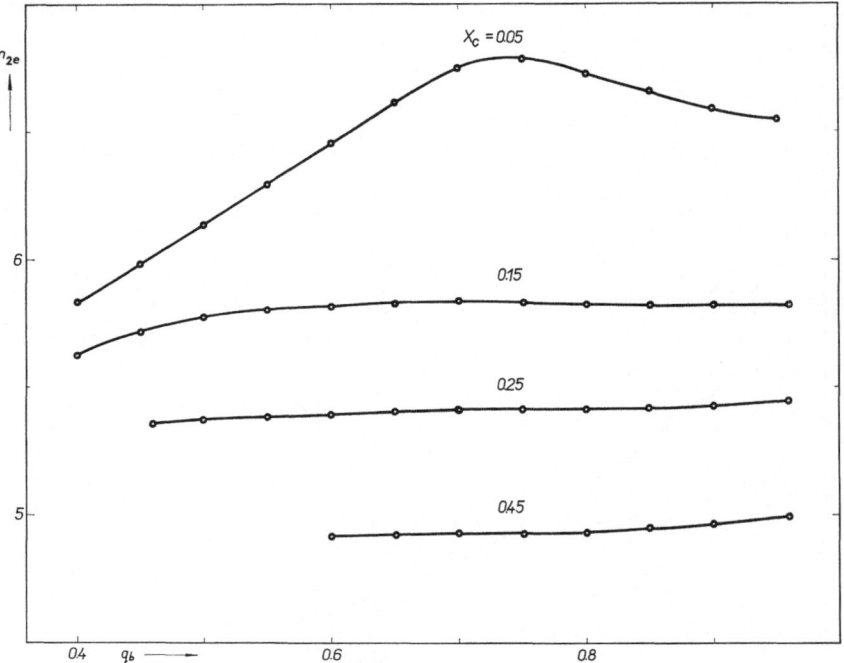

Fig. 6.   Final mass of the secondary component at the end of rapid mass loss.

3. The Non-Stationary Models

The investigation into the process of mass exchange itself must be made by computing evolutionary sequence of models deviating from thermal equilibrium (*non-stationary models*). For this purpose, Henyey's method for integrating the time-dependent equations of stellar structure (HENYEY *et al.*, 1964) was adapted by our Ondřejov group (PLAVEC *et al.*, 1968; KŘÍŽ, 1968).

Figure 7 is a typical example of the theoretical H-R diagram for the mass-losing component ($m_{1b} = 5\ M_\odot$, $q_b = 0.6$, $X_{cb} = 0.25$). At the beginning, the diminishing mass, radius as well as absorption of up to 95% of the luminous flux in the expanding outer layers cause a rapid decrease in luminosity and effective temperature. When the radius begins to increase again (at point d) and the mass loss is slowed down, the star partly restores its luminosity. In six cases computed by us ($X_c = 0.45$, 0.25 and 0.15, respectively, $q_b = 0.6$ and 0.8, respectively), the rapid phase lasted between 5 and 10×10^5 years, the average rate of mass loss was of the order of $10^{-6}\ M_\odot$/year, and the maximum rate was $2 \times 10^{-5}\ M_\odot$/year.

The star returns to thermal equilibrium at e; the parameters for this stage derived from non-stationary model calculations agree very well with the stationary models. Even for the case $X_{cb} = 0.05$ the differences are not great, although the assumption of a negligible change in chemical composition is not fulfilled.

Further evolution proceeds again on the nuclear time-scale, but the star adheres

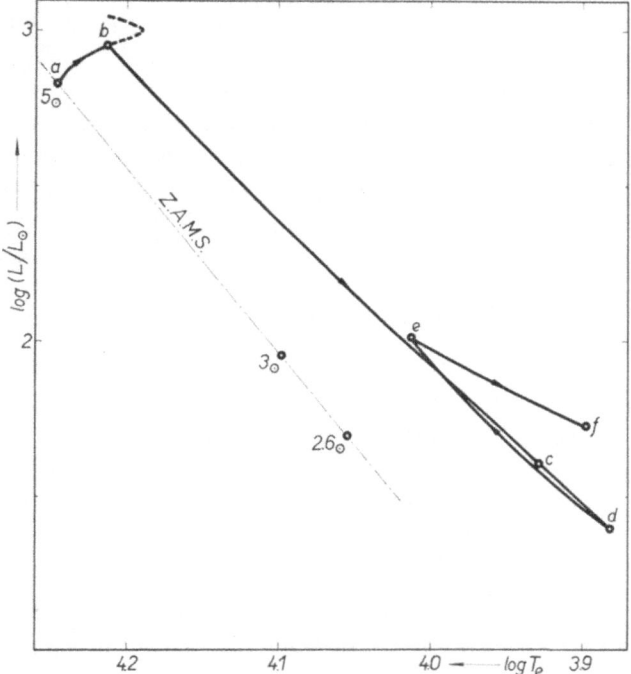

Fig. 7. Typical H-R diagram for non-stationary models representing the process of mass loss
($X_{cb} = 0.25$, $q_b = 0.6$).

to the Roche limit and a *slow mass loss* goes on; the rate is of the order of 10^{-8} $M_\odot$/year. This is the phase at which the semi-detached binaries are observed. In some cases this phase is quite long and does not end until hydrogen is very nearly exhausted in the core of the mass-losing star, an overall contraction sets in and the star detaches itself from the Roche limit. This appears to be the case for higher initial mass ratios, represented here by $q_b = 0.8$. But the effect of X_{cb} introduces differences. For $X_{cb} = 0.05$ the *Algol phase* lasts but 3.7×10^6 years and the system changes but negligibly during that period. With higher X_{cb}, the Algol phases are longer. For $X_{cb} = 0.25$ the duration is 36×10^6 years and the mass loss amounts to 1 $M_\odot$. As a consequence, the period increases from 1.2 to 2.0 days, and the primary changes from B9 to A3 and its magnitude decreases by 1^m. For $X_{cb} = 0.45$ the Algol phase would last 85×10^6 years, but in fact it ends already after about 19×10^6 years since by that time the secondary fills its critical Roche lobe, too, and becomes unstable.

Similarly, for $X_{cb} = 0.25$, $q_b = 0.6$ the phase of slow mass loss of the primary would last 66×10^6 years, but in fact the Algol phase is terminated already about after 46×10^6 years. The tendency toward formation of a contact binary owing to a faster evolution of the secondary is more pronounced for lower q_b, since the disparity of masses after the rapid phase is greater.

Nevertheless, two unusual results were obtained for $q = 0.6$. With $X_{cb} = 0.45$, the primary contracts at the end of the rapid phase and mass loss ceases. The decrease of radius lasts 0.55×10^6 years (segment e–e' on the track in Figure 8) and is followed by a normal slow nuclear evolution (e'–e'' in Figure 8), lasting 18.8×10^6 years. At the end of this phase the star reaches again its Roche limit and a phase of slow mass loss would begin, but at about the same time the secondary becomes unstable, too.

For $X_{cb} = 0.05$ and $q_b = 0.6$, the rapid phase as defined by stationary models is unusually long, 5.4×10^6 years, but in fact for most of this period it has the character of a slow phase (Figure 9). This is because with this very low hydrogen abundance the change in chemical composition due to nuclear processes is no longer negligible.

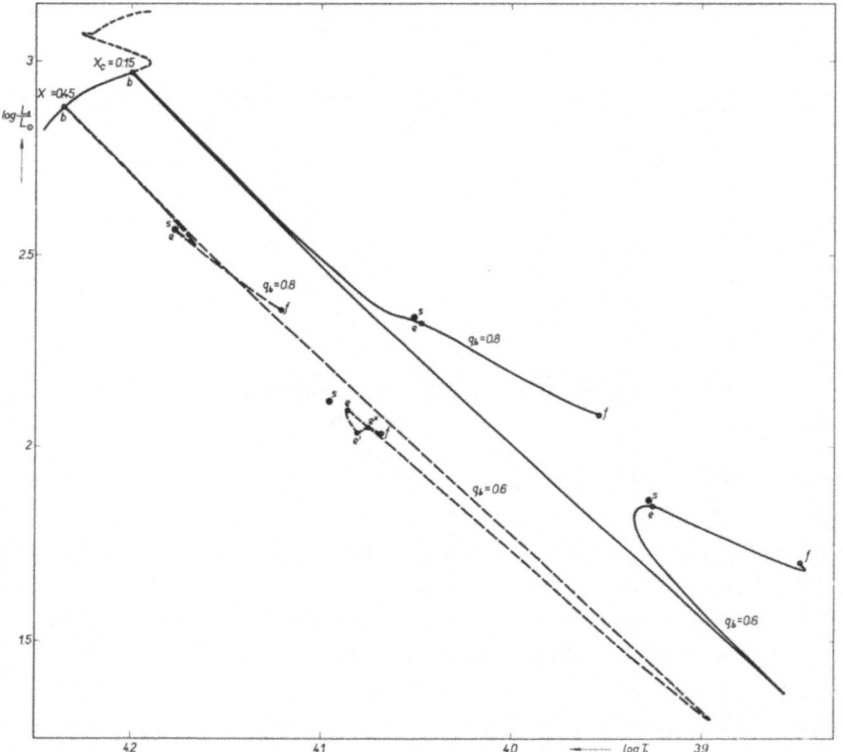

Fig. 8.   Theoretical H-R diagram for the process of mass loss with $X_{cb} = 0.45$ and 0.15.

At e', $X_c = 0.032$. The regular slow phase that follows represents actually a transition into case 'B' of mass loss, since the convective core disappears after another 9.6×10^6 years. The computation was discontinued at an age 27×10^6 years after the beginning of the mass exchange; by that time the mass loss was going on and it was in fact accelerating. The mass was $1.0\ M_\odot$ and the hydrogen-deficient intermediate zone already reached right to the surface.

It is just this great variety in the character, duration and importance of the Algol phase that makes the comparison with observed systems very difficult, since many time-consuming sequences of non-stationary models must be computed.

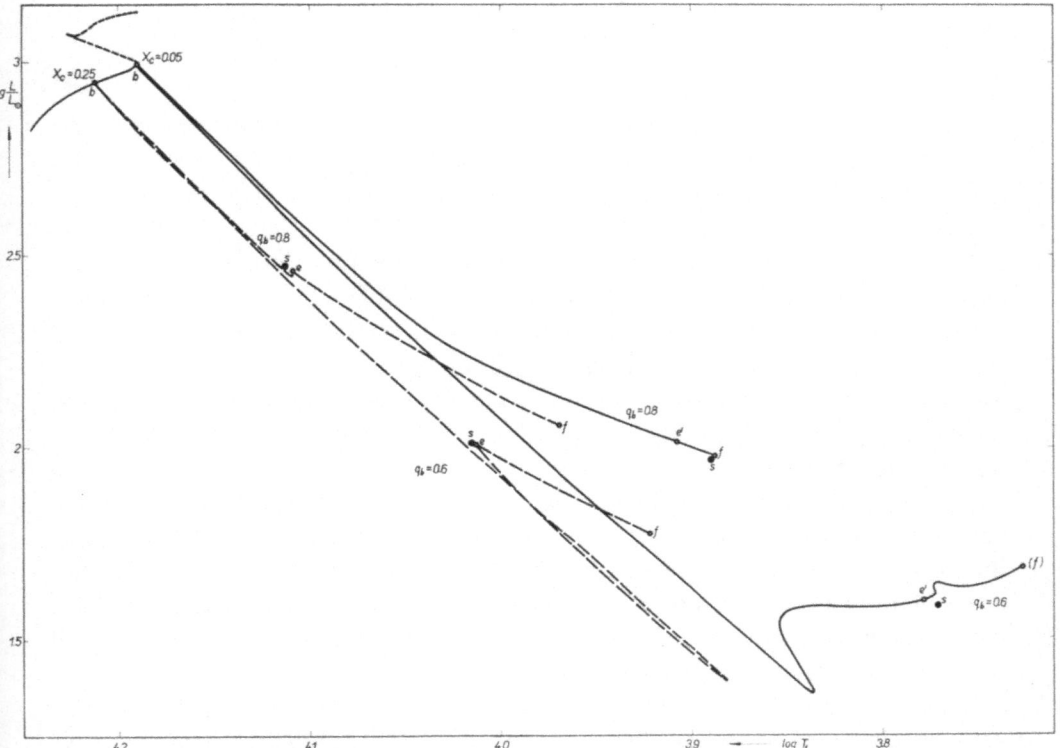

Fig. 9. Theoretical H-R diagram for the process of mass loss with $X_{cb} = 0.25$ and 0.05.

All the known semi-detached Algol-like eclipsing binaries are probably in the phase of slow mass loss; this conclusion is supported by the well-known phenomena of gaseous streams and period fluctuations. These phenomena must be much more conspicuous in systems undergoing the rapid mass exchange. Several short-period binaries (like GO Cyg, DM Del, DO Cas) might be on the verge of this phase. Among them is also the shell star o And. Recent discovery of the duplicity of 17 Lep leads us to suspect that the phenomenon of the shell stars (and of some other peculiar objects) may be connected with possible rotational instability of the mass-gaining component at the phase of rapid mass exchange.

References

BATTEN, A. H.: 1968, *Publ. Dom. Astrophys. Obs., Victoria* **13**, 119.

HENYEY, L. G., FORBES, J. E. and GOULD, N. L.: 1964, *Astrophys. J.* **143**, 483.

KIPPENHAHN, R. and WEIGERT, A.: 1967, *Z. Astrophys.* **65**, 251.

KOCH, R. H., SOBIESKI, S., and WOOD, F. B.: 1963, *Publ. Univ. Pa. Astron. Ser.* **9**.

KŘIŽ, S.: 1968, *Bull. Astron. Inst. Czech.* **19**, 248.

PACZYŃSKI, B.: 1966, *Acta Astron.* **16**, 231.

PLAVEC, M.: 1968, in *Highlights of Astronomy* (ed. by L. Perek), D. Reidel Publ. Co., Dordrecht, p. 396.

PLAVEC, M., KŘíž, S., HARMANEC, P., and HORN, J.: 1968, *Bull. Astron. Inst. Czech.* **19**, 24.

Discussion

Wood: Would you like to say anything at all about β Lyrae?

Plavec: I think that β Lyrae and V 367 Cygni are two systems observed at the phase of rapid mass exchange. Their period indicate that they enter under case 'B'. A little more about it is said in my recent paper (*Advan. Astron. Astrophys.* **6**).

Herczeg: There may be a sort of 'statistical difficulty' raised against this interpretation of Algol stars based essentially on the evolution during the hydrogen-burning phase (case 'A'). The change of the radii is, in the case of both components, rather limited and this means that the progenitors, stars in the previous phase of evolution must be also eclipsing binaries, with the ratio of radii not differing *fundamentally* from that of the Algol binaries. Thus the chances of discovery should be roughly comparable. My question is: why don't we observe a considerable number of these (detached) progenitors?

Plavec: I am working on a statistical discussion of this problem, and I am optimistic about the result. The Algol systems are overabundant, say, 2.5–3 times. But do not forget that for systems with dissimilar masses the mass exchange creates much more favorable conditions for observation; on the main sequence, we have very shallow eclipses. It should not be forgotten, however, that in systems of small mass, case 'B', too, produces semi-detached binaries.

Sahade: I agree with Dr. Plavec in that β Lyrae is undergoing rapid mass loss, now I would like to know how do you explain the system within the scheme you have described?

Plavec: I do not think that such a complicated system like β Lyrae could be interpreted in any detail by the present – still crude – theory of mass exchange. My opinion is based only on a very rough consideration of the problem as described in my article in the *Advan. Astron. Astrophys.* **6** (1968).

EVOLUTION THROUGH MASS EXCHANGE IN
CLOSE BINARY SYSTEMS OF TOTAL MASS 2.5 $M_\odot$

S. REFSDAL* and A. WEIGERT**

Behlen Laboratory of Physics, University of Nebraska, Lincoln, Neb., U.S.A.

Abstract. Calculations show that the observed semi-detached binary systems of mass $M_1 + M_2 \approx 2.5$ $M_\odot$ can be explained in terms of a mass exchange which starts after the central hydrogen in the original primary has been exhausted.

It was shown earlier (KIPPENHAHN and WEIGERT, 1967; PACZYŃSKI, 1966, 1967) that a mass exchange of 'case A' in a close binary system can give rise to the typical features observed in semi-detached systems; the now more massive star is on the main sequence while the now less massive component is above the main sequence and has higher luminosity than a main-sequence star of the same mass. ('Case A' means that the mass exchange starts during central hydrogen burning of the original primary; if it starts after central hydrogen burning, we speak of 'case B'.) Calculations of such mass exchange are available only for rather large total masses. When extrapolating the results to smaller total masses $(M_1 + M_2 \approx 2.5\,M_\odot)$, one has difficulties to explain the corresponding observed semi-detached systems (GIANNONE *et al.*, 1968). In this mass range, the observed systems show very small mass ratios and the secondaries have extremely high over-luminosities (compared with main-sequence stars of the same mass).

To investigate how one can explain these observed features, the evolution through

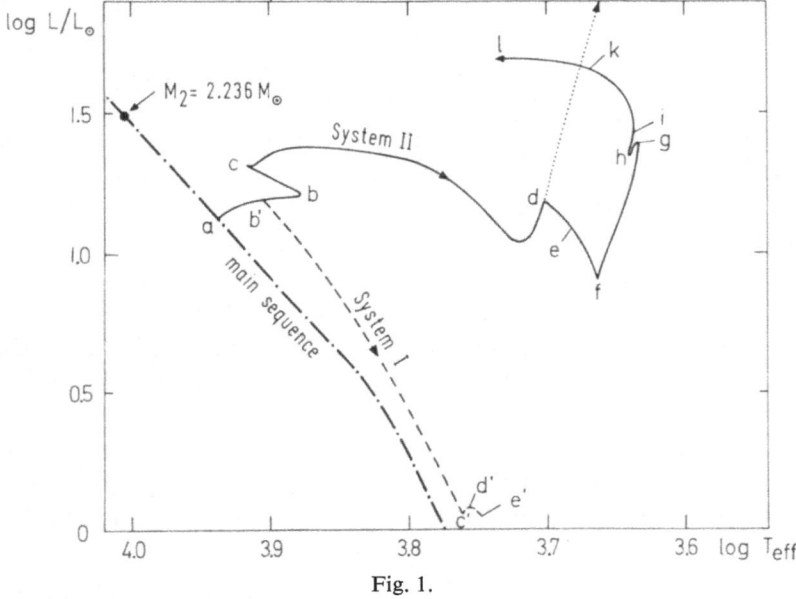

Fig. 1.

* On leave of absence from the University of Oslo.
** On leave of absence from the University of Göttingen.

M. Hack (ed.), Mass Loss from Stars. All rights reserved

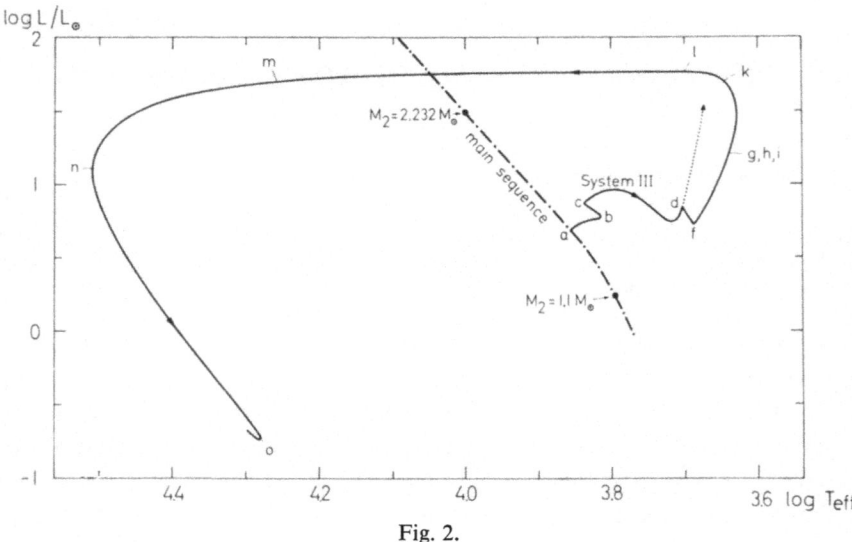

Fig. 2.

mass exchange was calculated for three systems of total mass $M_1 + M_2 = 2.5\,M_\odot$. The systems started as pairs of main-sequence stars of different mass ratio μ and of different separation A of the components. The evolutionary tracks of the original primaries are given in Figures 1 and 2.

System I started with $M_1 = 1.8\,M_\odot$, $M_2 = 0.7\,M_\odot$, and $A = 4.35\,R_\odot$. Mass loss from star M_1 starts during central hydrogen burning (point b' in Figure 1), i.e. the system follows case A. In a rapid mass loss ($b'-c'$), the star transfers about $0.7\,M_\odot$ to its companion. In the following slow phase of further mass exchange ($c'-e'$), roughly $0.2\,M_\odot$ are transferred to the now more massive star. In this slow phase, the system is semi-detached (sd) for a long time; however, the star M_1 has then rather small over-luminosities δM_{bol} and the mass ratio μ is not far from 1. These calculated quantities are not sufficient to account for the values observed in such sd-systems (Figure 3). Therefore, the observed sd-systems of $M_1 + M_2 \approx 2.5\,M_\odot$ cannot have been created by a mass exchange of case A.

Systems II and III both follow case B. They start with main-sequence stars of $M_1 = 1.8\,M_\odot$, $M_2 = 0.7\,M_\odot$, $A = 11\,R_\odot$ (System II) and $M_1 = 1.4\,M_\odot$, $M_2 = 1.1\,M_\odot$, $A = 8.6\,R_\odot$ (System III). Central hydrogen burning is finished in the original primary at point c (Figures 1 and 2). This star fills its critical Roche volume for the first time at point d, when it moves upwards in the HR diagram near the Hayashi line. A very rapid phase of mass exchange follows ($d-f$), in which the mass ratio is more than inversed. At point f, the stars have $M_1 = 0.51\,M_\odot$ (System II) and $M_1 = 0.93\,M_\odot$ (System III). In the ensuing phase ($f-k$), the system is still semi-detached since further mass of the original primary (now secondary) is slowly shifted over its critical Roche lobe. This slow phase of mass exchange lasts 10^8 years and 3×10^8 years in the two systems considered. This time interval appears to be long enough to give a reasonable probability for observing such systems. For this phase, the calculated over-luminosities δM_{bol} (compared with a main-sequence star of the same mass) are plotted over

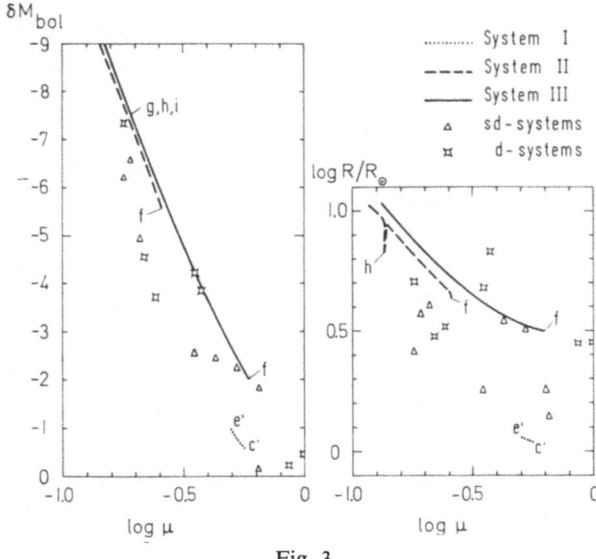

Fig. 3.

the mass ratio μ in Figure 3. The observed semi-detached systems of $M_1 + M_2 \approx 2.5\,M_\odot$ are indicated by open triangles. The right part of Figure 3 gives the radii R of the same stars. The calculated systems of case B agree sufficiently with the observed systems in the values of δM_{bol} and μ, and they have large enough radii. We may conclude, that the observed sd-systems have originated from main-sequence pairs and are now in a slow phase of mass exchange following case B.

At the end of mass exchange, the now secondary stars of the Systems II and III have masses of $M_1 = 0.265 M_\odot$ and $0.268\,M_\odot$, respectively. Since nearly their whole hydrogen-rich envelopes are stripped off, the stars consist mainly of a very condensed helium core containing about 96% of the star's mass. These stars move then in about 10^7 years over to the left of the main sequence (to point n in Figure 2) and become white dwarfs (point o in Figure 2).

During the slow phase of mass exchange (from f to k in Figures 1 and 2), the star's radius depends mainly on the mass contained in its degenerate helium core. The mass of this core grows, since the surrounding hydrogen-shell source burns further outwards, and thus the radius of the star increases. This increase is nearly independent on the mass in the outer envelope. The mass loss normally stops when the mass left in the hydrogen-rich envelope drops below 4% of the mass of the star. Therefore, the maximum radius of the star attained at the end of mass loss is dependent only on the final mass M_{1f} of this star. These facts make it possible to derive predictions how the final system (after mass loss) will change if one changes the initial system. For instance, if only the initial separation A_0 and the total mass M_t are changed, then the final mass of the original primary varies according to

$$M_{1f} \sim A_0^{0.15} M_t^{0.25}.$$

A special situation must show up, when during the slow phase of mass loss the helium core increases to $0.45 M_\odot$. Then central helium burning will start and the radius of the star will shrink. The system thus will become detached and the mass exchange stops. There are detached systems observed which have an evolved, over-luminous secondary. These systems are also plotted in Figure 3 (open stars). One might consider their secondaries to be such stars with hydrogen-burning shell sources and with central helium burning which has started during mass exchange. Then, however, their periods would have to be much larger than the observed periods. There seems to be another possibility for explaining at least some of the observed d_{es}-systems (i.e. detached systems with evolved secondaries). In the evolutionary calculations reported here, the semi-detached phase of slow mass transfer was interrupted for some time. In this intermediate phase, the now secondary star contracts and is detached from its critical Roche lobe. This can be most clearly seen from the evolutionary track of System II in Figure 1 (from point g to point i). Such an intermediate contraction seems to be quite common in corresponding phases of evolution. It is due to a shell source moving through a chemical discontinuity which has been left at the bottom of an outer convective zone at its deepest extension.

References

GIANNONE, P., KOHL, K. and WEIGERT, A.: 1968, *Z. Astrophys.* **68**, 107.
KIPPENHAHN, R. and WEIGERT, A.: 1967, *Z. Astrophys.* **65**, 251.
PACZYŃSKI, B.: 1966, 1967, *Acta Astron.* **16**, 231; **17**, 1.

Discussion

Underhill: You showed a diagram of $\log P$ versus $\log$ mass ratio predicted for systems with total mass $2.5 M_\odot$ and you plotted observed points. The scatter of points around this line was on the average, equivalent to an error of 0.3 in the log. Since the period of eclipsing binaries are well known, the 'error' must be attributed to the observed mass ratio. I cannot concord with you that the material which is presented is a good confirmation of the theory. I presume all the observed systems have a total mass of $2.5 M_\odot$. As I recall the diagram, a systematic correction was indicated to the observed mass ratio in the direction of increasing the observed mass ratio. Confirmation that the total mass of all the observed system is $2.5 M_\odot$ and that a correction of the observed mass ratio on the systematic direction indicated are reasonable, is needed before one can accept these data as proof of the correctness of the mass exchange theories.

Refsdal: The period of a system depends on the mass ratio and the initial parameters of the system (total mass, initial mass ratio, and initial distance). Scatter in these initial parameters can easily explain the scatter in the period-mass ratio diagram.

Plavec: We should not forget that in fact a grid of models with varied initial conditions would be necessary before we are able to say with certainty that the theory can or cannot explain the observations. In my opinion, Weigert and Refsdal's models do give the correct qualitative explanation of overluminosity of the subgiant secondaries in Algol systems of low mass.

MASS EXCHANGE IN CLOSE BINARIES OF
MODERATE PERIOD AND MASS

S. KŘÍŽ

Astronomical Institute of the Czechoslovak Academy of Sciences, Ondřejov, Czechoslovakia

Abstract. Results of numerical computations are presented for the evolution of a binary consisting of components 13.8 $R_\odot$ apart having the masses 5 $M_\odot$ + 4 $M_\odot$. The mass exchange starts after the exhaustion of the central hydrogen (case 'B'). A possible connection between shell stars and products of mass exchange is indicated.

Till now, the mass exchange in case 'B' was numerically studied only for systems with small masses (2 $M_\odot$ + 1 $M_\odot$) and large masses (9 $M_\odot$ + 3.13 $M_\odot$). But there are many systems of moderate mass, such as σ Aql, U Oph, AG Per, ζ Phe.

In order to study the mass transfer in such binaries, the evolution of the primary component of 5 $M_\odot$ was computed for a binary in which the initial mass of the secondary was chosen to be 4 $M_\odot$, the distance 13.8 $R_\odot$ and the initial chemical composition $X = 0.602$, $Y = 0.354$, $X_{CN} = 0.015$. The corresponding initial orbital period is 1.98 days. The method of computation was published by Kříž (1968).

Within 52×10^6 years after the zero-age main sequence, the convective core of the primary disappears, the nuclear-energy production occurs in a thick shell and a rapid expansion of the envelope begins. At the age of 53.69×10^6 years, the star fills the Roche lobe and the mass exchange starts. At this time the radius of the primary is two times greater than the initial radius of the homogeneous star.

Figure 1 shows the change of the mass of the primary with time and the variation of the rate of mass loss, $-dm_1/dt$. It is clearly visible that the mass exchange consists of a rapid and slow phase, similarly as in case 'A'. In the course of the rapid phase, the rate of the mass loss attains two local maxima. The first maximum is 2×10^{-5} $M_\odot$/year, the second maximum is 1.1×10^{-5} $M_\odot$/year. The locations of both maxima are indicated by points 3 and 5. The second maximum may be connected with the occurrence of the outer convective zone, which exists approximately between points 4 and 6. During 4.2×10^5 years of the rapid mass exchange the mass of the primary is decreased to 0.94 $M_\odot$, therefore the average rate of the mass loss is 0.97×10^{-5} $M_\odot$/year.

Near point 6 the process of the mass loss is slowed down considerably, but goes on for another 6.6×10^5 years. At that phase the average rate of the mass loss is 3.8×10^{-7} $M_\odot$/year only. The mass transfer is terminated at point 7; this corresponds to the ignition of the helium at the star's centre. The final mass of the primary is 0.69 $M_\odot$, the mass of the secondary is 8.31 $M_\odot$. It means that the primary has lost 86% of its initial mass. The distance between components has increased to 167 $R_\odot$ and the orbital period increased to 83.4 days. The hydrogen content at the surface of the primary component decreased from $X = 0.60$ to $X = 0.26$.

M. Hack (ed.), Mass Loss from Stars. All rights reserved

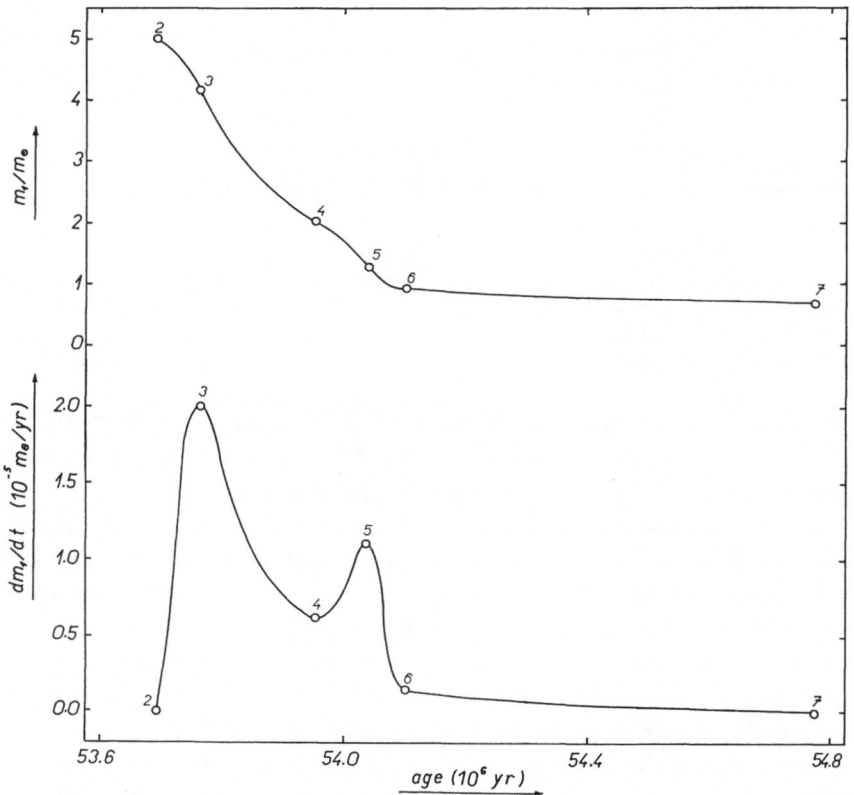

Fig. 1.   Change of mass of the primary and variation of the rate of mass loss; in the latter case, the units on the ordinate are 10^{-5} $M_{\odot}$ per year.

Our computations were terminated when the primary component detached itself from the Roche lobe, began to contract rapidly and was evolving towards the helium main sequence.

The evolutionary path on the H-R diagram is shown in Figure 2. The track of the secondary star before the mass transfer is also indicated. Because it is not clear whether and how the secondary accommodates the great amount of the material carrying with it also a great amount of angular momentum, it was impossible to compute the further evolution of the secondary. Tentatively it can be assumed that the final stage of the secondary component is roughly a one of a main-sequence star with mass of 8.3 $M_{\odot}$. The final remnant of the primary component is likely to be a helium star on the 'helium main sequence'. This fact is in good agreement with the result obtained by KIPPENHAHN and WEIGERT (1967) for a close binary 9 $M_{\odot}$ + 3 $M_{\odot}$. The corresponding final points for both components are indicated in Figure 2, too.

Some parameters of described binary at some points of its evolutionary track are given in Tables I and II. Point 1 means the main-sequence stage of the binary.

The results can be generalized for binaries of moderate masses and periods,

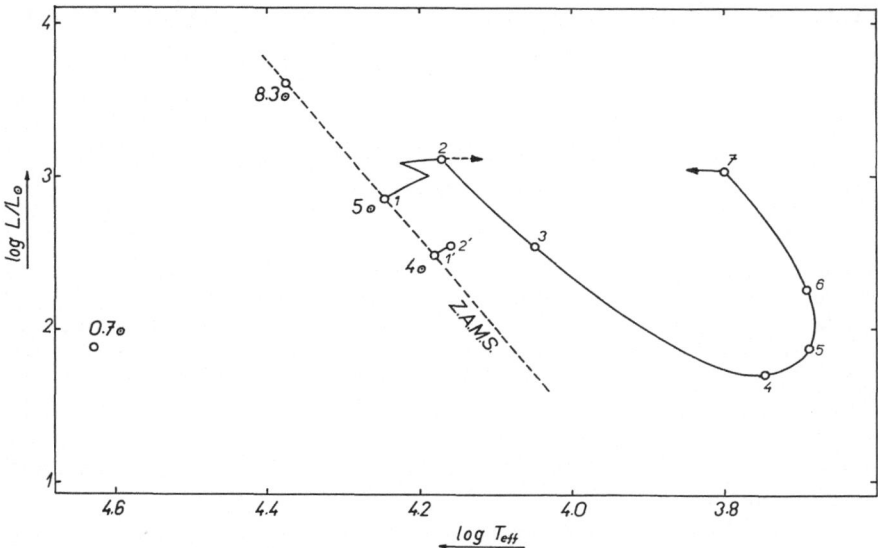

Fig. 2.   The theoretical H-R diagram for the system 5 $M_\odot$ + 4 $M_\odot$.

similar to the studied case, as follows: As soon as the primary fills the Roche lobe, the
large-scale rapid mass exchange begins. That phase is probably caused by the instability against a material outflow of the envelope above the hydrogen-burning shell.
When the mass decreases, the envelope expands and the star's radius tends to be larger
than the Roche limit. Consequently, the material must leave the primary, the radius
increases, and so on. This self-accelerating process is interrupted, when the hydrogen-
rich envelope is removed and the helium-rich layers come near the star's surface.
Further slow mass outflow is caused by the slower expansion of the envelope because
of the changes in the star's interior. However, the duration of this Algol-like phase
is so short that our chances to observe such a system are very small. Slow phase is
terminated when helium is ignited in the core of the star, as already pointed out by
PACZYŃSKI (1966).

TABLE I

Parameters of the system 5 $M_\odot$ + 4 $M_\odot$

Point	Age (10⁶ years)	m_1 ($M_\odot$)	m_2 ($M_\odot$)	A ($R_\odot$)	P (days)
1	0	5.00	4.00	13.8	1.98
2	53.693	5.00	4.00	13.8	1.98
3	53.765	4.14	4.86	13.6	1.94
4	53.955	1.99	7.01	28.2	5.78
5	54.035	1.32	7.68	53.7	15.2
6	54.113	0.94	8.06	96.1	36.4
7	54.772	0.69	8.31	167	83.4

WHITE-DWARF PRODUCTION IN BINARY SYSTEMS OF LARGE SEPARATION

DIETMAR LAUTERBORN

Göttingen, Germany

Abstract. Numerical calculations are carried out for the evolution of a binary system with a primary of 5 $M_\odot$ and a secondary of 2 $M_\odot$, revolving round another in a circular orbit of 300 $R_\odot$. After finishing central helium burning, the primary starts to transfer mass to its companion. After the mass loss, the star of originally 5 $M_\odot$ has become a star of 1 $M_\odot$. This star has a carbon-oxygen core which is a well-developed white dwarf, and a very extended hydrogen shell. The final system has a distance of 815 $R_\odot$; the period is 2.3 years.

If a star of – say – 5 $M_\odot$ is losing mass as member of a close binary system there are clearly three possibilities:

(a) the mass loss may begin during central hydrogen burning (so-called *case A*)

(b) the mass loss may begin during the expansion phase leading to helium burning (so-called *case B*)

(c) the mass loss may begin after the star has finished central helium burning and is evolving along or near the Hayashi track (*case C*).

The three cases A, B, C are well different concerning the order of magnitude of the corresponding radii. This can be seen in Figure 1, which shows the change of radius with time for a star of 5 $M_\odot$ calculated by Hofmeister in Munich in 1965. Look at Figure 1: first there is a slow increase of radius during hydrogen burning (if the mass loss starts during this phase we then have case A; the radii are relatively small); then at time $t \approx 5.2 \times 10^7$ a one has a sudden increase of radius during expansion to helium burning (if the mass loss starts during this phase we then have case B; the

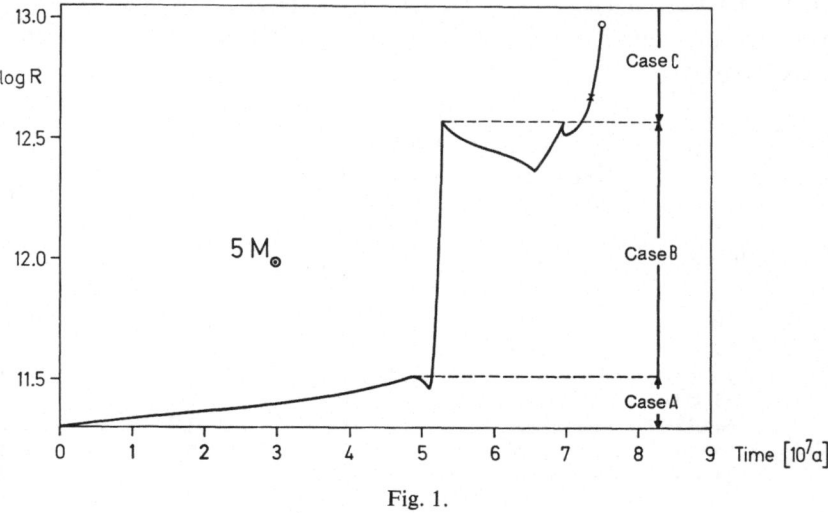

Fig. 1.

glers. A possible suggestion is that they represent the final stage of a sequence like that computed by Kippenhahn and colleagues. Recently I have been able to find that at least one of these stars is a spectroscopic binary; the period is 4^d, the amplitude is small, the mass function low. It is quite possible that this object is composed from an A5 main sequence and a white dwarf. I think, with respect to the other stars among the blue stragglers, I cannot yet say that I know all to be double, but I would not be surprised if most of them result to be double as observations increase. The characteristic of these stars is that they very nearly match the prototype system worked out by Kippenhahn and coll., except that it could not adapt the present configuration in an evolutionary process where angular momentum and mass were conserved. The present separation and masses would prove that there has been some angular momentum and mass loss.

Another aspect of this problem is this: all these arguments suggest that there must be a good many of these processes in the solar neighbourhood. Calculations of how many stars among the A stars must have been produced by mass exchange suggest that you can account for most of the A0 type stars; I don't believe this of course.

Plavec: Well, I am extremely pleased to hear from Dr. Deutsch that he finds out one or two of the blue stragglers are binaries, because I believe that mass exchange is the easiest way to get these stars. Mass exchange tends to create a primary star more massive and brighter than the original primary. In this case a star of 5 $M_\odot$ becomes of 8 $M_\odot$ and therefore is shifted up on the main sequence.

Underhill: It has been proposed that mass transfer in systems similar to that described will account for 'blue stragglers', Am and Ap stars. These suggestions (Renson, Van der Heuvel) imply that nuclear reactions producing heavy elements have occurred in the parts of the star which are transferred. Do the calculations of your case indicate that such nuclear reactions will have occurred and that mass exchange is a possible explanation of the unusual abundances which are deduced to be present in the atmosphere of Ap and Am stars?

Plavec: No, there is no evidence in favour of additional nuclear reaction at the early phases of stellar evolution at which mass exchange in case 'A' or 'B' occurs. I like the idea that blue stragglers are transformed binaries, but I must say that the explanation of the Am or Ap stars requires either an additional hypothesis explaining the anomalous abundances, or it must be looked for in another direction. A mere mass exchange does not seem sufficient.

Nariai: Observationally we know two systems, v Sgr and HD 30353, in which the helium-rich component is brighter and possibly less massive. Can we expect such a phase before the system reaches the final stage? The abundance of hydrogen at the surface of these stars is less than 1%.

Plavec: I think yes, although the phase when the helium star is large and bright is quite short – the star begins to shrink very soon.

WHITE-DWARF PRODUCTION IN BINARY SYSTEMS OF LARGE SEPARATION

DIETMAR LAUTERBORN

Göttingen, Germany

Abstract. Numerical calculations are carried out for the evolution of a binary system with a primary of 5 $M_\odot$ and a secondary of 2 $M_\odot$, revolving round another in a circular orbit of 300 $R_\odot$. After finishing central helium burning, the primary starts to transfer mass to its companion. After the mass loss, the star of originally 5 $M_\odot$ has become a star of 1 $M_\odot$. This star has a carbon-oxygen core which is a well-developed white dwarf, and a very extended hydrogen shell. The final system has a distance of 815 $R_\odot$; the period is 2.3 years.

If a star of – say – 5 $M_\odot$ is losing mass as member of a close binary system there are clearly three possibilities:

(a) the mass loss may begin during central hydrogen burning (so-called *case A*)

(b) the mass loss may begin during the expansion phase leading to helium burning (so-called *case B*)

(c) the mass loss may begin after the star has finished central helium burning and is evolving along or near the Hayashi track (*case C*).

The three cases A, B, C are well different concerning the order of magnitude of the corresponding radii. This can be seen in Figure 1, which shows the change of radius with time for a star of 5 $M_\odot$ calculated by Hofmeister in Munich in 1965. Look at Figure 1: first there is a slow increase of radius during hydrogen burning (if the mass loss starts during this phase we then have case A; the radii are relatively small); then at time $t \approx 5.2 \times 10^7$ *a* one has a sudden increase of radius during expansion to helium burning (if the mass loss starts during this phase we then have case B; the

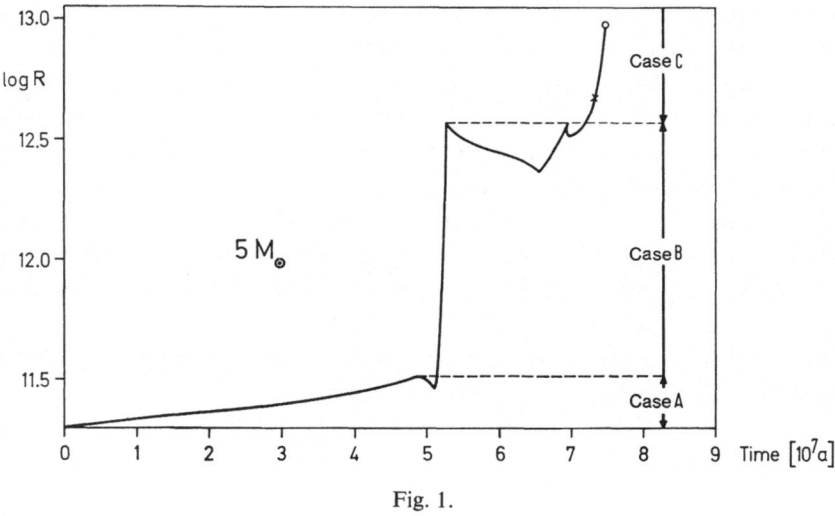

Fig. 1.

corresponding radii are already larger); then at time $t \approx 7.3 \times 10^7$ a there is another expansion phase along or near the Hayashi track (if the mass loss starts during this phase we then have case C; the radius is now relatively large).

From now on I shall only speak about *case C*; I shall give a report of evolutionary calculations carried out for the evolution of a special system of case C.

The special system I will speak of is characterized by the following parameters: there is a primary of 5 $M_\odot$, accompanied by a secondary of 2 $M_\odot$, revolving round one another at a distance of 300 $R_\odot$. The corresponding period then is 230 days.

I start to look at this system at the moment hydrogen begins to burn in the center of the star of $5M_\odot$. As it is, I did not calculate the evolution of the 5 $M_\odot$-star through the phases of hydrogen and central helium burning; I could very effectively make use of evolutionary calculations carried out by Hofmeister in Munich. I just took the last model of the mentioned evolutionary sequence of Hofmeister (distribution of chemical elements: $X = 0.6020$, $Y = 0.3540$, $Z = 0.044$) as the first model of my own computations.

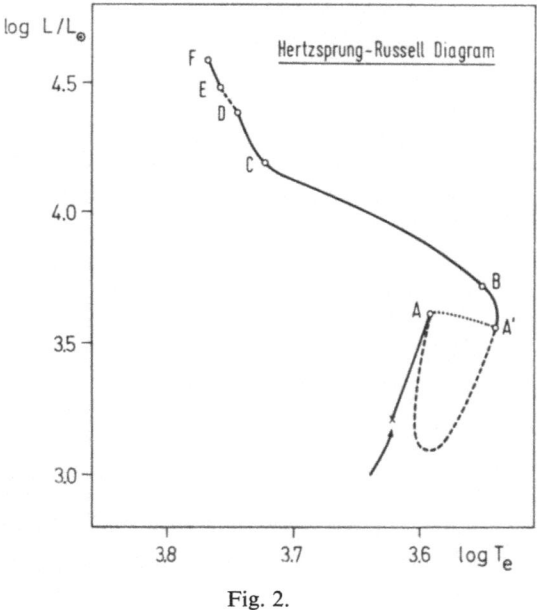

Fig. 2.

Figure 2 shows the evolution of the primary star of 5 $M_\odot$ in the Hertzsprung-Russell diagram. The evolution up to the point named A is the well-known evolution along the Hayashi track. At A the mass loss sets in. It turned out that the mass loss can be divided into two phases: a very quick phase leading from A to A' and a some-what slower phase from A' to B. At B the mass loss has come to an end; the star of 5 $M_\odot$ has lost as much as 4 $M_\odot$ – all this mass has been transferred to the secondary. At B the star begins to contract and to move to the left in the HRD. At the point C the old hydrogen shell is reignited; this stops the contraction, and the further evolution of the star is determined by the hydrogen and the – somewhat less effective – helium shell which are burning outwards and increasing the radius of the star. At F the volume of the star has become so large that again the star begins to lose mass. The calculations have led just up to the point F. The calculations go on.

At F the originally primary star of 5 $M_\odot$ has become already nearly a white dwarf. It may seem surprising that I speak of a white dwarf though the star, at F, has a luminosity of some 50000 $L_\odot$ and a radius of some 200 $R_\odot$. But let us look at Figure 3, which shows the structure of the star just before and immediately after the mass loss.

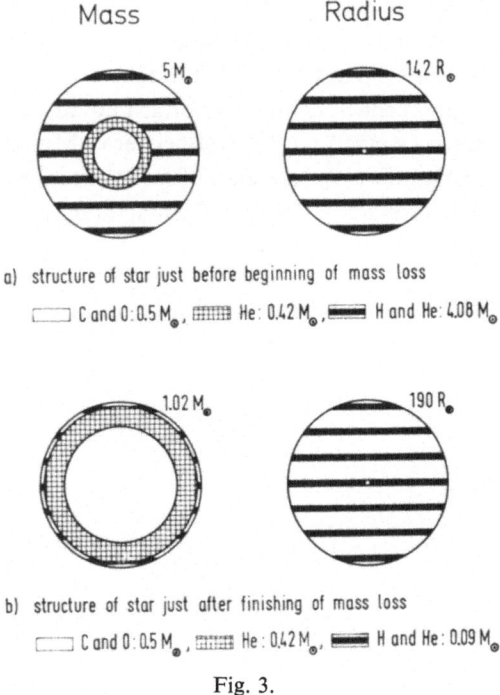

Mass   Radius

a) structure of star just before beginning of mass loss

C and O: 0.5 $M_\odot$, He: 0.42 $M_\odot$, H and He: 4.08 $M_\odot$

b) structure of star just after finishing of mass loss

C and O: 0.5 $M_\odot$, He: 0.42 $M_\odot$, H and He: 0.09 $M_\odot$

Fig. 3.

At the left of Figure 3 one sees the distribution of chemical elements over the mass, at the right the distribution of chemical elements over the radius of the star. The upper part shows the situation just before the onset of mass loss, the lower part shows the situation immediately after the end of the mass loss.

Let us look first at the *upper part*: situation before the mass loss starts. There is a small core of carbon and oxygen, surrounded by a small shell of pure helium and a large shell of mixed hydrogen and helium. At the right, for the radius, you see that only this outer region of H and He is contributing to the radius. The core gives only a very small, nearly invisible point in the middle of the circle in Figure 3, upper part, right hand.

Let us look then at the *lower part*: situation after the mass loss has come to an end. At left, for the mass: nearly the whole shell of hydrogen and helium has vanished. It has remained a mass of not more than 1 $M_\odot$. The core has now a large fractional mass, but this large fraction of mass even now gives only a small fraction in radius –

just the small nearly invisible point in the middle of the circle in Figure 3, lower part, right hand.

Figure 3 shows the situation immediately after the end of the mass loss. During the evolution up to F (Figure 2) there is no great change of the general structure of the star. The structure of the last computed model is the following: there is a large core of carbon and oxygen of 0.95 $M_\odot$, but this large mass gives only a small contribution to the radius. Then there is a very small shell of He, and above it a shell of mixed H and He, both shells making up for only a very small fraction of mass of 0.06 $M_\odot$; nevertheless, this small fraction of mass gives a large fraction of radius.

Now the further evolution of the 1.01 $M_\odot$ star is evident from the following three facts:

(1) It is well known from related calculations that the radius of a star of 1 $M_\odot$ consisting only of carbon and oxygen will be much smaller. Thus, once the small shell has vanished, there is no possibility for the star to maintain such a large radius.

(2) We have found: *If* the shell is removed, the star's radius will be very much smaller. Now it is certain that the shell *will* be removed – either by shell burning or by mass loss.

(3) Further on, it is known that a star of 1.01 $M_\odot$ will not reach the carbon burning stage (carbon burning is the next nuclear burning to be taken into account).

Thus the star has no other possibility than to become a white dwarf.

And indeed, if one goes into the interior of the star and inspects the situation one finds a well-developed white dwarf. The interior of the red giant star *is* already a white dwarf.

Now then we can write down the parameters for the final system: there is now a white dwarf of 1 $M_\odot$, accompanied by a main-sequence star of 6 $M_\odot$, revolving round one another at a distance of now 815 $R_\odot$. The corresponding period is 2.3 years.

The distance of 815 $R_\odot$ is a rather large one for a system which has had mass exchange. The distance of this system is already in the order of magnitude of the perihelion distance of the Sirius system.

I'd like to make three final remarks concerning the question why I spent my time with studying just this system – one could invent a lot of other systems. There were three points:

(a) Up to now, no calculations have been carried out for a system of case C.

(b) More important: I was interested in finding a system of relatively large distance which nevertheless has undergone mass exchange. For a system with a period of 2.3 years one would not think, at first sight, that it has had mass exchange.

(c) Up to now only white dwarfs of relatively small mass (0.25 $M_\odot$ in the investigations of Kippenhahn *et al.* and of Refsdahl and Weigert; and 0.6 $M_\odot$ in a paper on which Prof. Plavec reported in his paper) have been found in direct calculation from the main-sequence into the white-dwarf stage with mass loss in an adequately chosen binary system.

I hoped then to find a white dwarf of higher mass, and indeed it turned out that the final star had a mass of 1 $M_\odot$.

Discussion

Morton: What is the surface temperature of your white dwarf with the extended envelope?

Lauterborn: Rather low. The star is a red giant. – To the question what I think of Ptolemy I've two remarks:

(1) The question whether or not Sirius was a red star in Greek and Roman times seems to me to be definitely settled by the paper of See. There are so many explicit remarks about the colour of Sirius in the Greek and Roman literature that Sirius necessarily must be said to have been a red star some 2000 years ago.

(2) It is not sure with the same definiteness that the red colour resulted from the fact that the star was in the same condition as my 1 $M_\odot$ white dwarf, just in the phase of second mass loss. Theoretical arguments and comparison with similar calculations make it probable that the time-scale for the contraction from the red-giant into white-dwarf stage is somewhat larger (though not very much larger) than the some 1000 years required for Sirius B.

But there is another possibility. It is well known that white dwarfs with so very thin shells generally can have a thermal runaway which will give a rather large change of the radius. The time-scale for these thermal runaways is known and is in the range of the 1000 years of Sirius B. So I think it more probable that it was such a thermal runaway which caused Sirius to be red in Greek and Roman times.

But the question remains unsettled up to now. It can be answered by continuing the present calculations for the evolution of the 1 $M_\odot$ white dwarf with extended shell.

Underhill: J. R. W. Heintze at Utrecht has been studying the Sirius system attempting to fit the continuous energy distribution of the A star and the Hγ profile with a model atmosphere theoretical spectrum. Heintze finds that published photometry on Sirius indicates extra light at long wavelengths (excess red light). And he finds that the suggestion which occurs from time to time that Sirius is a triple system requires further investigation. Heintze's conclusions are given in *Bull. Astron. Inst. Neth.* **19** (1968).

Lauterborn: I know nothing of the energy distribution in the Sirius system. But let me say that I did not mean that the system I got as final system (white dwarf of 1 $M_\odot$, main-sequence star of 6 $M_\odot$, distance 815 $R_\odot$) is the Sirius system. I simply took the first system of case C that, from theoretical considerations, could be expected to give a white dwarf of relatively large mass. It then turned out that the distance of the two stars after mass loss was already in the order of magnitude of the perihelion distance of Sirius and that the white dwarf had very precisely the mass of Sirius B. As it is, one can say now that in some years for the theoreticians it will be possible to get, by mass exchange in a binary system, a final system which is very much closer to the Sirius system. I did reflect upon this question and got some results, but I do not dare to present them to you at this moment.

POSSIBLE EFFECTS OF THE RESONANCE IN
CLOSE BINARIES ON THEIR MASS EXCHANGES

J.-P. ZAHN

Laboratoire d'Astrophysique de l'Observatoire de Nice, France

Abstract. Present calculations on the forced non-radial oscillations of main-sequence stars show that the resonances in close binaries are much more probable than was assumed up to now. Some predictions are made about the behaviour of a star entering such a resonance.

In the papers which have been published up to now on the mass exchange in close binaries, it has been assumed that the stars are in perfect hydrostatic equilibrium. Their surface is then an equipotential surface, and they lose matter when they reach the Roche limit. Proceeding in this way, one does not take into account a phenomenon which is not proved to be negligible: the possible resonance of a low-frequency non-radial oscillation in one of the stars, due to the orbital motion of its companion which induces a variable gravitational field. This can in fact occur when the rotation of the star is not synchronized with the orbital motion; such lack of synchronism is observed in a few cases, and it is also expected on theoretical grounds for Cowling-type stars, because the time-scale of rotational braking of these stars in detached binaries is in general larger than the nuclear time-scale (ZAHN, 1966).

First of all: what is the probability of such a resonance? Unfortunately, the present state of our knowledge concerning high modes of oscillation is limited to the stars which do not rotate, and this, of course, is not the case of the components of an actual close binary. But it is likely that the properties of the oscillations of a rotating star do not differ, qualitatively, from those of a non-rotating star. It is well known that such a star has an infinity of low-frequency modes of free non-radial oscillations; these are the so-called g-modes (COWLING, 1941). Recently, TASSOUL and TASSOUL (1968) have shown how to determine the periods of the high-order modes of a star which is described by a realistic model of internal structure. Being interested in the forced oscillations, we have estimated the ability of these modes to enter a resonance when the star is submitted to the attraction of a revolving companion. Only the most important contribution to the tide, which is described by a spherical harmonic of second order, has been considered.

The results will be published elsewhere in their complete form; here we shall just mention the most interesting points. First, we must define what we call the strength of a given resonance. Let us take as a measure of the distortion of the star the amplitude a of the tide on the line joining the centers of the binary. In the hydrostatic approximation, a is fixed at a certain level a_0 which depends on the masses and the distances of the two components, and also a little on the mass-concentration. But, in fact a is a function of the period of the exciting gravitational field. In the framework of the linear theory, and neglecting any kind of damping, it becomes infinite for the period of each free mode.

M. Hack (ed.), Mass Loss from Stars. All rights reserved

We shall examine later which process limits the amplitude there. Let us define now the strength of a resonance, which is a measure of the probability, for the star, to enter this resonance. Calling ΔP the distance between two successive resonances (which tends to a finite limit with increasing order), and δP the quantity from which the star must leave the resonance-period to recover the spherical shape (Figure 1),

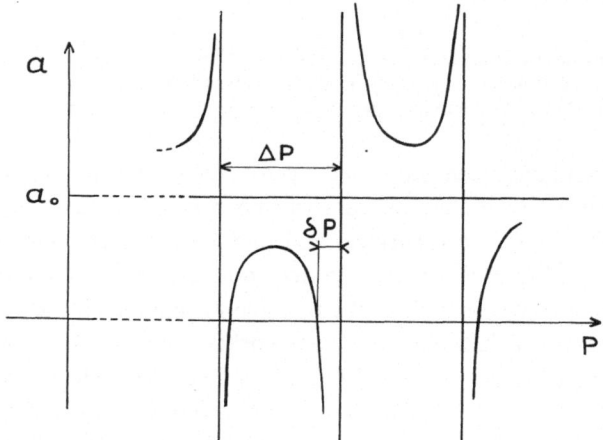

Fig. 1.   A detail of the resonance spectrum of a star; it shows the amplitude a of the tide vs. the period P of the exciting potential. The intervals δP and ΔP are used to define the strength of the resonance: $\Lambda = \delta P/\Delta P$.

we define the strength of the resonance by:

$$\Lambda = \delta P/\Delta P. \tag{1}$$

This quantity can be calculated directly, for a given model of internal structure. Our first results concern only main-sequence stars having an outer radiative zone. Main-sequence stars of low mass, which have a deep outer convective zone, are of no interest here: they are known to synchronize very rapidly their rotation with the orbital motion in a close binary, and thus are not subject to tides (ZAHN, 1966). The models which have been used are those of KUSHWAHA (1957), of 10 and 2.5 $M_\odot$.

The results are surprising in the sense that the strength of the resonances depends very much on the model, and thus on the mass, of the star. For example, we give here the Λ's for one of the highest modes calculated, the 24th harmonic (Table I) of

TABLE I

Strength of the resonance for the 24th harmonic

Mode	Model (mass in solar units)	Strength (Λ)	Period (days)	$(2\pi/P)^2 R^3/GM$
g 24	2.5	0.126	0.78	0.0358
	10	0.0420	2.20	0.0133
g 14	10	0.0149	1.33	0.0363

each model. But it is more significant to compare two modes belonging to the same value of the non-dimensional parameter $(2\pi/P)^2 R^3/GM$, which measures in the present case the intensity of the outer exciting potential; to the 24th harmonic of the 2.5 $M_\odot$ model corresponds then the 14th harmonic of the 10 $M_\odot$ model.

Anyway, one sees that the less massive star is much more able to enter a resonance. If the periods were distributed in random, they would have about one chance out of four to fall in the interval of width δP surrounding the resonance-period, in the typical example given above. Thus one can conclude that, for main-sequence stars of 2 or 3 $M_\odot$, the resonances are quite a common phenomenon.

Now, we have to ask ourselves what actually happens when a star, during its evolution, meets such a resonance. It is clear that the amplitude cannot grow to infinity, but the viscosity and the radiative damping are by themselves far too weak to limit it. Thus one must look for some non-linear effects, combined with turbulent dissipation taking place when the amplitude reaches a level which is not permitted by the linear approximation. An upper limit of the amplitude is certainly given by the simple condition that two successive layers, in the star, cannot cross each other. The most severe restriction applies to the radial displacement, and it can be written:

$$\partial\,\delta r/\partial r > -1\,. \tag{2}$$

From the shape of the oscillation, one can see that near the surface it is the most difficult to fulfil this condition. This difficulty increases also with the order of the mode because the distance between two successive modes decreases. For a high enough order, the above condition can be shown to be equivalent to:

$$\frac{a - a_0}{R} < \frac{1}{6}\left(\frac{2\pi}{P}\right)^2 \frac{R^3}{GM}, \tag{3}$$

where $(a - a_0)$ is the part of the tide due to the dynamical effects.

It is tempting to try to predict what happens just before the amplitude reaches this very upper limit. When the star approaches the resonance, somewhere near the surface, in the tidal bulge or at right angles, the lines of flow are pinched together. There, the continuity equation can be satisfied only in two ways: either by an increase of the density, which would lead to an inversion of the density gradient, or by an acceleration of the horizontal motions depending very sharply on the depth. In both cases, one may expect the appearance of an instability, of the Rayleigh-Taylor or Kelvin-Helmholtz type; the turbulent motions which are then produced may contribute more or less to the energy transport and thus react on the local structure of the star. Two conditions for the limitation of the amplitude are then fulfilled: the introduction of an effective damping by the turbulence, and a shift of the resonance-period due to the modification of the structure of the star.

It is unlikely that such a resonance would deeply modify the shape of the star except, perhaps, when it approaches the Roche limit. Then, in particular, the hydro-dynamical flow near the Lagrangian point could be quite different from that which was assumed, for example, by KRUSZEWSKI (1964) when he calculated the path of mass-particles leaving this region. One can expect also that the resonances may introduce a kind of modulation in the rate of mass loss, for a star has to cross a few of them during its post main-sequence evolution.

But, of course, one has to go farther in the investigation of this phenomenon, and especially take into account its non-linear character, before risking any further predictions. However, it seems that the resonances can no longer be neglected in the study of close binaries of moderate mass.

References

COWLING, T. G.: 1941, *Monthly Notices Roy. Astron. Soc.* **101**, 367.
KRUSZEWSKI, A.: 1964, *Acta Astron.* **14**, 231.
KUSHWAHA, R. S.: 1957, *Astrophys. J.* **125**, 242.
TASSOUL, M. and TASSOUL, J.-L.: 1968, *Ann. Astrophys.* **31**, 251.
ZAHN, J.-P.: 1966, *Ann. Astrophys.* **29**, 313, 489, and 565.

Discussion

Plavec: Are these resonance and oscillation due to the fact that stars do not rotate in synchronism with the orbital motion? Or because the orbit is eccentric?

Zahn: No. The cause is lack of synchronism between the orbital motion and rotation.

Plavec: So, if these stars should have time enough on the main-sequence stage to synchronize their rotation there would be no such tremendous oscillations.

Zahn: Yes, quite.

SOME PROBLEMS CONCERNING GAS FLOWS IN
CLOSE BINARY SYSTEMS OF DWARF STARS

V. G. GORBATZKY

Leningrad University Observatory, Leningrad, U.S.S.R.

Abstract. The influence of the stream flowing from the secondary on the disk-like envelope of the main star is briefly considered. It is found that the lifetime of the envelope after the flow ceases must be less than 10^6 sec.

1. The disk-like envelopes of stars in close binary systems discovered by Struve are the subject of intense study by both observers and theoreticians. The clearing-up of the kinematics and dynamics of these formations may be useful for understanding the nature of processes in stars. The role which the disk-like envelopes play in close binary systems consisting of dwarf stars (such as U Gem, DQ Her, WZ Sgt, and similar systems) is especially important. The radiation of the envelope makes an important contribution to the total observed brightness of such a system. Sometimes it may even give the major part of the total radiation. The formation of the disk-like envelopes is connected probably with the activity of these stars manifested in the outbursts. The study of the envelopes is especially interesting for this reason.

In the present brief report we shall review some problems which emerged in the theoretical study of these objects in Leningrad University.

2. PRENDERGAST (1960) has shown that the motion of the gas particles in the disk-like envelope is circular Keplerian and consequently, the azimuthal component of velocity is $V_\varphi \sim r^{-1/2}$, where r is the distance from the centre of the star. Therefore in the envelope there is a non-zero azimuthal velocity gradient along the radius. There is also the developed turbulence. The turbulent viscosity causes the gas flow towards the surface of the star. The dynamics of the process have been considered by GORBATZKY (1965). The steady state of the envelope can exist only if there is a permanent supply of gas. If this supply ceases the envelope will vanish because the gas it consists of will join the star.

The only imaginable mechanism of the mass supply of the envelope is the gaseous stream flowing from the secondary to the main star. Spectroscopic observations by KRZEMINSKY and KRAFT (1964) give us convincing evidence of the presence of such a stream ('jet') in system WZ Sgt. The velocity V of the gas in the jet is of the order of 10^8 cm sec^{-1} and its kinetic energy must be great enough. When the jet collides with the envelope its energy transforms partially into thermal energy and eventually into radiation. At the place where the jet encounters the envelope the high-temperature region will be formed – the so-called 'hot spot'.

The heated gas from this hot spot will be distributed along the envelope by rotation. The essential part of the continuum and line radiation of the envelope is the

M. Hack (ed.), Mass Loss from Stars. All rights reserved

transformed kinetic energy of the jet. This conclusion is confirmed by the direct determination of the electron temperature T_e in the envelope of RW Tri. This temperature determination is based on the UBV light-curves near to the occultation phase. It was found that in the outer part of the envelope ($r \approx 0.9 \, r_0$) T_e is near 2×10^4 K, while in the inner parts ($r \approx 0.5 \, r_0$) T_e is several times smaller (here r_0 is the radius of the disk). The conclusion is that the source of radiation mainly affects the outer regions of the envelope. Heating by the gaseous stream is the only possible external source of such a kind.

3. Some of the effects that may be caused by the extra radiation are discussed by GORBATZKY (1967). In particular, it was noted that the shoulder on the light-curve before the main light minimum marks the position of the best visibility of the hot spot. The computations were also made to find the gas motions in the streams. They have shown that at the moment of the light maximum of the shoulder (the phase $\approx 0.9 \, P$) the position of the hot spot is opposite to the observer.

The height of the shoulder is usually several tenths of magnitude. Consequently the energy ΔE radiated by the hot spot is only several times less than the total radiation of the system (10^{31}–10^{32} erg sec^{-1}). The conclusion is that the mass Q flowing into the envelope per second is of the order

$$Q \approx \frac{\Delta E}{V^2} \approx 10^{15} - 10^{16} \text{ g sec}^{-1}. \tag{1}$$

The total gas flow in the envelope also is Q and if we assume the envelope to be a cylinder of height Δz_0 and radius r_0

$$2\pi r_0 \Delta z_0 \rho_0 V_r^0 = Q, \tag{2}$$

where V_r^0 is the radial gas velocity at the external boundary of the cylinder and ρ_0 is the corresponding density of the gas. The estimates of r_0, Δz_0 and ρ_0 inferred from the observational data are $r_0 \approx 10^{10}$ cm, $\Delta z_0 \approx \frac{1}{10} \rho_0$, $\rho_0 \approx 10^{-11}$ g cm^{-3}. Now (2) gives $V_r^0 \approx 10^6$ cm sec^{-1}. The azimuthal component V_φ^0 has been given as $V_\varphi^0 \approx 10^8$ cm sec^{-1}, so that $V_r^0 \approx 0.01 \, V_\varphi^0$. It is easily shown that the lifetime of the envelope, after the supply of the gas ceases, is

$$t_r \approx r_0^2 / R V_r^0, \tag{3}$$

where R is the radius of the main star (or of the inner boundary of the envelope). As $R \gtrsim 10^8$ cm, one gets $t_r \lesssim 10^6$ sec. This value is very small when compared with the observed lifetimes of the envelopes in all known binary systems. The conclusion is that the envelopes are in the quasi-steady state.

In VV Pup the radiation of the hot spot makes the main contribution to the total light of the system. As it is known, in 1948–49 the brightness of the star was 2^m–3^m less than usual. One may speculate that the star was lacking the envelope at this period and consequently, there was no stream from the secondary to the main star. If this is so, the wide spread opinion that the filling up of the critical Roche surface by the secondary star is the main cause of the outflow of matter, becomes doubtful.

If we adopt this point of view, we must conclude that for 2 years the secondary star in VV Pup was much smaller in size than the critical Roche surface, and then its size increased again. It is very difficult to imagine such a situation without coming to serious contradictions with the current theories of stellar structure and evolution. It seems to be useful to investigate the other factors that may cause the outflow of gas from the secondary star.

4. There is one fact concerning the system WZ Sgt which indicates that the picture of the outflow of gas from the L_1 point is more complex than usually supposed. The mass ratio in this system is very high: $M_1/M_2 \approx 20$. The position of the hot spot is also known. The corresponding phase is $0.9 P$. The gas-dynamical equations describing the gas flow in this system were solved numerically. The expansion of the gas was taken into account in the computations.

The computations have shown that the observed position of the hot spot can be obtained only if one assumes $V_r^0 \approx 100$ km sec^{-1} and $V_\varphi^0 \approx 100$ km sec^{-1}, where V_r^0 is the velocity at the point L_1 in the direction of the main star, and V_φ is the velocity in the perpendicular direction. This result may be considered as evidence of the presence of some non-gravitational force facilitating the outflow of matter. As an alternative, one can suppose that the rotation of the secondary is not synchronized with the orbital motion. The understanding of the nature of the gas outflow would be of great importance, as it might throw some light on the causes of the nova outbursts.

References

GORBATZKY, V. G.: 1965, *Trudy astr. Obs. Leningr. gos. Univ.* **22**, 16.
GORBATZKY, V. G.: 1967, *Astrophys.* **3**, 245.
KRZEMINSKY, W. and KRAFT, R.: 1964, *Astrophys. J.* **140**, 921.
PRENDERGAST, K.: 1960, *Astrophys. J.* **132**, 162.

LES ÉTOILES DOUBLES, INDICATRICES D'UNE PERTE DE MASSE SÉCULAIRE DES ÉTOILES

J. DOMMANGET

Observatoire Royal de Belgique

1. Eléments orbitaux et fréquence des couples stellaires

Les valeurs des éléments orbitaux physiques des couples stellaires (demi-grand axe, période et excentricité) couvrent des intervalles particulièrement étendus.

Tout d'abord, en ce qui concerne la distance de séparation des composantes des binaires, on sait qu'elle varie depuis environ un diamètre solaire (0.01 UA) jusqu'à quelque 0.01 parsec (DOMMANGET, 1967a). Ceci confère au demi-grand axe A exprimé en unités astronomiques, un domaine de variation défini approximativement par les conditions:

$$8.00 < \log A < 3.30 . \tag{1}$$

Par ailleurs, les périodes et les excentricités présentent des domaines de variation approchés, limités respectivement comme suit:

$$\text{plusieurs heures} < P < 10\,000 \text{ ans} \tag{2}$$

et

$$0 \leqslant e < 1 . \tag{3}$$

Quant à la fréquence des étoiles doubles, elle présente les caractéristiques suivantes (DOMMANGET, 1969):

(a) Plus de 50% des étoiles sont doubles (c'est-à-dire que plus des 2/3 des étoiles font partie d'un système binaire);

(b) La fréquence des valeurs de $\log A$ est quasi-constante dans le domaine (1) sauf sans doute au voisinage de ses bornes: elle est de 13% environ par intervalle de variation d'une unité en $\log A$.

2. Option parmi diverses hypothèses sur la formation des étoiles doubles

Il semble que les constatations faites ci-dessus permettent de faire un choix parmi les théories proposées pour la formation des couples stellaires.

On peut distinguer les trois hypothèses suivantes:

(a) Formation des couples comme conséquence de la distribution au hasard des centres de condensation dans le milieu galactique;

(b) Capture mutuelle de deux étoiles passant au voisinage l'une de l'autre;

(c) Origine commune des composantes (scission d'un astre unique, présence de deux noyaux dans une atmosphère commune, etc.).

M. Hack (ed.), Mass Loss from Stars. All rights reserved

De ces trois hypothèses, seule la dernière est à retenir, mais assortie d'une condition que nous allons préciser.

En effet, une distribution au hasard des étoiles ne conduit qu'à une très infime proportion d'étoiles doubles contrairement à ce que l'on observe. Par ailleurs, elle conduit à une fréquence de binaires décroissant avec la distance de séparation alors que l'observation indique que cette fréquence croît très sensiblement quand A décroit (fréquence constante pour un intervalle $\Delta \log A =$ constante).

Quant à la seconde hypothèse, elle est à rejeter pour les raisons suivantes:

(1) La fréquence de rencontre de deux étoiles dans l'espace par le hasard des circonstances et suite à leurs mouvements spatiaux propres, est très faible. Par exemple (DOMMANGET, 1967a), des rapprochements d'étoiles à moins de 0.1 parsec = = 2000 UA n'ont lieu qu'au rythme statistique d'un rapprochement tous les 5×10^9 années, tandis que si la distance descend à moins de 0.01 parsec = 200 UA, ce rythme est d'un rapprochement tous les 5×10^{11} années. Une fois de plus, on ne peut expliquer par cette hypothèse, la grande fréquence des couples que l'on observe.

(2) Le rapprochement de deux étoiles venant de directions différentes ne peut se faire que suivant une orbite relative hyperbolique (voire parabolique). La capture ne peut donc avoir lieu que si un mécanisme quelconque (milieu résistant, marées, échange de matière) transforme l'hyperbole en ellipse et cela dès le premier passage périastral. Un mécanisme d'une telle ampleur paraît donc si pas impossible, en tous cas extrêmement rare.

Il reste alors la troisième hypothèse, celle accordant aux composantes d'un couple, une origine commune. Cette hypothèse explique sans difficulté majeure, la grande fréquence des couples stellaires. Mais elle ne peut, à elle seule, expliquer la grande dispersion des valeurs de leurs éléments orbitaux physiques à l'intérieur des domaines (1), (2) et (3), sans devoir faire appel à un (ou des) mécanisme évolutif complémentaire.

Or de tels mécanismes sont connus et permettent d'expliquer aisément les grandes valeurs de P, de A et de e. Plusieurs phases sont néanmoins à distinguer dans l'hypothèse considérée.

Tout d'abord, lorsque le couple est serré, soit immédiatement après sa naissance, les éléments P, A et e peuvent subir des variations importantes dans des intervalles de temps relativement courts par suite des mécanismes déjà cités (échange de masse entre les composantes, marées, freinage dans une atmosphère commune, etc.).

Toutefois, au delà d'une certaine limite de la distance de séparation, par exemple au delà de 0.1 ou 1 UA, ces phénomènes s'évanouissent complètement. Par conséquent, si leurs effets respectifs peuvent expliquer dans certains cas au moins un accroissement des éléments A et P lorsque les composantes sont rapprochées, ils ne peuvent expliquer les grandes distances de séparation (10000 ou 1000 UA) ni les grandes périodes observées parmi les couples visuels.

Seule une perte de masse séculaire, semble-t-il, pourrait expliquer ces grandes orbites et devrait être prise en considération également tout au long de la vie d'un couple stellaire, soit donc aussi, parmi les mécanismes produisant une évolution des éléments des couples serrés.

On peut ainsi calculer, simplement à titre d'indication, qu'un couple tel que γ Geminorum, dont les éléments orbitaux physiques sont actuellement:

$$\begin{cases} P = 6 \text{ ans} \\ A = 8 \text{ UA} \\ e = 0.30, \end{cases}$$

et dont la masse totale est de 15 $\mathfrak{M}_\odot$, peut présenter une évolution orbitale telle qu'après un intervalle de temps de l'ordre du milliard à dix milliard d'années, ses éléments deviennent:

$$\begin{cases} P = 1400 \text{ ans} \\ A = 120 \text{ UA} \\ e = 0.30 \end{cases}$$

pour autant que sa masse décroisse jusqu'à une masse solaire. Une telle évolution nécessite une perte de masse environ 1000 fois plus importante que celle due au rayonnement lumineux.

Par ailleurs, on sait que la perte de masse peut être fonction de la distance de séparation des composantes et se trouver sensiblement accrue lors des passages périastraux. Dans ces cas, on constate alors que l'excentricité augmente dans des proportions sensibles, ce qui peut expliquer les orbites à fortes excentricités parmi les couples écartés. De plus, l'orbite finale présente alors non seulement une grande excentricité, mais encore de plus grandes valeurs de P et de A que celles trouvées plus haut. Pour e passant de 0.30 à 0.90 par exemple, on trouverait finalement des valeurs de l'ordre de:

$$\begin{cases} P = 14\,000 \text{ ans} \\ A = 600 \text{ UA} \\ e = 0.90 \end{cases}$$

3. Conclusions

L'existence même des étoiles doubles, surtout celle des étoiles doubles visuelles, serait donc un indice d'une perte de masse séculaire des étoiles.

La variabilité de l'importance du phénomène d'un couple à l'autre (et sans doute d'une composante à l'autre) expliquerait aisément la dispersion constatée pour les caractéristiques physiques et orbitales des binaires. La perte de masse pourrait être importante chez certaines d'entre elles, même pour des classes spectrales relativement avancées. Elle pourrait être très faible chez d'autres qui ne subiraient dès lors qu'une évolution orbitale peu importante ce qui serait en accord avec la décroissance de la fréquence des couples pour des distances de séparation croissantes.

Il reste cependant en particulier à mettre en évidence par l'observation, que certains couples présentent une perte de masse variable suivant la distance de séparation des composantes afin de rendre cohérent, le point de vue exposé ici. Or, à la suite d'une recherche statistique portant sur l'évolution possible des couples stellaires par perte de masse séculaire (DOMMANGET, 1963), nous avons établi une liste de couples ayant le plus de chance de présenter un tel phénomène périastral (DOMMANGET, 1964, p. 21) et, parmi eux, nous en avons retenu surtout un, β Arietis, pour sa faible période et la facilité consécutive d'observer ce couple à ses passages périastraux successifs (DOMMANGET, 1967b). Un examen spectroscopique et photométrique de ce couple paraît souhaitable dans le cadre des recherches sur les mécanismes de perte de masse des étoiles.

Bibliographie

DOMMANGET, J.: 1963, *Ann. Observ. Roy. Belgique*, troisième série, **9**, fasc. 5.

DOMMANGET, J.: 1964, *Communication de l'Observatoire Royal de Belgique*, no. 232.

DOMMANGET, J.: 1967a, Colloque 'On the Evolution of Double Stars'. *Communication de l'Observatoire Royal de Belgique*, Série B, no. 17, p. 25.

DOMMANGET, J.: 1967b, Colloque 'On the Evolution of Double Stars'. *Communication de l'Observatoire Royal de Belgique*, Série B, no. 17, p. 168.

DOMMANGET, J.: 1969, *Publications Scientifiques du Ministère de l'Air*, Notes Techniques, Paris; et *Communications de l'Observatoire Royal de Belgique* (en cours de publication).

MASS LOSS FROM UNSTABLE STARS

INFRARED SPECTRA OF NOVA DELPHINI 1967

YVETTE ANDRILLAT

Laboratoire d'Astronomie, Faculté des Sciences, Montpellier, France

and

LÉO HOUZIAUX

Département d'Astrophysique, Faculté des Sciences, Mons, Belgium

Abstract. Infrared spectra at 39 Å/mm of Nova Delphini 1967 have been obtained in the wavelength range λ 6500–λ 8800. Typical microphotometer tracings are displayed and an estimate of the received flux is given for the most prominent lines at nebular stage. Loss of matter is very likely to occur.

Since July 10, 1967, Nova Delphini 1967 has been extensively observed at the Observatoire de Haute-Provence, and short descriptions of the spectra have been given by FEHRENBACH *et al.* (1967, 1968). This paper reports on the infrared spectra obtained from July 13, 1967 until the beginning of September 1968. The grating spectrograms, obtained with the coudé spectrograph of the 193-cm telescope on ammoniated IN plates cover the region from Hα to the sensitivity limit of the emulsion, i.e. about 8800 Å. The dispersion is 39 Å/mm.

It is not our intention to give here a detailed description of the spectral evolution of the nova, but to show the most typical aspects of the infrared features. Figure 1 shows a microphotometer tracing of the spectrum obtained on July 28, 1967. It shows strong lines of O I (at 7772 and 8446 Å) and of Ca II with P Cygni type profiles. The absorption components are wide and displaced towards the violet. Paschen lines are somewhat fainter and display also P Cygni profiles. Multiplets 1, 3 and 8 of N I appear as rather narrow emissions; this is also the case for the Fe II lines at 7516 and 7712 Å, as well as for the Mg II doublet at 7877 and 7896 Å.

An important outburst occurred on Dec. 12, 1967, which brought the nova's visual magnitude around 3.5. Figure 2 exhibits the aspect of the infrared spectrum a few days later (Dec. 20, 1967). O I lines are very broad and strong, and a violet emission component has appeared. The same type of profile is observed for Ca II lines, whose strength is now much larger than the Paschen lines, broad and ill-defined. A violet absorption flanks both the O I and Ca II lines. Fe II lines appear as diffuse bands, as do the N I components. A wide emission superimposed on the atmospheric Z band may be attributed to He II 8236 Å. Diffuse lines appear also at 7000, 7112 and 8330 Å. They are tentatively identified as transitions respectively due to O I, C II and C III.

At the nebular phase, while Hα is still very strong, Ca II, N I and Fe II lines have considerably weakened or even disappeared. On the other hand, He I λλ 6678, 7065 and 7281 show up as wide bands with triple structure. He II λ 8236 is still present. The most conspicuous features are now [O II] λλ 7319–7330 and O I λ 8446. It is interesting to note that, while this last line became stronger and stronger, the triplet at 7772 Å is just noticeable. Other nebular lines include [A III] λ 7136, and C II λ 7236. Paschen

M. Hack (ed.), Mass Loss from Stars. All rights reserved

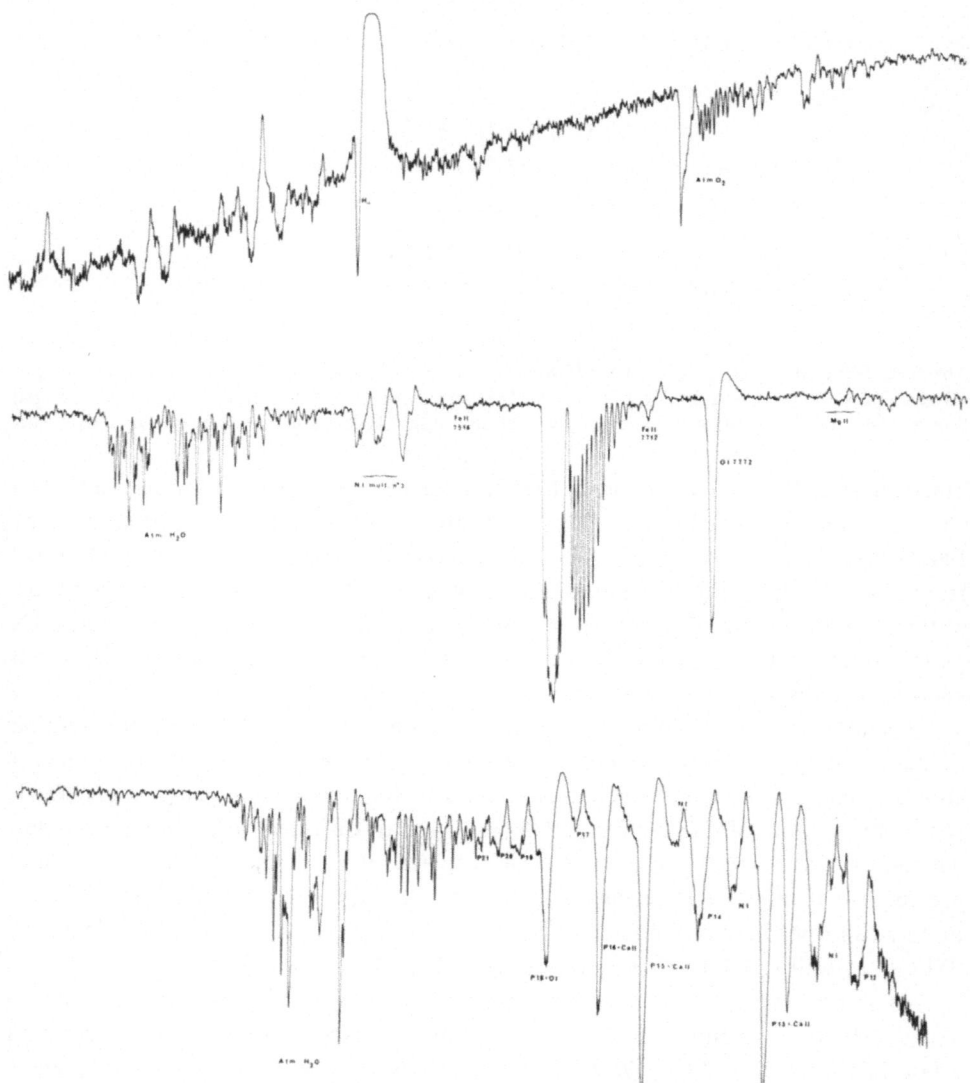

Fig. 1.   Microphotometer tracing of the spectrum taken on July 28, 1967. Original dispersion:
39 Å/mm.

lines are not easily recognized, being very broad and blended with the Ca II triplet.
Figures 3–7 reproduce line profiles of some of the transitions observed during the
nebular phase. Some of the lines have been obtained at a dispersion of 19 Å/mm.

These profiles show the large width and the complicated structure of the lines.
Some of the lines as C II 7236 Å and [O II] 7330 Å are doublets with rather widely
spaced components.

It has been possible to compare the continuous spectrum with the radiation emitted
by the B3 III star ε Cas, observed at the same zenith distance. Admitting a red gradient

of 1.3 for the comparison star, it is found that the nova gradient on August 7, 1968, in the wavelength range 5500–7500 Å, is 1.71, which means that the slope of the energy curve $d \log I/d(1/\lambda)_\mu$ equals 0.66. This value is compatible with the one found for the gradient of Nova Herculis 1963 at the same evolutionary stage (ANDRILLAT, 1963). Interstellar reddening is so far not known.

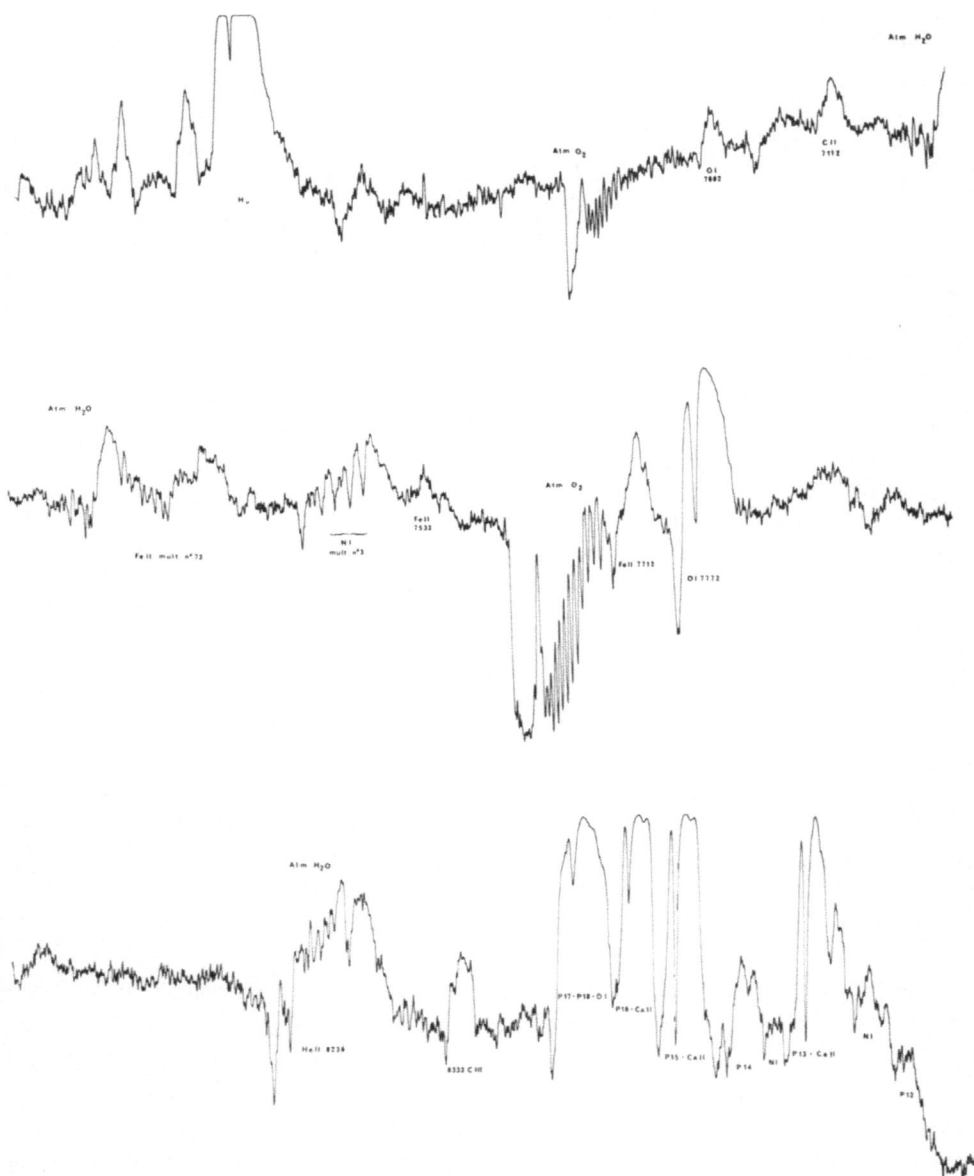

Fig. 2. Microphotometer tracing of the spectrum taken on Dec. 20, 1967. Original dispersion: 39 Å/mm.

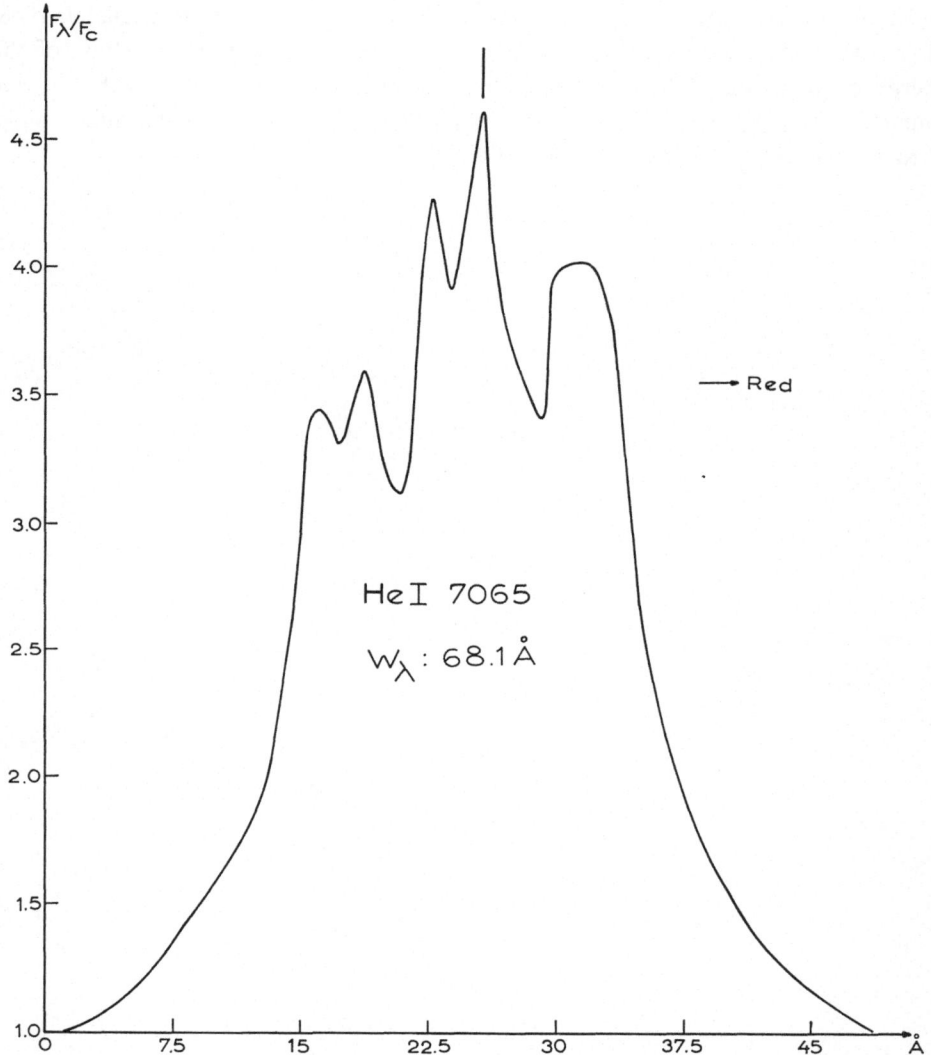

Fig. 3.   Intensity profile of He I λ 7065 observed on August 7, 1968. The vertical mark indicates the observed wavelength 7069.7 Å. Original dispersion: 39 Å/mm.

As the continuous spectrum was measurable on some of our plates, it has been possible to derive equivalent widths for the lines. They are given on the figures for some of the lines. Other values include He I λ 6678 (34.6 Å) and O I λ 7772 (21 Å), on August 7, 1968. At the time of observation the visual magnitude in the continuous spectrum was 7.0 ± 0.2. Knowing that a $V = 0.0$, $B-V = O$ star emits 3.8×10^{-9} ergs cm^{-2} sec^{-1} A^{-1} at 5400 Å, it is possible to infer the energy received from the nova in the various lines; these energies are given in Table I.

Further analysis requires data on the star's distance, on the geometry of the shell,

TABLE I

Line	$E(\text{ergs cm}^{-2} \text{ sec}^{-1})$
Heɪ 6678	8.4×10^{-11}
7065	1.3×10^{-10}
[Oɪ] 7772	2.8×10^{-11}
8446	1.8×10^{-10}
[Oɪɪ] 7319	2.6×10^{-10}
7330	
Aɪɪɪ 7136	6.1×10^{-11}
Cɪɪ 7236	5.0×10^{-11}
Feɪɪ 7712	8.0×10^{-12}

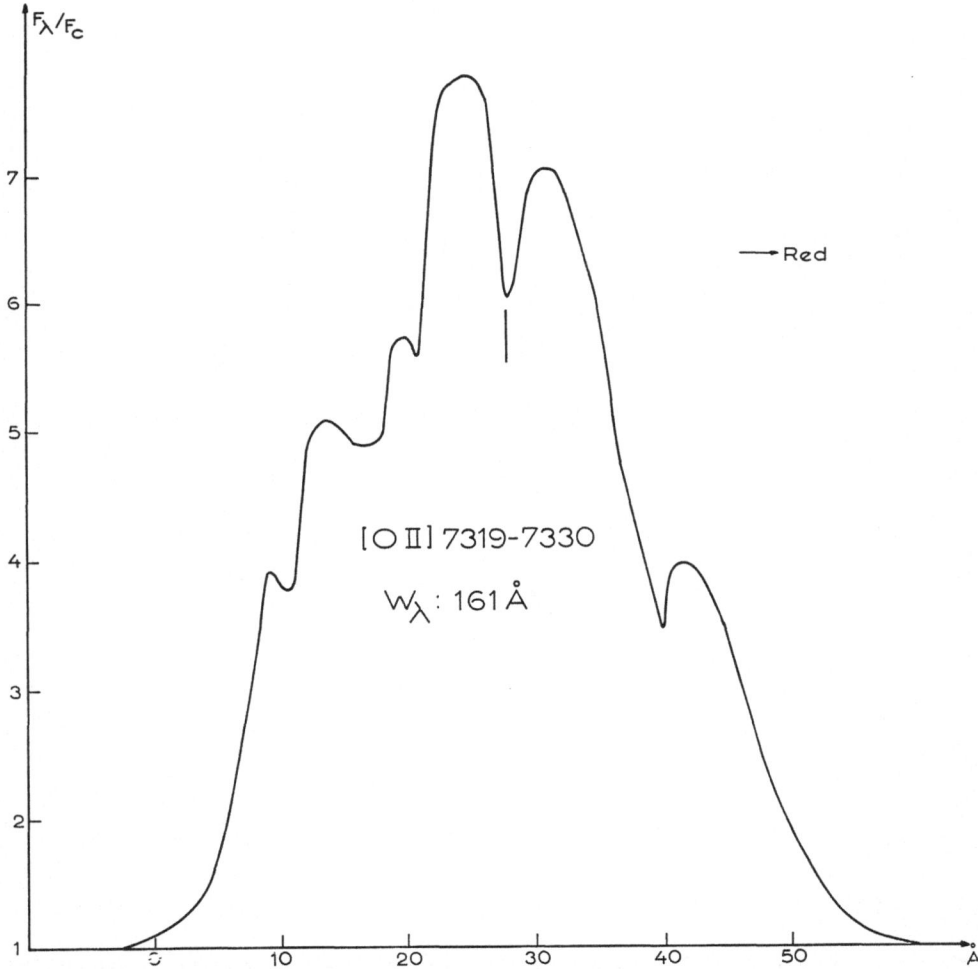

Fig. 4. Intensity profile of [Oɪɪ] $\lambda\lambda$ 7319–7330 observed on August 7, 1968. The vertical mark indicates the observed wavelength 7325.7 Å. Original dispersion: 39 Å/mm.

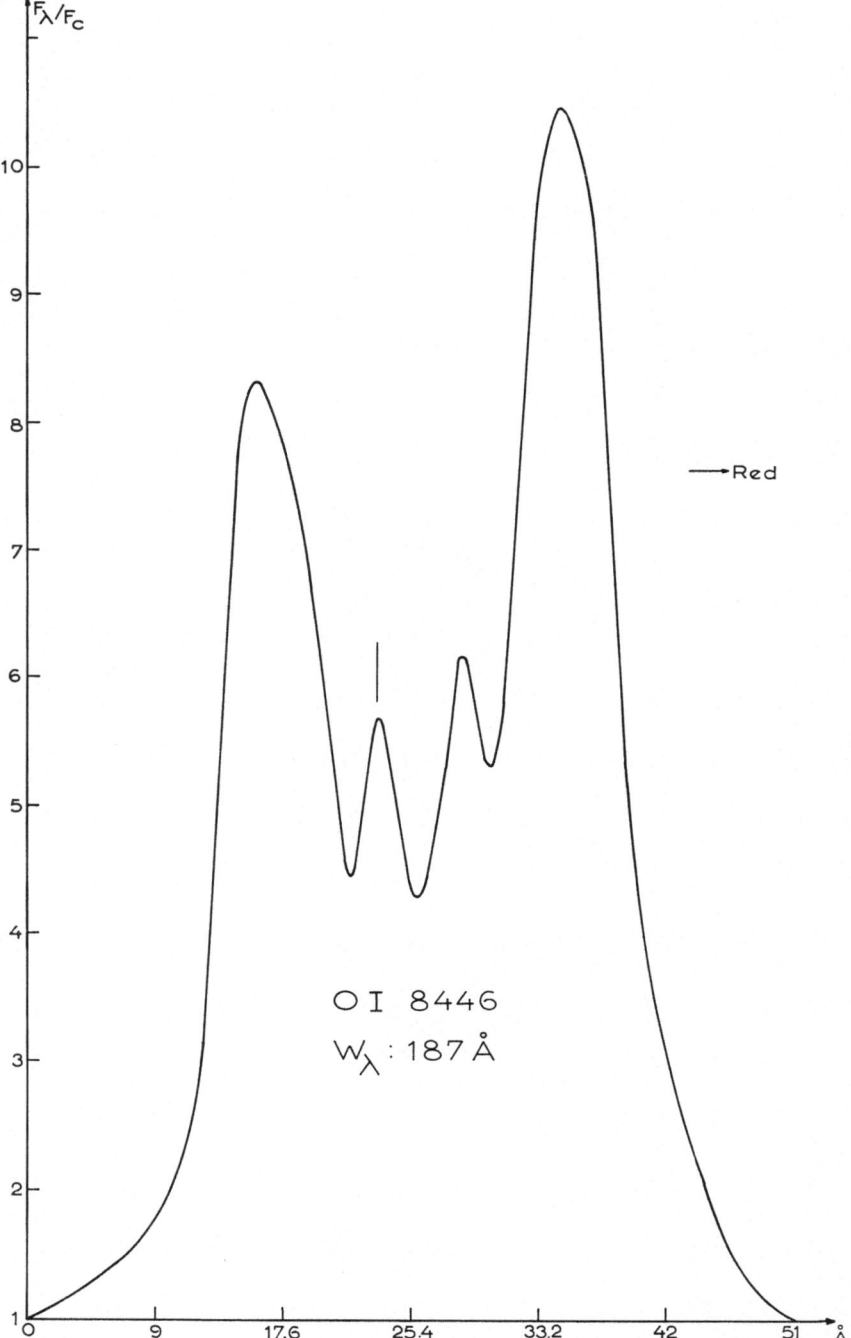

Fig. 5.   Intensity profile of O I λ 8446 observed on August 12, 1968. The vertical mark indicates the
observed wavelength 8443.3 Å. Original dispersion: 19 Å/mm.

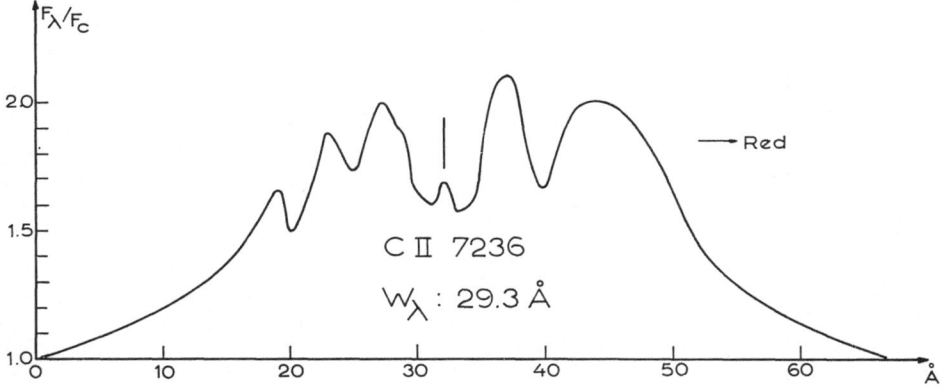

Fig. 6.   Intensity profile of C II λ 7236 observed on August 7, 1968. The vertical mark indicates the observed wavelength 7234 Å. Original dispersion: 39 Å/mm.

and the physical conditions prevailing at the time of observation, which cannot be obtained from infrared data alone.

However, qualitative information on the oxygen lines may lead to interesting conclusions regarding the excitation mechanism of the infrared O I lines. As seen in Figure 1, shortly after maximum, both λ 8446 and λ 7772 show strong emission of comparable strength; this occurs also after the Dec. 12 outburst. However, on August 7, 1968 Table I shows that $I(8446)$ is about six times $I(7772)$. This change in relative intensities may be interpreted in the following way: at maximum light, the shell is not too extended and the electron density is relatively high, both lines are then formed by recombination of O^+ ions. However, as the shell expands, the density lowers considerably as well as the recombination rate, while the kinetic temperature gets higher and higher, as shown by the appearance of forbidden lines due to O III. The intensity of λ 7772 decreases, but, in the case of λ 8446, another mechanism takes over, namely the cascade from the 3d ^{3}D level, populated through fluorescence with the Ly β line, for which the shell has become more transparent, due to its lower density.

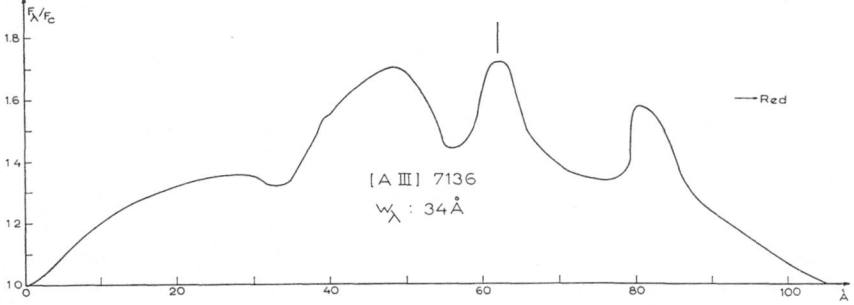

Fig. 7.   Intensity profile of [A III] λ 7136 observed on August 7, 1968. The vertical mark indicates the observed wavelength 7145.8 Å. Original dispersion: 39 Å/mm.

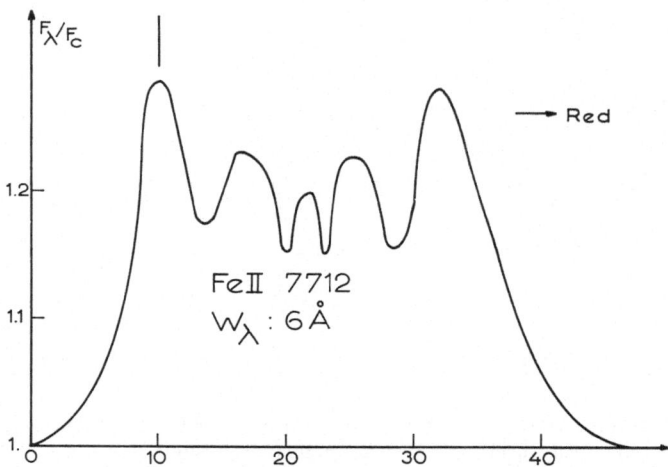

Fig. 8.   Intensity profile of Fe II 7712 observed on August 7, 1968. The vertical mark indicates the
observed wavelength 7681.6 Å. Original dispersion: 39 Å/mm.

At some stage of the evolution the two mechanisms must be competitive for the formation of λ 8446. Due to a long interruption in the observations in the spring of 1968,
this epoch could not be determined.

As it is the case in planetary nebulae, the intensity ratio $I(8446)/I(6300+6363)$
may be used in order to determine the energy density at Ly β, and then the hydrogen
density, without any determination of distance (ANDRILLAT and HOUZIAUX, 1967).

During the nebular phase, it is most probable that matter escapes from the nova.
The large widths of the observed lines lead to expanding velocities of the order of
1000 km sec^{-1}. The velocity of escape may be written as 618 $(M/R)^{-1/2}$ km sec^{-1},
with M and R in solar units. For the nova $(M/R)^{-1/2}$ is very likely to be smaller than 1.
An evaluation of the rate of mass loss requires however a good determination of the
electron density, as most of the material is fully ionized hydrogen and singly ionized
helium.

References

ANDRILLAT, Y.: 1963, Colloque sur les novae, les novoïdes et les supernovae, Paris, CNRS, 1964,
    p. 136.
ANDRILLAT, Y. and HOUZIAUX, L.: 1967, in IAU Symposium No. 34: *Planetary Nebulae* (ed. by
    D. E. Osterbrock), D. Reidel Publ. Co., Dordrecht, p. 68.
FEHRENBACH, C., ANDRILLAT, Y. and BLOCH, M.: 1967, *C. R. Acad. Sc. Paris* **266**, 106; 1968, **265**, 583.

Discussion

McCarthy: I wish to congratulate Dr. Houziaux and Mme Andrillat on the excellent
series of infrared spectra. And I should like to ask what were the observational
circumstances under which the data shown in the final slide were obtained. I note that
the observations extend beyond the limits of IN plates.

Houziaux: The last spectrum has been obtained with a spectrograph which gives 230 Å/mm; it is opened at $f/2.4$ and ammoniated Kodak IM emulsion has been used.

De Groot: I would like to know if it is already possible to give some more information on the numerical results about the mass loss and the density of Lyβ for Nova Delphini.

Houziaux: So far we have not worked out any numerical result.

NOVA DELPHINI 1967-8

J. B. HUTCHINGS

Dominion Astrophysical Observatory

Abstract. A discussion of the peculiar, slow nova Delphini 1967 is given, based on a large amount of data obtained in Victoria. A summary of the important changes in the spectrum is given, and consideration of the line profiles and line velocity curves leads to the conclusion that the photosphere attained a diameter in excess of 10^9 km for an extended period of time. After June 1968 no further matter was ejected and collapse of the photosphere left an auroral type spectrum with complex structure in its emission lines.

1. Introduction

Having been a 5th magnitude or brighter object in the Northern sky for over a year, the phenomenon of nova Delphini should need little introduction. Many descriptions of its spectacular changes are already in circulation, and its brightness, position, and longevity have provided an excellent opportunity for systematic study of a star undergoing considerable mass loss.

It seems well established that in general the duration of a nova outburst is related to the initial increase in brightness of the star. It is less certain but still quite likely that the duration of the phenomenon is related to absolute brightness at maximum, so that the mass loss may be caused largely by radiation pressure. However, there are large uncertainties in the estimates of absolute magnitude as well as an intrinsic spread caused by differences between novae of gravitational field, rotation, effects of companion stars, and mass and luminosity of the prenova star. The general rule seems to be followed in the case of nova Delphini, whose prenova state has been identified (STEPHENSON, 1967) as a 12.0 mag blue star, making the rise to the September maximum about 7.0 mag, which is amongst the lowest on record. This is consistent with the slowness of the nova and its unusual behaviour.

I now wish to run briefly through the data collected at the Dominion Astrophysical Observatory (full description in preparation for D.A.O. publications), describing the main features of the spectrum, and then sketch a preliminary analysis. We were fortunate in having a new grating spectrograph installed just before the outburst in July 1967, and we have obtained a series of spectra of the nova at 15 Å/mm, weather permitting, over 11 months of the year. A number of spectra were obtained at 60 Å/mm to fit in with the requirements of other observers, and I should acknowledge my gratitude to the many D.A.O. observers, who together took most of the spectra. At present there are some 160 spectra, over 100 of which are on the 15 Å/mm dispersion. A few spectra were taken at 2.5 Å/mm, which yielded some important information.

While it has as yet been impossible to do justice to the data available, detailed analyses have been made of good-quality plates at intervals of about 1 month over the past year. The discussion is now divided into three sections, covering the three main spectral phases of the nova.

M. Hack (ed.), Mass Loss from Stars. All rights reserved

2. Phase 1

The first phase of the nova lasted from outburst until December 1967. During this time the spectrum consisted of similar P Cygni type profiles for all lines, mainly attributable to H, Fe II, Cr II, and Ti II.

Figure 1 shows the profiles (photographic darkening) of Hγ over this period, which are seen for the most part to be broad in absorption, with a sharp and relatively

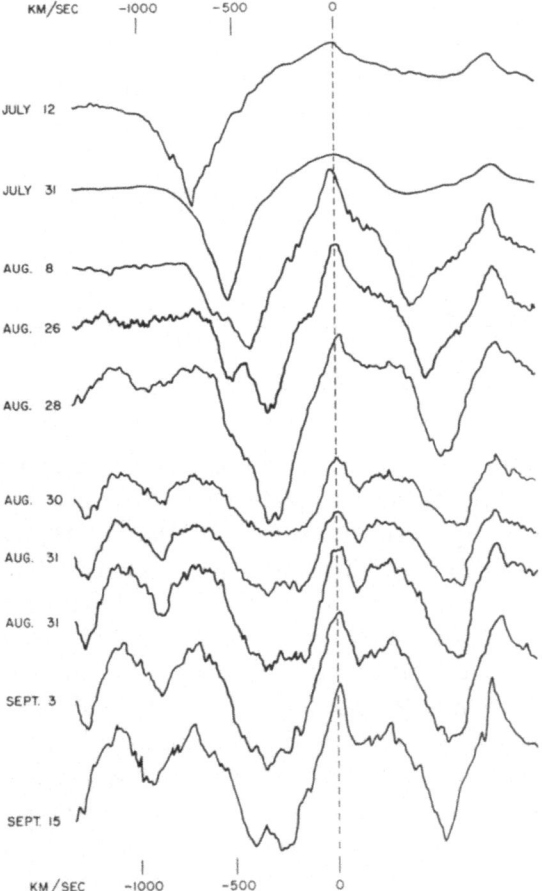

Fig. 1.   Tracings of Hγ during phase 1.

weak emission component. This type of profile is very typical of a continuous atmosphere extending perhaps 1 or 2 radii from the photosphere (see below), rather than a detached shell moving away from the star. The earliest two profiles do have a sharper minimum, which is more suggestive of a detached shell. This suggests that a shell-like structure was ejected in July and that the photosphere itself later expanded below the shell at approximately the same rate.

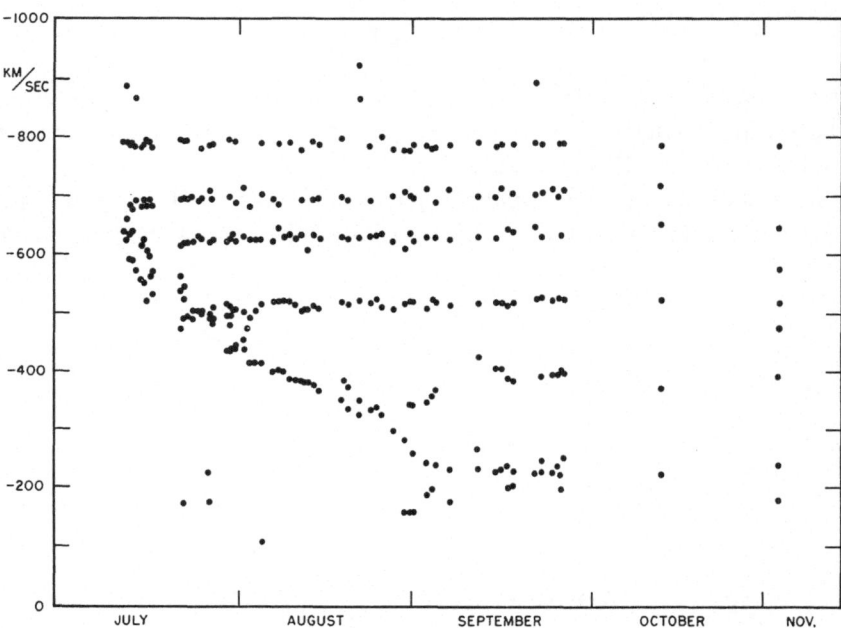

Fig. 2. Absorption-line velocities during phase 1.

Figure 2 shows the velocity curves for this phase. The bottom curve represents the mean velocity of the Balmer and CaII absorption lines, while the four upper horizontal curves are very narrow components of the CaII H and K lines, which are found in addition to the main components. These four narrow lines are seen gradually weakening throughout phase 1, and the velocity of the lower three may show a slight increase of velocity with time. However, there is little if any indication of interstellar damping to their outward motion, and we may make the assumption that their expansion is ballistic.* In order that the shells reach a constant velocity as soon as July 12th, and have so low a velocity dispersion, it is necessary that they be ejected at least 100 km/sec above $V_{esc.}$ from a star of relatively small radius. The most likely values of g and R for the star give $V_{esc.}$ about 1500 km/sec and the maximum radius at 'CaII shell' ejection some 10^7 km. It thus seems likely that these shells were ejected in the nova outburst some time before the main explosion, while the central star was still relatively small. It is possible to match the main absorption curve with a ballistic for the same $V_{esc.}$ if we assume the photosphere then to be $100-200 \times 10^6$ km in radius. If we assume that the photosphere itself then expanded at about the same rate, we obtain a radius of some 1000×10^6 km by mid-September. In this connection we note that a maximum in the light curve was not reached until mid-September and that the initial rise of a nova is generally associated with expansion of the photosphere.

* There is an alternative assumption that these lines represent shock waves, which will not be discussed in detail here.

A further notable feature of the sharp emission lines in phase 1 was a systematic difference of some 10 km/sec between the Balmer and metallic lines (in the sense Balmer more positive). This difference was also observed in 1968 but only when new, sharp emission peaks were observed. The correlation with emission peak width does not necessarily reflect only measuring accuracy, as the velocities agree well when the emission lines are less sharp. The phenomenon is possibly connected with the extent of the atmosphere and the relative opacities of Balmer and metallic lines. Balmer lines are formed in the less dense outer layers of an extended atmosphere, which are more transparent to the receding (positive velocity) matter. This argument adds support to the continuous atmosphere (as opposed to separate shell) hypothesis, but it is not clear why the velocity difference should always be close to 10 km/sec.

It is not intended to give a catalogue of the elements present in the spectrum. It is, however, worth noting that as phase 1 progressed, the excitation potentials of the lines observed corresponded to a gradually decreasing temperature of the envelope.

3. Phase 2

This phase began with the outburst in December 1967 when the star brightened to 3.5 mag. No observations were secured until early in January, when a very changed spectrum was seen. Phase 2 was characterized by a spectrum with rapidly changing

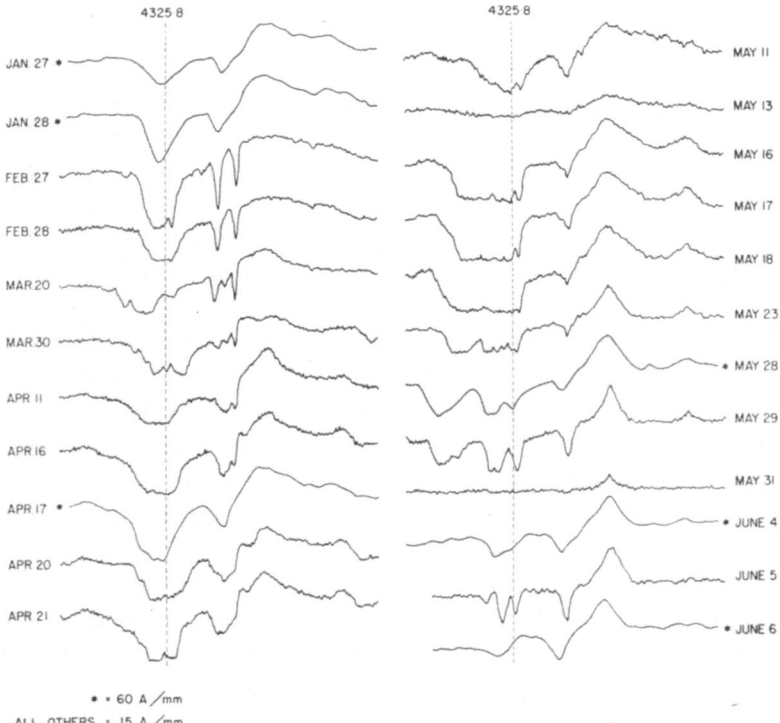

Fig. 3. Tracings of Hγ during phase 2.

multiple absorption and emission lines, accompanied by rapid changes in the brightness of the star, extending to the end of May 1968.

Figure 3 shows a series of plate density tracings of Hγ and the changes can readily
be seen. The profiles during this phase are different from those of phase 1, having
sharp absorption components which weaken with time, and initially sharp emission
components which spread in time. These are typical of distinct shells which detach
themselves from the photosphere and expand from it, as illustrated by the computed
profiles in Figure 4.

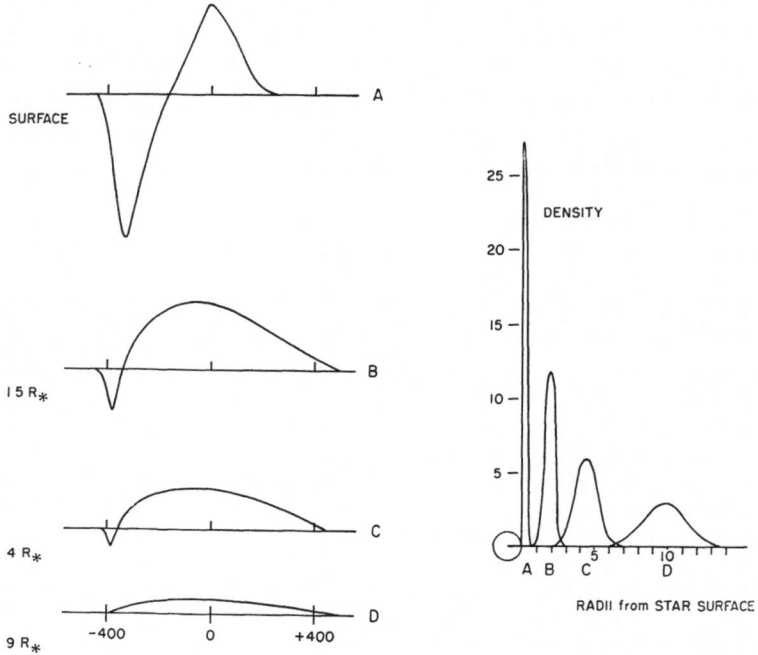

Fig. 4.   Computed profiles of Hγ for shell ejected from star at 400 km/sec.

It therefore appears that during phase 2 the photosphere is more or less stationary
and repeatedly ejecting shells at velocities up to 1500 km/sec. It is possible to trace
the history of some of the shell profiles and by matching profiles with those computed,
knowing the expansion velocity, to obtain a photospheric radius. Such values come
out to be in the region of 1000×10^6 km, which is in good agreement with the figures
estimated for phase 1. Figure 5 shows the velocities of the absorption components
during 1968.

The shells ejected during phase 2 show definite differences in temperature which
may be estimated by noting the excitation potentials of the lines found in the various
absorption components. These form a coherent picture of temperature differences, with
the exception of a few lines, which may very possibly be sensitive to departures from
LTE, and are being investigated with this in mind.

At the end of March the emission lines became stronger and sharper and sharp

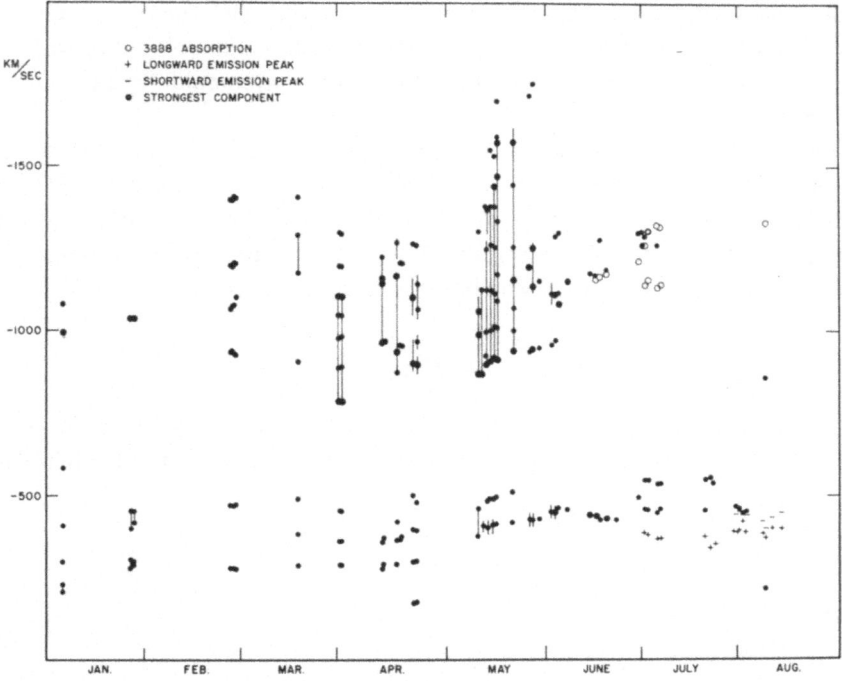

Fig. 5.   Line velocities during 1968.

weak emission components of high-temperature lines (He I, Mg II, C II, Si II) were seen. These died away, but reappeared in June in a similar way, which we shall deal with shortly. The high-temperature lines vanished in April and Cr II lines reappeared, evidently in a new shell of lower temperature. This shell had the strongest and sharpest lines during April, while other high-velocity absorption lines weakened and merged.

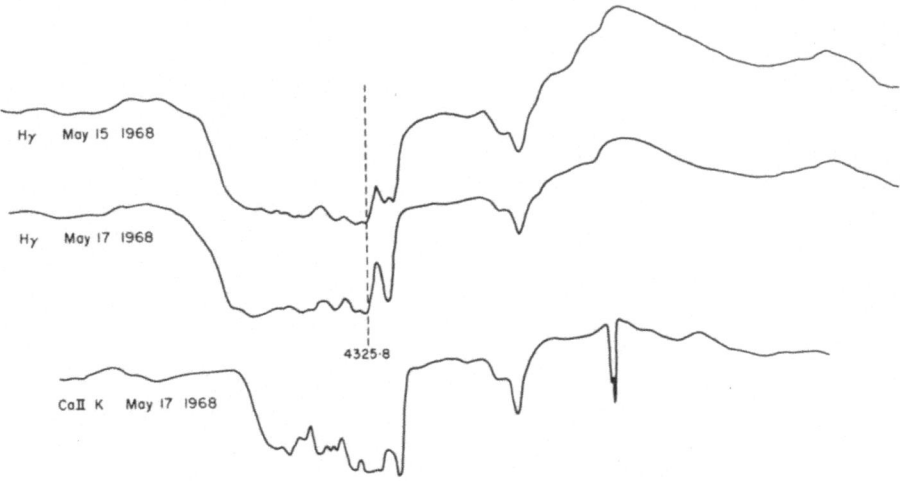

Fig. 6.   Smoothed tracings from high-dispersion plates (2.3 Å/mm).

Finally, May brought a spectacular end to phase 2 with the ejection of some 10 new shells of high velocity, most of which blended, or obliterated each other within a few days. Only the strongest Fe II lines and one Cr II absorption could be seen in the new spectrum.

Figure 6 shows tracings from high-dispersion plates (2.3 Å/mm) taken 2 days apart, showing the changes and blending. The main activity was in the Balmer and Ca II lines and the absence of new features in the remainder of the spectrum is taken to reflect an abundance difference rather than a temperature difference here. This is borne out by events in phase 3, which followed these outbursts.

4. Phase 3

This phase followed the end of shell ejection and saw many drastic changes in the spectrum. The Balmer emission lines became sharp and strong and the high-tempera-

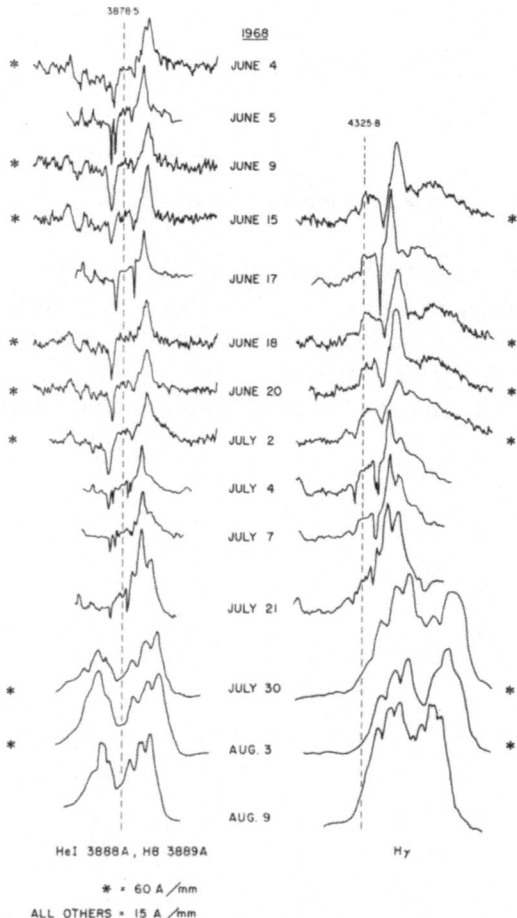

Fig. 7.   He I 3888 and Hγ during phase 3.

ture lines reappeared strongly in emission, some 20 km/sec more positive than the Balmer and old Fe II lines. In the ultraviolet, the region 3600–3100, previously heavily line blanketed, appeared line-free and very strong. Broad absorption components appeared at 4060–4070, possibly due to N III or Si IV absorption. Simultaneously, the He I lines 3888 and 3187 appeared in absorption, very sharp and strong. These are from the metastable 2^3S level and are the only line which may appear in absorption at high dilution. Figure 7 shows tracings of He I 3888 and Hγ for this phase.

The following hypothesis is therefore proposed. After the May outbursts, which may have been metal-poor, the cessation of ejection, and the expansion of the older metallic shells, the ultraviolet line blanketing was removed. This allowed hot ionising radiation to reach the inner edge of the last high-velocity shells. This produced the B-type emission lines and the He I 2^3S absorption. Similar events may have occurred in March but died due to later metallic ejection. Comparison of the continua in January and June (from Hα down to 3100 Å) shows a great strengthening in the UV and weakening in the red with time; that is, the central star in June may be shrinking and heating up again. Since the beginning of phase 3 the luminosity of the nova has been decreasing steadily so that there has been less radiation pressure to support the photosphere at its maximum extension.

From mid-June into July a further development was the building up of a second emission component with positive velocity, in the Balmer lines and a few other strong lines. The velocity is some 400 km/sec, possibly variable by 50 km/sec. Figure 8 shows direct intensity profiles of Hγ obtained with a high-resolution photoelectric scanner developed at the D.A.O. The strength of Hγ emission is seen to be several times that of the local continuum, and the building-up of the positive emission component and

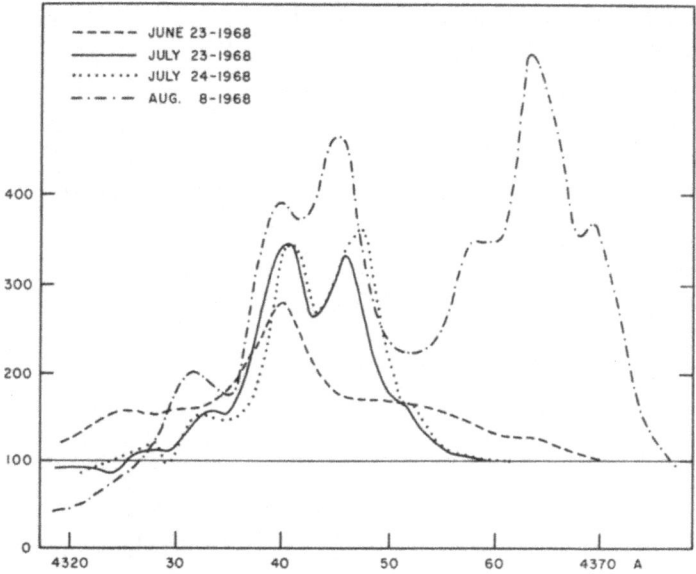

Fig. 8.   Photoelectric scanner observations of Hγ.

then a negative velocity component is clearly seen. Such fragmentation of the emission spectrum has been recorded at similar stages of other novae and taken as evidence for the non-singularity of novae. While it is very difficult to explain such structure geometrically on a single star hypothesis, it is not certain that the phenomenon is unconnected with collapse of the photosphere or fragmentation of ejecta.

The B star lines faded by the end of July and were rapidly replaced by the forbidden auroral lines of O III and Ne III, showing the same triple structure, but with weaker outer components. In early August these lines began to fade and the original (central) component of the Balmer lines weakened. This is as far as observation extends at present and only very tentative conclusions can be drawn. The very rapid development of the high temperature and then the auroral spectrum seem to indicate a rapid collapse of the photosphere, thus increasing the radiation dilution.

5. Interstellar Ca II

These lines show up sharply on all of the spectra of 15 Å/mm or better dispersion, and indicate an interstellar velocity of $-2.9 \pm .1$ km/sec. However, the very high-dispersion plates show the lines to be double (Figure 6). A high-dispersion plate of the 5.5 mag B 1.5 supergiant HD 190603, some 10° away in the sky, shows interstellar Ca II with four components, so that there is no close comparison of distance. However, the ratio of equivalent widths of Ca H and K in the two stars is 1.36 and 1.35 respectively, so that we may compare the stars as follows:

The supergiant is found to have $M_v = -7.6$ and interstellar absorption of 2.1 mag. If the nova is as reddened as the B star then at nova $m_v = 5$, its $M_v = -7.5$ and its distance some 1000 parsecs. If the nova is unreddened, the figures are $M_v = -5.3$ and distance 350 parsecs. The nova lies at galactic latitude 14° while the B star is almost in the galactic plane. Therefore it seems likely that

(a) the nova is less reddened than the B star,

(b) the distance is less than 400 parsecs, to keep it in the galactic plane.

Also, the slowness of the nova should give it a maximal M_v close to -6, from comparison with other novae. Finally the calibration by BEALS (1937) of Ca K equivalent width indicates a distance of some 500 parsecs.

References

BEALS, C. S.: 1937, *Publ. Dom. Astron. Obs.* **6**, 336.
STEPHENSON, C. B.: 1967, *Publ. Astron. Soc. Pacific* **79**, 584.

Discussion

Friedjung: The profiles of the emission lines indicate for other novae an asymmetry in the ejection. For Nova Aquilae the profiles were needle-shaped, and this has been well correlated with two masses of gas, one observed to be approaching the observer, one receding. Did I understand that the photosphere continued to expand up to the end of phase 1? The light curve seems to have flattened before the end of phase 1, and the photosphere could not have expanded.

Hutchings: The asymmetry of the emission lines in the nebular stage may indeed indicate non-spherical symmetry of the ejected matter. However, the interpretation of the line profiles formed closer than 10 photospheric radii suggests that here the matter *is* distributed fairly symmetrically. The photoelectric light curve does indicate that the luminosity of the star increased slowly until about mid-September. From this time until December the velocities of recession are low so that even if the photosphere ceased to expand below the ejecta, the approximation of continuous atmosphere is not seriously in error. We do not have many observations in the period October–December, so that I cannot be sure about this point.

Larsson-Leander: You mentioned in your paper that the various maxima of the emission profiles might be due to the multiplicity of the system. Surely you did not mean that this has anything to do with the possibility of the star being a close binary, but that it is due to irregularities in the expanding principal envelope.

Hutchings: Structure in the nebular emission lines has been suggested as indicating duplicity of the stellar system. However, it appears more likely to me that the structure in Nova Delphini is caused by irregular structure of the emitted matter. I hope future observations of the nebular mass throw more light on this.

Dallaporta: I have been told that Nova Delphini has presented a quite anomalous light curve, with a very long stay to the maximum and a very slow increase. Which could be the physical reason for this anomalous behaviour in respect to more standard novae?

Hutchings: As mentioned in my opening remarks (see above) the correlation between rapidity of nova development and absolute magnitude at maximum suggests that the rapidity of the ejection mechanism was controlled primarily by radiation pressure from the central star.

POLARIZATION OF NOVA DELPHINI 1967

BEN ZELLNER

Lunar and Planetary Laboratory, The University of Arizona,
Tucson, Ariz., U.S.A.

Abstract. The light of Nova Delphini 1967 has intrinsic linear polarization, with both a small-amplitude fluctuation with a time-scale of days and a secular variation with a time-scale of months. Rotation of the plane of polarization with time, arising from the interaction of intrinsic and interstellar polarization, is present, but too small to allow a confident evaluation of the interstellar component. A few observations of Nova Vulpeculae 1968 suggest that it also has variable polarization.

Intrinsic polarization in the light of an active nova was first reported by EGGEN *et al.* (1967), who found the recurrent nova T Pyxidis to be variable in the degree, position angle, and wavelength-dependence of polarization. The degree of polarization was found to change by more than 0.5% in 2 hours and more than 2% in 2 weeks, while rotations of position angle as large as 32° per day were reported.

The linear polarization of Nova Delphini 1967 was measured by the author in August 1967 and April–August 1968. Measurements were taken in seven filters from 0.33 to 0.96 μ with the 154-cm Catalina reflector of the Lunar and Planetary Laboratory. The equipment and observing procedure have been described by COYNE and GEHRELS (1967).

The nova was found to exhibit variable polarization, though not so violently as T Pyxidis. A rapid, irregular variation of up to 0.3% per day is present in all filters but most active in the red and green. There is also a slow secular increase in all colors except the deep ultraviolet; the greatest increase is in visible light, and has amounted to more than 0.8% over the span of a year.

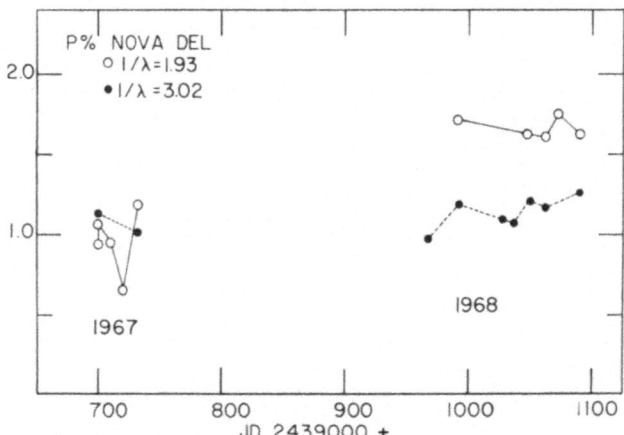

Fig. 1.   Polarization of Nova Delphini 1967 as a function of time, in green (open circles) and ultraviolet light (filled circles).

M. Hack (ed.), Mass Loss from Stars. All rights reserved

Figure 1 shows the time variation of polarization for the filters at wave numbers of 1.93 and 3.02 μ^{-1} (green and ultraviolet). The radius of the open circles gives the typical probable error of a single observation. The secular increase in the green filter is about 0.06% per month, while that in the ultraviolet filter is very small or absent.

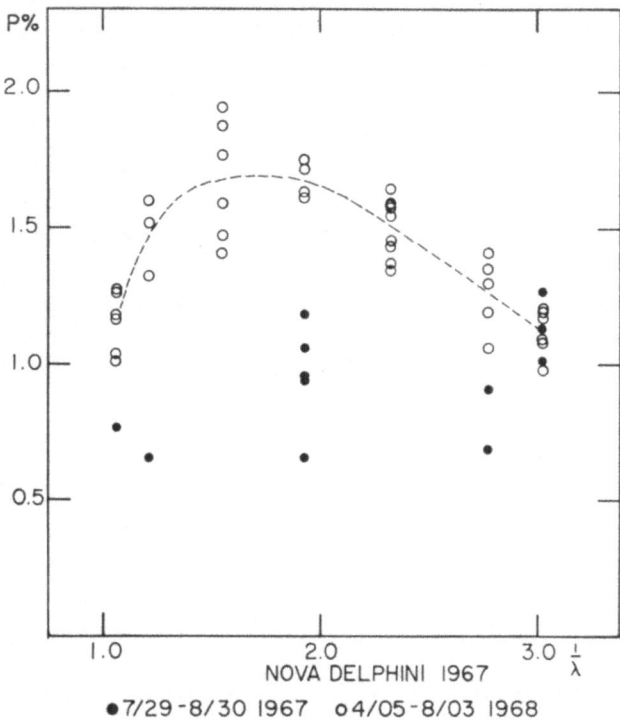

Fig. 2. Polarizations of Nova Delphini 1967 vs. wave number in inverse microns. Filled circles are 1967 observations, open circles 1968 observations. The scatter is real, reflecting both secular and short-term variations.

Figure 2 gives all polarizations observed through August 3, 1968, plotted as a function of wave number in inverse microns. While the wavelength-dependence in 1967 was in general rather flat, with a slight rise into the ultraviolet, the mean of the 1968 observations (dashed line) shows a peak at about 0.57 μ with a slow decline into the ultraviolet and a rapid drop into the infrared. Observations in late August 1968 (not shown) show no change from this pattern.

It is possible in principle to sort out a constant interstellar polarization from an intrinsic polarization that is fixed in position angle but variable in degree, by the way in which the observed position angle changes as the observed polarization increases. Such a systematic rotation of the plane of polarization is seen in the Nova Del data, but it is small, amounting to only 3° or 4° over a year. Since this is only a few times larger than the probable error of a good observation of position angle, an attempt to find the interstellar component by this method would be of doubtful validity.

In view of the large change in polarization, this small rotation requires that the interstellar and intrinsic components be nearly parallel or perpendicular, or that the interstellar polarization be small.

The mean curve through the 1968 observations shows a wavelength-dependence similar to that for the characteristic interstellar polarization (COYNE and GEHRELS, 1967). Thus one might suspect that the interstellar polarization is dominant in 1968, and that the reduced polarization in 1967 and the short-term variations reflect the activity of an intrinsic component which is nearly orthogonal to the interstellar polarization. However, this is by no means a firm conclusion.

A few observations of Nova Vulpeculae 1968 in blue light show a steady increase from 1.08% polarization at the end of May to 1.42% in late August. There is no indication of a short-term variability. The wavelength-dependence resembles that of interstellar polarization, so the intrinsic component may be small.

Intrinsic polarization of novae is a natural consequence of light scattering in a circumstellar ring or shell. Until a case arises where the interstellar component can be evaluated with some confidence, however, observations of polarization will not give much quantitative information about the properties of the shell.

References

COYNE, G. V. and GEHRELS, T.: 1967, *Astron. J.* **72**, 887.
EGGEN, O. J., MATHEWSON, D. S., and SERKOWSKI, K.: 1967, *Nature* **213**, 1261.

Discussion

Treanor: Zellner's observational results are very interesting: as regards to short-term variations, it would be valuable to have in some subsequent work of this kind, independent mutual confirmations by observers working simultaneously with different equipment. Zellner's arguments regarding the separation of the intrinsic and interstellar polarization are ingenious, but seem to me to be pushing the observational results rather hard. It is not quite true that the two types of polarization can always in principle be separated by studying the variations of position angle. At best it is only the variable part which can be separated to some extent; also the use of the mean interstellar wavelength-dependence curve, while it is the only procedure possible, is a weak step in this argument.

FRESH EVIDENCE CONCERNING THE TYPE
OF EJECTION OF NOVAE

M. FRIEDJUNG

Institut d'Astrophysique, Paris, France

Abstract. Reasons are firstly summarized for believing that ejection of matter by novae continues long after maximum light. Preliminary conclusions of a study of Balmer absorption-line profiles of nova Herculis 1963 are also given, the methods used being capable of indicating the nature of absorption systems.

I shall talk today about fresh evidence concerning the nature of the ejection process in novae after maximum light. I shall firstly discuss fresh arguments based mainly on old observations of nova V 603 Aquilae to be shortly published in *Astrophysical Letters*, and then some interesting results for nova Herculis 1963 just obtained by me. The latter are unfortunately only of a preliminary nature; I hoped to have more results to present but much time was lost by me for various reasons in the past few months.

Reasons were previously given by me (FRIEDJUNG, 1966a) based on the light curves of various novae, for believing that ejection of most material after maximum light was continuous rather than instantaneous. In this piece of reasoning it was supposed that the continuous radiation of an expanding nova envelope producing all the light seen, was mainly due to free-bound and free-free transitions of hydrogen, supposed mainly ionized. In such a case the observed rate of fading of the continuous spectrum of several novae after maximum light, would be too slow for ejection to be instantaneous. The assumptions behind this reasoning can be criticized as they have been by POTTASCH (1967), but fresh, more rigorous reasons based mainly on the observations of V 603 Aquilae can be given. Such arguments will be based on V 603 Aquilae, for not only was this nova well observed, but also quantities for it varied in such a way that the interpretation is much less ambiguous than for more recent novae.

One means of testing whether the hydrogen of a nova envelope is almost completely ionized is to study the absorption spectrum. As is well known, the absorption lines of a nova occur in absorption systems, each system consisting of lines of a certain range of excitation with a certain Doppler shift. The systems that appear after maximum light have been classified by McLaughlin into the broad categories Principal, Diffuse Enhanced, and Orion. In practice the situation is generally more complex than this simple classification would suggest; there can be more than one system of a certain category, the systems of this category having similar Doppler shifts. Reasons have already been given by me (1966a) for supposing that systems classified as Principal and Diffuse Enhanced are formed in parts of a nova envelope where hydrogen is mainly neutral. These systems show, beside lines of hydrogen, lines of neutral and ionized metals. For V 603 Aquilae they showed at certain phases

M. Hack (ed.), Mass Loss from Stars. All rights reserved

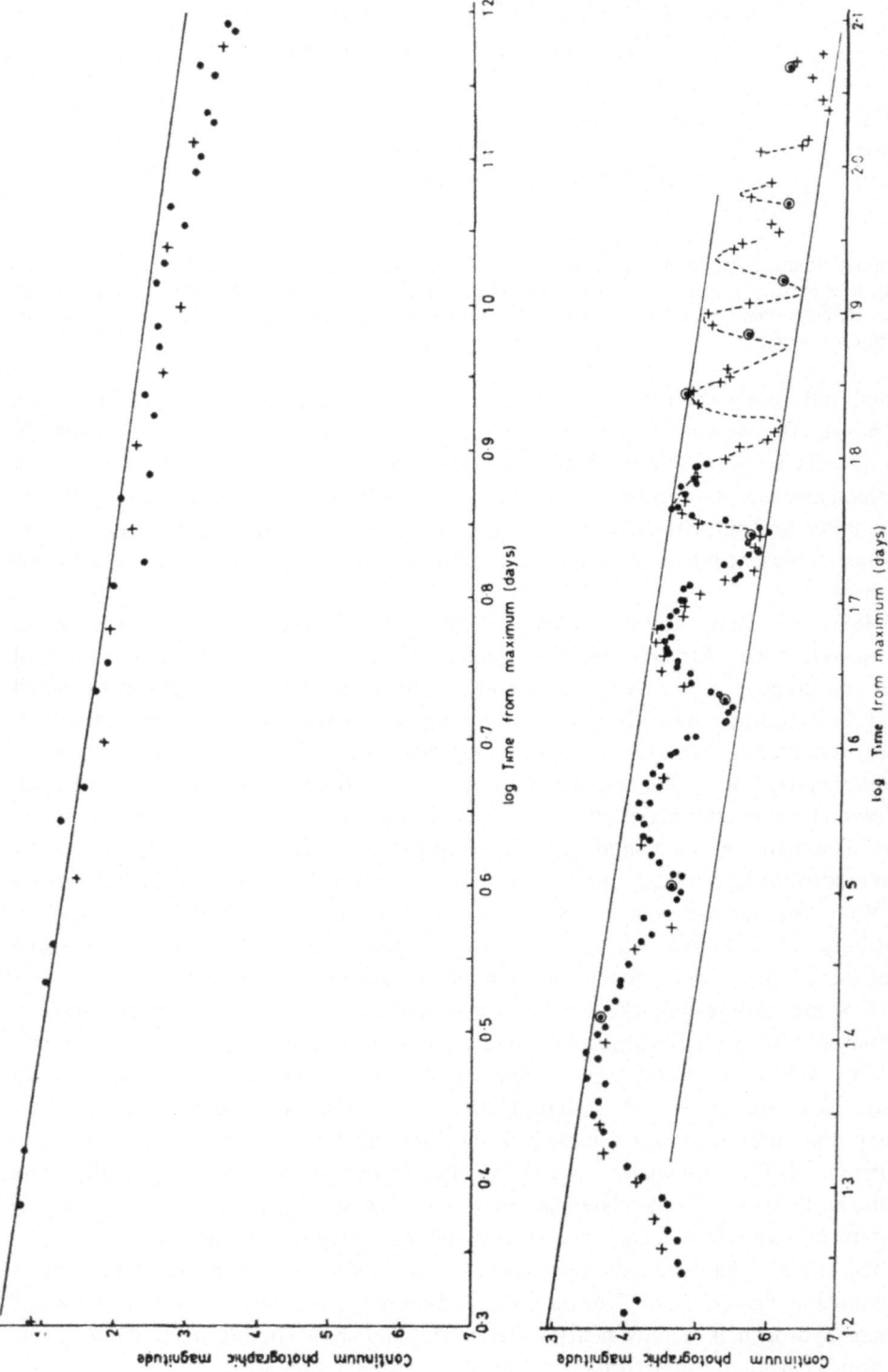

Fig. 1.   The continuum magnitude of V 603 Aquilae against log time from maximum

the sodium D lines even, which definitely suggests formation in a region where hydrogen was predominantly neutral.

At phases when the Principal and Diffuse Enhanced absorption systems of V 603 Aquilae disappeared, it is reasonable to assume that hydrogen was almost completely ionized in the envelope, and the previous light curve arguments can be applied. Figure 1 is a graph of the magnitude of the continuous spectrum of V 603 Aquilae against log time from maximum. The light oscillations took place between lines in the diagram which are almost parallel, and at the minima one can use the argument concerning the absence of the Principal and Diffuse Enhanced systems to deduce that the hydrogen was almost completely ionized. The rate of fading from one minimum to the next is still too slow for instantaneous ejection. However, the fading now considered is over a smaller time interval than in the original reasoning (FRIEDJUNG, 1966a), and hence the discrepancies between the amounts of fading predicted by instantaneous ejection models and those observed, are less. They could be partly removed by supposing a change in temperature of the envelope. over the interval considered and the double ionization of helium over this interval. However, this removal would be only partial, and also if the line joining the minima is extrapolated back to maximum light, the time interval considered becomes much larger.

Besides the arguments from the light curve of V 603 Aquilae which can be extended further, the absorption lines provide further reasons against instantaneous ejection. The lines of the Principal system were extremely narrow. This must be considered as showing them formed in a region much larger than that giving rise to emission, if as one generally does, one interprets the violet shifts of nova absorption lines as due to expansion at high velocity of the material ejected by a nova. These narrow lines were superposed on blended Balmer emission lines near the Balmer limit, and geometrical considerations indicate that both line and continuous emission came from a much smaller radius than the line absorption. Thus all the material could not have been in the same instantaneously ejected thin shell. Forms of instantaneous ejection model without a thin shell also contradict the observations, as they would require smaller expansion velocities near the centre of a nova envelope, the material in such cases near the centre having been ejected at the same time, but expanded much less than that further out. This would contradict observations of the widths of emission lines, which can be used to give the expansion velocity of the line-emitting region.

A strong argument against instantaneous ejection is finally provided by the Orion-system absorption lines of V 603 Aquilae. This system was for reasons given by McLAUGHLIN (1965), almost certainly formed beneath the layers producing the other absorption systems. The material producing it cannot have all been ejected at a similar time to that of the material further out, as it had a higher velocity and would have collided with the latter long before it finally disappeared. It seems that the persistence of this system can only be explained by continued ejection, if one wishes also to account for the observed velocity variations of it. Moreover, at some phases absorption lines belonging to it absorbed much of the underlying continuous spectrum, showing that a large part of this was formed in still lower layers.

All the arguments given suggest that most of the emission of V 603 Aquilae came from near the centre of the envelope. The spectrum showed that all layers observed were in expansion, thus one cannot suggest that the emission came directly from the central star. Continued ejection fits the observations best.

As shown previously by me (FRIEDJUNG, 1966a) simple continued ejection models tend to have two light emission and density maxima, one near the outer edge, one near the centre. The one near the centre seems to have produced most of the light for V 603 Aquilae as just shown, and probably does for many other novae. The problem then remains how to explain the absorption systems. It was previously suggested by me (FRIEDJUNG, 1966a, c) that the system or systems belonging to the category 'Principal' of McLaughlin were formed at the outer edge, the others near the centre of a nova envelope. A velocity gradient in the envelope was required. There were, however, serious difficulties and a new interpretation was tentatively suggested by me recently (FRIEDJUNG, 1968), in which the Orion and Diffuse Enhanced systems are formed by high-velocity material immersed in material slowly expanding outwards, at lower velocities characteristic of the system or systems referred to as Principal in McLaughlin's terminology. More data are required before a final decision concerning the formation of the absorption systems can be made.

In the current investigation the idea has been to study the optical thickness of the Balmer absorption lines at different positions, corresponding to different Doppler shifts and hence radial velocities. Such a study is not only able to yield the optical thickness but, what is more interesting, the proportion of light that the line can absorb, which gives information about the geometry of line formation.

For the purpose of the study spectra of nova Herculis 1963 taken at the Haute-Provence Observatory, were kindly lent by Dr Fehrenbach. The spectra are of high dispersion (9.7 Å/mm for the blue-sensitive ones, 12.4 Å/mm for the red plates). Up to now only one spectrum, for March 4, 1963, has been studied, but even this gives some results of interest.

For Balmer absorption lines broadened mainly by the Doppler effect, the optical thickness of a particular line is proportional to $f\lambda$, where f is the oscillator strength, and λ the wavelength. In Figure 2 a graph is shown for positions in the different Balmer lines, corresponding to one radial velocity. The abscissa of the graph is the optical thickness of each line relative to that for $H\delta$; the ordinate the difference between the logarithm of light before and that after line absorption at a position with the relevant radial velocity, for the relevant Balmer line. These measures, it was shown, were not very sensitive to errors arising from the finite resolution of the spectrograph. To find the light intensity before line absorption, two different assumptions were used. According to the first assumption the Balmer emission-line profiles were symmetrical, so the light before absorption at a negative radial velocity equalled that at the corresponding positive radial velocity. This may not be valid because of occultation of some parts of the envelope by the central parts, and so a second assumption was also tried. According to this, the emission-line intensity was constant for negative radial velocities, the profiles hence being asymmetric.

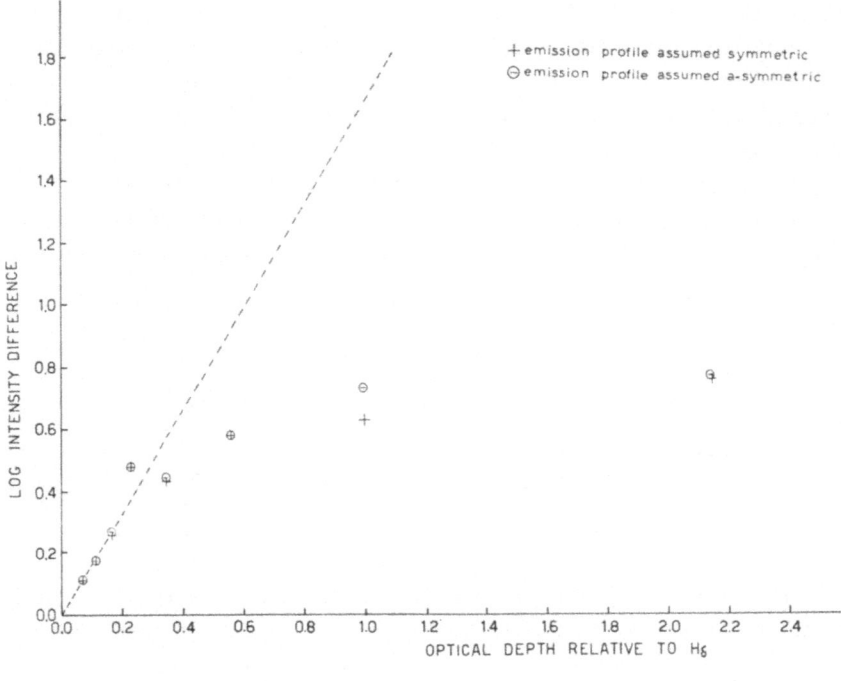

Fig. 2.

The graph of Figure 2 is for a radial velocity of 1094 km/sec, very near the velocity corresponding to the centres of the Principal system absorption lines; the Balmer lines for this nova having Principal and Diffuse Enhanced components. As one can see from the diagram, the graph is curved, while one would expect it to be straight if all light at the relevant radial velocity could be equally absorbed. (Note that the values for H 12 were not plotted in the diagram, as visual inspection shows the line obviously blended.) One can conclude that all light at the relevant radial velocity is not in a position to suffer absorption to the same extent.

Similar results are obtained for 4 other radial velocities corresponding to positions inside a Principal absorption system line, the range of velocities for such results being 1245–1018 km/sec. The results plotted in Figure 2 and in the graphs for the other radial velocities show that the graph points corresponding to Balmer lines far from the start of the Balmer series appear to fall near a straight line for each graph, this being quite noticeable in Figure 2. This would appear to indicate that for lines far from the start of the Balmer series, all the light of a given wavelength was absorbed to the same extent. The straight lines for the graphs corresponding to each radial velocity gave optical thicknesses for each Balmer line, and these in their turn gave for Hβ, Hγ, Hδ, the proportion of light that was not for presumably geometric reasons, absorbable by the material producing absorption far from the start of the Balmer series. There are clearly uncertainties in the optical thickness determination, evident if one considers the scatter of points in Figure 2 and the other graphs. However, the optical thicknesses are so large for Hβ and Hγ, that the proportions of light calculated as unabsorbable are quite reliable (errors due to optical thickness uncertainty $\leqslant 10\%$ except for Hγ at velocity of 1018 km/sec). For Hδ the same type of errors are greater, being near 20% for velocities of 1169 and 1120 km/sec, and even larger for the other velocities, excepting 1094 km/sec, where the error is below 10%.

Results for the proportions of light unabsorbable, calculated assuming a symmetric emission-line profile at various radial velocities, are given in Table I. All these velocities correspond to positions inside a line of the Principal absorption system.

TABLE I

Radial velocities

Line	1245 km/sec	1169 km/sec	1120 km/sec	1094 km/sec	1018 km/sec
Hβ	0.35	0.20	0.15	0.17	0.35
Hγ	0.29	0.22	0.16	0.17	0.33
Hδ	0.13	0.17	0.17	0.21	0.36

In general these values and values obtained assuming an asymmetric profile are sufficiently small, so as to indicate that both continuous and line emission were absorbable (at 1018 km/sec this is not true for Hβ and Hδ). This sets an important condition for a model of the formation of the Principal absorption system. For a nova continuously ejecting, one expects most continuous emission to be formed either near the centre of the envelope or near the outer edge (FRIEDJUNG, 1966a). Now for this nova at least about 9/10 of the continous radiation was absorbable for the Diffuse Enhanced system, which in novae, for reasons given by McLaughlin, is generally considered as formed nearer the centre of the envelope than the Principal system. This means that for nova Herculis 1963, most continuous emission was presumably formed near the centre of the envelope. If line absorption of the Principal system were produced by material at a constant velocity, continuous emission, being almost all formed near the centre of the envelope, would be only absorbable near one wavelength for geometric reasons. The velocity spread of absorption of continuous radiation by the Principal absorption system is much greater than expected, and shows there must have been a velocity gradient where the Principal system was formed, giving velocity differences in the line of sight of at least 150 km/sec, and probably more than 200 km/sec.

Another condition is set by the near-constancy of the velocity of maximum Principal absorption system absorption along the Balmer series. For the plate studied, this velocity seems only to have changed between 30 and 40 km/sec along the Balmer series. Now the profiles of the Hβ and Hγ Principal absorption system lines were determined by the light unabsorbable by the material that produced line absorption far along the Balmer series; this light suffering absorption it seems therefore by other material of a lower optical thickness. This material of lower optical thickness must have had a similar space and velocity distribution to produce maximum line absorption at a similar radial velocity to that of the material of higher optical thickness, hence one is naturally drawn to explain the observations in terms of small scale inhomogeneities of density. The ratio of optical thickness required is of the order of 100:1 near the Principal absorption system line centre, the exact value not being easy to determine because of observational uncertainties. An objection can be raised to the model of density inhomogeneities. These would be expected to affect absorption for the whole Balmer series, not just lines near the start. In such a case, graphs like that of Figure 2 would never become linear. The apparent linearity could, however, still be explained if line emission were formed at larger distances from the centre of the envelope than the continuous emission. Line emission is very weak far from the start of the Balmer series, and hence one might there observe only continuous radiation passing through a region of nearly constant optical thickness. It should be noted that even if the apparent linearity of the graphs far from the start of Balmer series were not real, and the true optical thicknesses for $H\beta$ and $H\gamma$ were much larger, the previous conclusions would not be changed much.

It must be noted that if there is a velocity gradient, and material at the outside can absorb line emission from further in, the velocities in the central regions cannot be less than those of material responsible for Principal system absorption. Thus velocity must at least for a certain range of distance decrease outwards.

Fewer results have been obtained for the range of radial velocities corresponding to absorption lines of the Diffuse Enhanced system. Most of the radiation of the same wavelength was absorbed at $H\beta$, and came from lower layers than the region of line formation. This includes line emission at these radial velocities, much less, however, than the line emission at radial velocities corresponding to the Principal absorption system. However, accurate quantitative calculations cannot be made without knowledge of the amount of light scattered by the spectrograph because of the low intensity of light not absorbed at $H\beta$ (about 10% of the continuum). No detailed measures for light scattered by the spectrograph used, have yet been published.

The results given for nova Herculis 1963 in this paper are only preliminary, and may have to be modified later. However, this approach may help to solve the problem of the nature of the absorption systems of novae.

References

FRIEDJUNG, M.: 1966a, *Monthly Notices Roy. Astron. Soc.* **131**, 447.
FRIEDJUNG, M.: 1966b, *Monthly Notices Roy. Astron. Soc.* **132**, 143.
FRIEDJUNG, M.: 1966c, *Monthly Notices Roy. Astron. Soc.* **132**, 317.
FRIEDJUNG, M.: 1968, *Astrophys. Letters* (in press).
McLAUGHLIN, D. B.: 1965, *Novae, Novoids et Supernovae*, Paris, p. 123.
POTTASCH, S. R.: 1967, *Bull. Astron. Inst. Neth.* **19**, 227.

Discussion

Larsson-Leander: I believe there is very strong evidence for the view that the principal spectrum of novae is formed in a more or less detached shell. In many instances, details in the contours of the principal emissions have been found to correspond to features in the visible expanding nebulae years afterwards. Also the principal spectrum does not react very much to fluctuations of light during the decline, while the Orion spectrum as to intensity and velocity is very sensible to light variations.

Friedjung: I believe the material of the principal absorption system consists of material ejected soon after maximum light, when the ejection rate is highest. The situation for the principal system emission is more complex. However, I agree as concerns the principal absorption.

A MECHANISM FOR THE PRODUCTION OF
PLANETARY NEBULAE

G. S. KUTTER

*Leander McCormick Observatory, University of Virginia,
Charlottesville, Va., U.S.A.*

and

M. P. SAVEDOFF and D. W. SCHUERMAN

*C. E. Kenneth Mees Observatory, University of Rochester,
Rochester, N.Y., U.S.A.*

Abstract. C^{12} stars in the range 1.04–1.55 $M_\odot$ are evolved to simulate the core evolution of the possible precursors of planetary nebulae. The nuclear shell burning in stars above 1.2 $M_\odot$ advances to within about 0.2 $M_\odot$ of the surface, where the intense radiation interacts with the surface matter and causes mass loss. Comparison between our theoretical results and observations suggests that this may be a mechanism by which planetary nebulae are formed.

1. Introduction

We seek a theoretical understanding of the mechanism responsible for the production of planetary nebulae. To this end we evolved simple, but representative, models of the possible parent stars.

According to observations these parent stars are evolved stars in the mass range from 1.0 to 1.5 $M_\odot$, and the formation of the nebulae represents a clean separation of the H/He envelope from the stellar core (O'DELL, 1963; GREENSTEIN and MINKOWSKI, 1964; OSTERBROCK, 1966; SEATON, 1966). The production of planetary nebulae is a common phenomenon with a rate of 0.1–1.1×10^{-12}/pc^3–year in the solar neighborhood (WEIDEMANN, 1968). Therefore, a commonly occurring mechanism in evolved stars of about 1 $M_\odot$ must be responsible. Such a mechanism will probably derive its energy from nuclear sources in the stellar interior to provide the 10^{49} ergs required to carry the envelope out of the gravitational well of the remaining stellar core.

We have chosen initially pure C^{12} stars in the mass range from 1.04 to 1.55 $M_\odot$ to approximate the cores of evolved stars and have followed their evolution. In this choice we were motivated by the possibility of nuclear ignition in an electron-degenerate shell, triggered by the T-inversion arising from the combined effect of a growing electron-degenerate core and neutrino emission. VILA (1965, 1966) and SAVEDOFF *et al.* (1968b) demonstrated that such a T-inversion occurs during the late evolutionary stages of solar mass stars with composition heavier than He^4. The addition of a H/He envelope should not drastically alter our results, except near the surface; in fact, it should only increase the likelihood of mass loss.

M. Hack (ed.), Mass Loss from Stars. All rights reserved

Model Nr.	1	2	3	4	5	6	7
Age (Years)	−139 500	−6676	−427.2	0.0	92.39	98.99	126.1
$\text{Log}(T_c)$	7.985	8.641	8.699	8.691	8.683	8.682	8.612
$\text{Log}(P_c)$	18.789	22.159	23.182	23.314	23.329	23.328	23.148
$\text{Log}(\rho_c)$	3.098	5.691	6.449	6.545	6.557	6.557	6.437
Ψ_c	−2.50	+1.98	5.88	6.80	7.02	7.04	7.16
R_*	10.537	9.780	9.669	9.655	9.653	9.653	9.675
$\text{Log}(T_{eff})$	4.724	5.203	5.326	5.344	5.347	5.347	5.329
$\text{Log}(L_{ph}/L_\odot)$	3.230	3.632	3.906	3.947	3.954	3.955	3.927
$\text{Log}(L_\nu/L_\odot)$	−1.652	4.137	5.218	5.351	5.379	5.381	5.541
$\text{Log}(L_{CC}/L_\odot)$	–	0.076	4.360	4.983	5.329	5.403	7.459
$\text{Log}\lvert L_g/L_\odot\rvert$	3.542	4.612	5.549	5.514	−∞	5.111	7.881
$-\Omega \times 10^{-50}$	0.212	1.35	2.34	2.53	2.56	2.56	2.28
$\text{Log}(\varepsilon_{CC})_{max}$	–	1.30	5.03	5.85	6.64	6.83	9.36
q	–	0.0	0.28	0.27	0.26	0.26	0.25
Ψ	–	1.98	2.39	2.77	2.69	2.50	1.09
$\text{Log}(\varepsilon_\nu)_{max}$	−0.42	+4.84	5.66	5.83	6.02	6.08	6.84
q	0.0	0.0	0.36	0.27	0.26	0.26	0.24
$q(\Psi = 0)$	–	0.40	0.63	0.65	0.66	0.66	0.66
Δq_{cv}	–	–	–	–	–	0.01	0.29
$x_{12}(cv)$	–	–	–	–	–	0.978	0.950

Model Nr. 1. Initial model
         2. T-inversion
         3. Max. T_c
         4. C^{12} ignition
         5. Max. $(-\Omega)$
         6. Onset of convection
         7. Max. L_{CC}
         8. Max. convection

         9. Min. $(-\Omega)$
      10. Min. L_{CC}
      11. $(\varepsilon_{CC})_{max}$ at $q = 0.30$
      12. $(\varepsilon_{CC})_{max}$ at $q = 0.51$
      13. $(\varepsilon_{CC})_{max}$ at $q = 0.71$
      14. Onset of surface convection
      15. Intense surface convection (results are very
              approximate, cf. discussion of Table III).

[a] Refers to convection just interior from surface; all other values of Δq_{cv} refer to convection just exterior to $q(\varepsilon_{CC, max})$.

2. Method of Computation

Our model computations assume hydrostatic equilibrium and are based on the HENYEY (1964) method of integration. At the surface we used the boundary condition

$$m = M: T = 0, P = c_1 T^{c_2},\qquad(1)$$

where the constants c_1 and c_2 are chosen to be consistent with the derivatives of the opacity, κ, and the ratio of the gas pressure to total pressure, β. Temperature-depend-

1.25 $M_\odot$ star

8	9	10	11	12	13	14	(15)
.6	516.6	1256	7985	18980	23420	25710	27270
.501	8.441	8.447	8.444	8.476	8.472	8.428	8.381
.854	22.700	22.725	22.832	23.263	23.818	24.391	24.706
.246	6.147	6.164	6.237	6.530	6.910	7.316	7.542
.24	7.31	7.38	8.19	11.1	17.5	29.0	39.4
.722	9.737	9.723	9.661	9.640	9.694	10.120	10.516
.287	5.254	5.258	5.267	5.295	5.330	5.205	5.069
.854	3.753	3.738	3.653	3.722	3.971	4.322	4.571
.294	5.016	5.018	5.164	5.416	5.633	5.528	5.484
.359	5.030	4.900	5.110	5.345	5.493	5.268	5.196
.745	$-\infty$	4.785	4.580	4.880	5.458	5.632	5.790
.85	1.65	1.69	1.81	2.27	3.19	4.64	5.77
.30	6.89	6.65	6.86	7.34	7.74	7.44	7.39
.24	0.25	0.26	0.30	0.51	0.71	0.84	0.88
.23	−0.11	−0.03	−0.02	−0.55	−1.06	−1.90	−1.88
.65	6.31	6.26	6.37	6.69	7.03	7.05	7.11
.24	0.25	0.26	0.30	0.51	0.71	0.84	0.88
.61	0.24	0.26	0.29	0.48	0.66	0.80	0.86
.40	0.37	0.17	0.25	0.21	0.09	0.02	0.02[a]
.920	0.898	0.874	0.661	0.360	0.279	0.635	1.000

ent electron degeneracy (GRASBERGER, 1961), electron conduction (COX and STEWART, 1965), nuclear reactions (TSUDA, 1963; REEVES, 1966), and energy loss by photo, plasma, and pair annihilation neutrinos (CHIU and STABLER, 1961; INMAN and RUDERMAN, 1964; ZAIDI, 1965; CHIU and MORRISON, 1960) are included. In convective zones we used the usual criterion of constant entropy and mass-averaged the composition.

The energy release rate of the C^{12} reactions, ε_{CC}, is highly T-sensitive. For example, at $T \simeq 10^{8.7}$ K, ε_{CC} is proportional to T^{35}, while ε_ν(photo) is proportional to T^8. Therefore, once ignition occurs the nuclear reactions will dominate the star's evolution until checked by expansive cooling.

In the description of the results in the following section, selected models are referred to by model numbers defined in the text and in Table I. Age 0 refers to the time of nuclear ignition, i.e. when ε_{CC} surpasses ε_ν at one or more shells. All quantities are measured in c.g.s. units unless otherwise indicated. Subscripts c, CC, cv, g, ph, s, and sd refer to the stellar center, C^{12} reactions, convection, gravitation, radiation, entropy, and sound, respectively. q is the fractional mass, Ω the gravitational potential energy of the star, ψ the electron degeneracy parameter (= Fermi energy/kT) and β

equals $P_{gas}/(P_{gas}+P_{ph})$. Further details can be found in KUTTER (1968) and KUTTER and SAVEDOFF (1967).

3. Results of Model Computation

A. GENERAL

We followed the evolution of the 1.04, 1.10, 1.25, 1.35, and 1.55 $M_\odot$ pure C^{12} stars, starting with initial models (Models 1) characterized by $\log(T_c) \simeq 8.0$ and $\Psi_c \simeq -2.5$ to assure that ν-losses and nuclear energy gains, as well as electron degeneracy effects, are negligible. The evolutionary tracks of these stars in the $\log(L_{ph}/L_\odot)$ vs. $\log T_{eff}$ diagram are illustrated in Figure 1. After an initial, nearly homologous contraction approximating a polytrope of index 3, the core becomes electron-degenerate (i.e. $\Psi_c > 0$) and the neutrinos become the dominant mode of energy loss. When $\log(T_c)$ and $\log(\rho_c)$ reach approximately 8.6 and 5.7, respectively, the temperature inverts near the center (Models 2). Except for the time scales, the evolution of these stars to the inversion of the temperature is essentially alike. However, as the peak of this temperature inversion grows and moves toward the surface, significant differences appear in the evolution of these stars because of their greatly varying degrees of electron degeneracy.

The 1.04 $M_\odot$ star misses carbon ignition by 0.06 in $\log(T)$ at $q=0.74$ (Model 4). After passing the points of peak luminosity (Model 5) and peak effective temperature (Model 6), the star evolves directly to the white dwarf stage on a time scale of nearly 4×10^8 years.

In contrast, the 1.10, 1.25, 1.35, and 1.55 $M_\odot$ stars ignite in electron-degenerate shells centered on $q=0.64$, 0.27, 0.16, and 0.01, respectively (Models 4). A C^{12} shell flash ensues (between models 5 and 9) during which the stars expand and the gravitational potential acts as an energy sink. This phenomenon is the same as the core He flash at the tip of the red giant branch. The shell flash occurs on a time scale of several hundred years, which is long relative to the time of free fall ($\simeq 10$ sec), but short relative to the heat diffusion time ($\simeq$ few 1000 years). Because of the sharp and rapid rise in the temperature to about $10^{8.9}$ K in the burning shell, a convective zone develops whose maximum outer boundary reaches $q=0.96$ and 0.65 in the 1.10 $M_\odot$ and 1.25 $M_\odot$ stars, respectively (Models 8). Hydrostatic equilibrium remains a valid approximation throughout the shell flash and the accompanying expansion. Contrary to our initial expectations, there is no indication of mass loss during this evolutionary phase, even though, for example, the stellar radius of the 1.10 $M_\odot$ star increases by a factor of 3.4 (but only by a factor of 1.2 in the 1.25 $M_\odot$ star). The gravitational wells are able to absorb the large energy release, which amounts to about 1/3 of the total pre-flash energy of the stars.

The post-flash evolutions of the 1.10 $M_\odot$ and heavier stars are marked by further significant differences. In the 1.10 $M_\odot$ star, the flash expansion is rapid enough ($v_{max}=34$ cm/sec) to extinguish the shell burning by its cooling effect, whereas it merely reduces the shell burning in the 1.25 $M_\odot$ star ($v_{max}=5$ cm/sec). Thus, when the steep temperature gradient produced by the shell burning in the 1.10 $M_\odot$ star

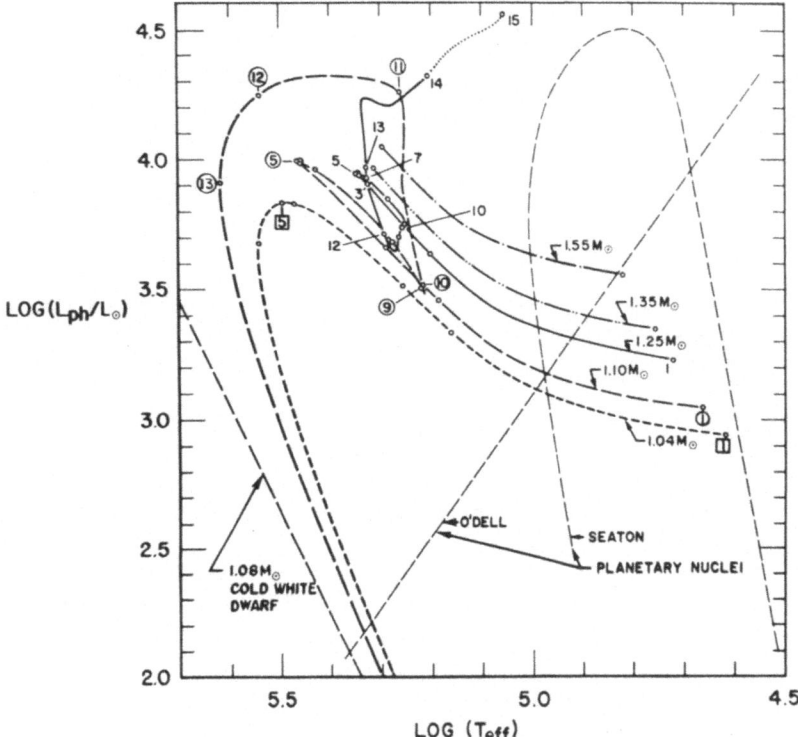

Fig. 1. Luminosity vs. effective temperature. The model numbers are defined in the text and in Table I. The dashed curves labelled 'Planetary Nuclei' refer to observational data by O'DELL (1963) and SEATON (1966). The track beyond Model 14 of the 1.25 $M_\odot$ star is dotted to emphasize the approximate nature of the results.

has dampened out on a time scale of about 5000 years, the star contracts to the white dwarf stage.

On the other hand, the post-flash evolution of the 1.25 $M_\odot$ star is characterized by a C^{12}-burning shell which advances from $q=0.26$ to 0.84 in about 25 000 years, leading to hydrodynamic surface conditions which we interpret as mass loss. The evolution of this star is summarized in Table I and in Figures 2 and 3.

The 1.35 and 1.55 $M_\odot$ stars were evolved only to the beginning of the shell flash in order to estimate the lower mass limit for central ignition. This limit is approximately 1.58 $M_\odot$. The evolution of stars below this limit, but above approximately 1.2 $M_\odot$ is expected to be similar to that of the 1.25 $M_\odot$ star.

B. MASS LOSS FROM 1.25 $M_\odot$ STAR

During the post-flash evolution to age 25 080 years, the 1.25 $M_\odot$ star adjusts itself so that the energy loss by ν-emission is approximately equal to the energy gains from the nuclear reactions, the small difference being balanced by the energy release from a slow gravitational contraction. The radiative energy loss is negligible. This near-

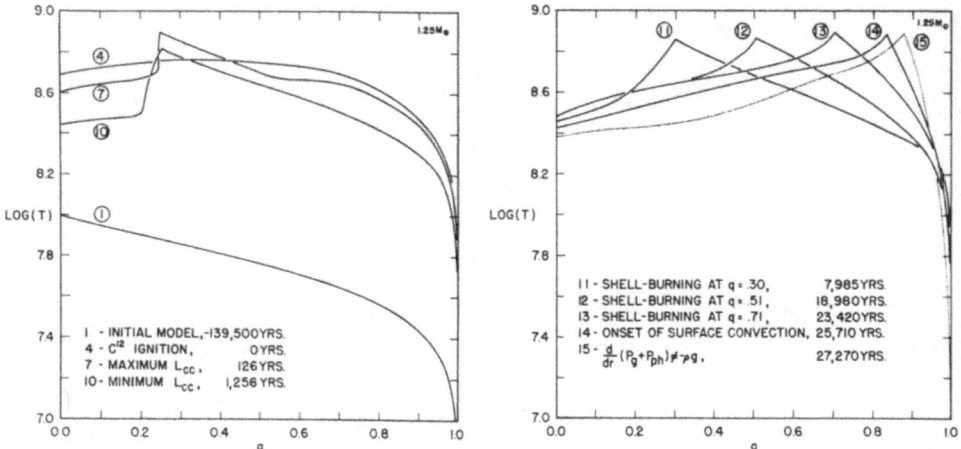

Fig. 2.   Temperature distribution of several models, 1.25 $M_\odot$ star. Model 15 is represented by a
dotted curve to emphasize the approximate nature of the results.

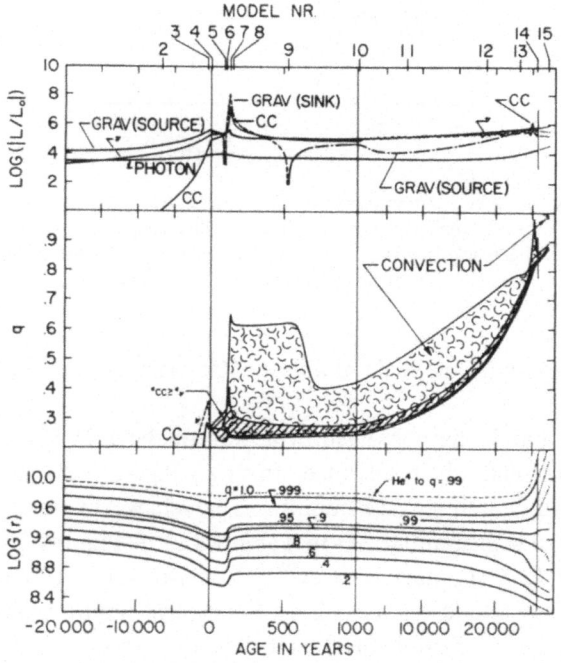

Fig. 3.   Log$|L/L_\odot|$, $q(\varepsilon_{CC}>\varepsilon_\nu)$, q(cv), and $r(q)$ as functions of age, 1.25 $M_\odot$. In the top graph,
labels refer to L_g, L_ν, L_{ph}, and L_{CC}. *Source* (solid line) and *sink* (dashed line) refer to gravitational
contraction and expansion, respectively. Only the variation of L_{CC} is drawn in detail from Model
9 on. The curve ------ between ages 600 and 25080 years represents time-averaged values of L_g.
Actually there exist fluctuations in L_g which accompany the oscillations in L_{CC}. In the center graph,
spatial extents of nuclear shell burning and convection are shown. CC and ν refer to locations of the
maxima of ε_{CC} and ε_ν. Dashed curve in bottom graph represents stellar radius if outer 1 % of carbon
were replaced by helium. Results beyond Model 14 are represented by dotted curves to emphasize
the approximate nature of the results.

balance between L_v and L_{CC} causes small oscillations in the physical parameters with periods of about 700 years. The oscillations in L_{CC} are illustrated in detail in Figure 3.

As the shell burning reaches $q=0.84$, the continuously steepening temperature gradient near the surface causes convection to start there (Model 14, described in Table II). The inner 90–95% of the star's mass is contracting (cf. lower part of Figure 3) and, in time, will follow the evolution to white dwarf conditions found in the 1.04 and 1.10 $M_\odot$ stars. However, the outer layers expand due to heating from the nuclear energy source, which explains the evolution to lower effective temperatures in Figure 1.

At age 27270 years (Model 15) the surface convection extends inward to $q=0.98$ and is highly superadiabatic close to the surface. Approximations of the physical conditions in this surface convective zone are listed in Table III at mass shells separated by about 2 density scale heights. The values of $\Delta\nabla T$, v_{cv}, and L_{cv} are based on the formulas given by SCHWARZSCHILD (1958, pp. 48–50) and the mixing length is equated to one density scale height. At the last mass shell, $q=1-1.8\times10^{-8}$, the actual T-gradient surpasses the adiabatic T-gradient by a factor of 2.6, the convective velocity comes to within a factor of 3 of the local sound velocity, and the divergence of the turbulent pressure comes to within 2 orders of magnitude of the divergence of $(P_{gas}+P_{ph})$. Any proper treatment of these surface convective layers should be based on a mixing-length theory and should include the effects of shockwaves, turbulent pressure, and the production of sound and gravity waves (cf. STEIN, 1967).

Thus, our final evolutionary models of the 1.25 $M_\odot$ star are very approximate and merely indicate the direction in which the star evolves. Most importantly, they indicate that, besides producing a growing surface convective zone, the march of the shell source toward the surface causes a continuous increase in L_{ph}. As L_{ph} approaches the value

$$L_0 = \frac{4\pi cGM}{\kappa} = 10^{4.11}\frac{M/M_\odot}{\kappa}L_\odot \text{ ergs/sec}, \qquad (2)$$

the force due to the divergence of P_{ph} becomes greater than the opposing force of gravitation and leads to a breakdown of the hydrostatic equilibrium equation. As we shall see in the following section, L_0 is actually an overestimate for hydrodynamic stability since it neglects the effect of P_{gas}.

For an electron-scattering atmosphere, L_{ph} surpasses L_0 about 3000 years after the onset of surface convection. We interpret this to mean mass loss. Since most, if not all, stars in a broad mass range with a lower limit near 1.2 $M_\odot$ would be expected to pass through this phase during a late stage in their evolution, this mass loss may be a mechanism by which planetary nebulae are formed. In the next two sections we will investigate further this possibility.

4. Analysis of Mass Loss

We now develop an accurate criterion for mass loss and estimate its rate in the late evolutionary stages of the 1.25 $M_\odot$ star. We assume spherical symmetry and steady-state flow and neglect magnetic and viscous forces. Further we assume that mass

TABLE II

Physical conditions of Model 14, 1.25 $M_\odot$ star

q	$\log(r)$	$\log(T)$	$\log(P)$	$\log(\rho)$	$\dfrac{L_{\mathrm{ph}}+L_{\mathrm{cv}}}{L_\odot}$	X_{12}	β	κ	ε_{cc}	ε_ν
$8.42(-5)^{b}$	7.128	8.428	24.391	7.316	$-1.72(-2)$	0.999	1.00	$3.0(-4)$	–	4.4(4)
0.0665	8.118	8.464	24.216	7.190	$-1.93(1)$	0.996	1.00	$6.7(-4)$	–	5.9(4)
0.465	8.499	8.653	23.469	6.664	$-6.99(1)$	0.138	1.00	0.031	1.1(1)	1.9(5)
0.747	8.689	8.766	22.439	5.874	$-7.67(2)$	0.042	0.99	0.055	9.6(2)	6.9(5)
0.809	8.774	8.836	21.862	5.297	$-1.14(4)$	0.079	0.92	0.054	1.4(5)	2.7(6)
0.827	8.822	8.873	21.603	4.957	$-2.03(4)$	0.185	0.80	0.055	4.8(6)	7.8(6)
0.836	8.852	8.881	21.476	4.785	1.09(4)	0.407	0.73	0.061	2.7(7)	1.1(7)
0.849	8.906	8.830	21.268	4.608	$6.56(4)^{a}$	0.641	0.71	0.070	9.8(5)	3.1(6)
0.862	8.958	8.778	21.067	4.465	$5.43(4)^{a}$	0.641	0.72	0.075	1.7(4)	9.4(5)
0.875	9.010	8.725	20.863	4.316	4.91(4)	0.682	0.73	0.081	2.7(2)	3.2(5)
0.889	9.064	8.667	20.641	4.157	4.54(4)	0.687	0.73	0.089	2.1	1.1(5)
0.904	9.121	8.604	20.398	3.982	4.21(4)	0.691	0.74	0.093	$8.0(-3)$	3.4(4)
0.919	9.180	8.535	20.133	3.789	3.90(4)	0.704	0.74	0.10	–	9.6(3)
0.934	9.242	8.462	19.847	3.579	3.57(4)	0.706	0.75	0.11	–	2.5(3)
0.954	9.336	8.344	19.378	3.228	3.07(4)	0.805	0.75	0.13	–	2.8(2)
0.971	9.430	8.215	18.869	2.847	2.65(4)	0.957	0.75	0.15	–	2.6(1)
0.989	9.569	8.000	18.004	2.193	2.26(4)	1.000	0.75	0.17	–	$5.0(-1)$
0.9969	9.700	7.755	16.997	1.422	2.13(4)	1.000	0.73	0.19	–	–
$1-6.2(-4)$	9.800	7.506	15.973	0.637	2.10(4)	1.000	0.72	0.21	–	–
$1-9.8(-5)$	9.871	7.254	14.933	-0.165	2.10(4)	1.000	0.70	0.22	–	–
$1-1.3(-5)$	9.920	7.006	13.879	-0.999	2.10(4)	1.000	0.65	0.24	–	–
$1-1.4(-6)$	9.955	6.768	12.817	-1.899	2.10(4)	1.000	0.55	0.31	–	–
$1-9.2(-8)$	9.987	6.477	11.521	-3.056	$2.10(4)^{*}$	1.000	0.39	0.44	–	–

a Convective mass shells.

b 8.42(−5) means 8.42×10^{-5}.

TABLE III

Physical conditions in surface convective zone of Model 15, 1.25 $M_\odot$ star

q	0.977	0.990	.9966	.99915	.99985	1-1.9(-5)	1-2.0(-6)	1-2.0(-7)	1-1.8(-8)
$\log(r)$	9.642	9.872	10.073	10.222	10.331	10.403	10.448	10.474	10.489
$\log(T)$	7.986	7.704	7.415	7.143	6.866	6.586	6.305	6.044	5.780
$\log(P)$	17.732	16.600	15.443	14.348	13.236	12.109	10.969	9.904	8.807
$\log(P_{cv})$	–	9.947	9.532	9.165	8.905	8.663	8.373	8.032	7.712
$\log(\rho)$	1.830	0.978	0.107	-0.178	-1.557	-2.411	-3.280	-4.104	-4.974
κ	0.17	0.19	0.21	0.22	0.28	0.44	0.80	0.77	1.3
l	1.3(9)	1.9(9)	2.4(9)	2.6(9)	2.3(9)	1.7(9)	1.1(9)	0.70(9)	0.45(9)
β	0.59	0.59	0.58	0.58	0.57	0.57	0.55	0.53	0.48
$-dT/dr|_{ls}$	2.55(-2)	8.89(-3)	3.53(-3)	1.77(-3)	1.06(-3)	7.46(-4)	5.87(-4)	4.95(-4)	4.14(-4)
$\Delta\nabla T$	1.47(-10)	3.53(-8)	8.14(-8)	2.10(-7)	8.91(-7)	4.89(-6)	2.85(-5)	1.32(-4)	6.62(-4)
$(L_{cv}+L_{ph})/L_\odot$	3.79(4)	3.73(4)	3.72(4)	3.72(4)	3.72(4)	3.72(4)	3.72(4)	3.72(4)	3.72(4)
$L_{ph}/(L_{cv}+L_{ph})$	1.00	0.921	0.866	0.808	0.655	0.422	0.238	0.256	0.169
v_{cv}	–	4.32(4)	7.30(4)	1.24(5)	2.41(5)	4.87(5)	9.49(5)	1.65(6)	3.12(6)
v_{sd}	1.07(8)	7.78(7)	5.60(7)	4.10(7)	2.99(7)	2.18(7)	1.60(7)	1.21(7)	9.27(6)

flow occurs only in the outer few per cent of the stellar mass and that the remainder of the star remains in hydrostatic equilibrium. The stellar mass, M, can then be taken to be constant in the analysis of the hydrodynamic envelope.

A. MASS-LOSS CRITERION

The four equations governing the mass flow are

$$\rho v r^2 = A/4\pi, \qquad \text{(continuity equation)} \qquad (3)$$

$$\rho \, d \left(\frac{v^2}{2} - \frac{GM}{r} \right) + d(P_{gas} + P_{ph}) = 0,$$
$$\text{(conservation of momentum)} \qquad (4)$$

$$\rho v \frac{d}{dr} \left(\frac{3}{2} \frac{P_{gas}}{\rho} + 3 \frac{P_{ph}}{\rho} \right) + (P_{gas} + P_{ph}) \, \nabla \cdot v = - \nabla \cdot \mathfrak{F},$$
$$\text{(conservation of energy)} \qquad (5)$$

$$\frac{d}{dr} P_{ph} = - \frac{\kappa \rho L_{ph}}{4\pi c r^2}, \qquad \text{(2nd moment of transfer equation)}, \qquad (6)$$

where v is the velocity, A is a constant measuring the rate of mass flow, and $\mathfrak{F}$ is the radiative flux.

We shall consider here only the electron-scattering case in the Thompson limit. The opacity, κ, as well as L_{ph} are then constant and $\nabla \cdot \mathfrak{F} = 0$ (SAMPSON, 1965). Physically, this means that there is no exchange of energy between the gas and photons (adiabatic flow) and, hence, that the temperature of the radiation field and the gas are not equivalent. Therefore, LTE cannot be assumed. A hydrodynamic envelope assuming LTE has been investigated by BISNOVATYJ-KOGAN (1967), although no mass-loss criterion applicable to a static model was found.

Using Equation (3) and $\nabla \cdot \mathfrak{F} = 0$, Equation (5) reads

$$\rho \, d \left(\frac{5}{2} \frac{P_{gas}}{\rho} + 4 \frac{P_{ph}}{\rho} \right) - d(P_{gas} + P_{ph}) = 0. \qquad (7)$$

The addition of Equations (4) and (7) yields

$$\frac{5}{2} \frac{P_{gas}}{\rho} + 4 \frac{P_{ph}}{\rho} + \frac{v^2}{2} - \frac{GM}{r} = B, \qquad (8)$$

where B is an integration constant representing the sum of the enthalpy, kinetic energy, and gravitational potential energy per gram. B must be positive for unbounded outward motion to occur. Employing Equations (3), (6), and (8) to eliminate ρ, P_{ph}, and P_{gas} in Equation (7), one finds

$$\frac{dy}{dx} = \frac{y[x(4-y) + (1-3\delta)]}{x[x(y-1)-1]}, \qquad (9)$$

where $x = Br/GM$ and $y = 2v^2/B$. The parameter δ equals L_{ph}/L_0, where L_0 is defined by Equation (2). When $\delta = 1$, the divergence of P_{ph} is equal to the gravitational body force.

Equation (9) forms the basis of our analysis. The parameter δ determines the nature of the flow, while B enters only as a scale factor. Mathematically, the singular point $(x, y) = (\delta, 1 + 1/\delta)$ is a saddle point, characterized by the intersection of two curves (MINORSKY, 1947), which we call the critical solutions. The slopes of the critical curves at the singular point are given by

$$\left.\frac{dy}{dx}\right|_{\pm} = \frac{-(2 + \delta) \pm \sqrt{13\delta^2 + 12\delta}}{2\delta^2}, \tag{10}$$

and the corresponding velocity by

$$v_{crit} = \frac{5}{3}\frac{P_{gas}}{\rho} + \frac{8}{3}\frac{P_{ph}}{\rho}. \tag{11}$$

The sign of $dy/dx|_{\pm}$ can be positive, negative, or zero according to whether δ is greater

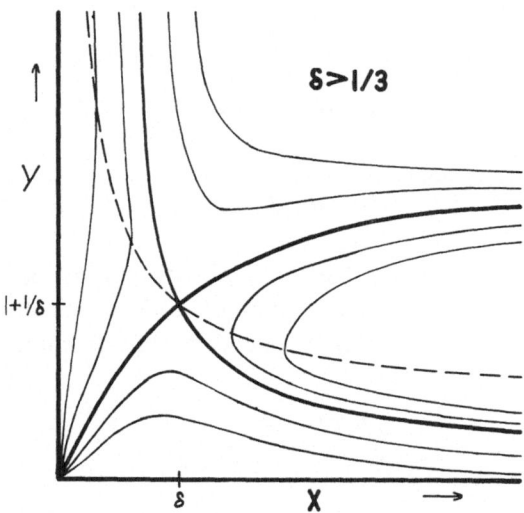

Fig. 4.   Critical solutions of Equation 9 for $\delta > 1/3$. The heavy curves represent the critical solutions; the dashed curve represents the locus of singular points.

than, equal to, or less than 1/3, respectively. In Figure 4 we have sketched several representative solutions of Equation (9) for the case $\delta > 1/3$, while Figures 5 and 6 show only critical curves for $\delta < 1/3$ and $\delta = 1/3$. The dashed curves, $y = 1 + 1/x$, are the loci of critical points.

We shall be interested in solutions with small velocities at the base of the envelope and with ρ vanishing for large r. When $\delta > 1/3$, the desired solution is the critical

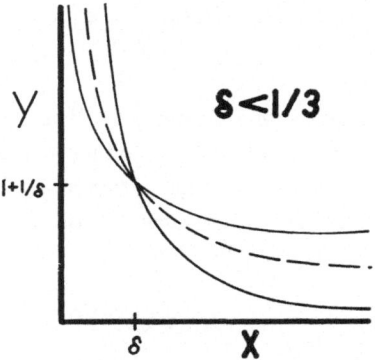

Fig. 5. Critical solutions of Equation (9) for $\delta < 1/3$. The dashed curve represents the locus of singular points.

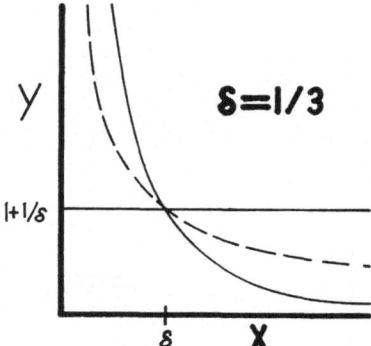

Fig. 6. Critical solutions of Equation (9) for $\delta = 1/3$. The dashed curve represents the locus of singular points.

(heavy) curve with positive slope. No solutions satisfying the above boundary conditions exist for $\delta < 1/3$. If $\delta = 1/3$, the solution is $v^2/2 = B$ for all r. Then the density decreases as r^{-2} so that the envelope is extended, and the enthalpy is equal to the gravitational energy everywhere, i.e.

$$\frac{5}{2}\frac{P_{gas}}{\rho} + 4\frac{P_{ph}}{\rho} = \frac{GM}{r}. \tag{12}$$

Thus we see that our static models can lose mass whenever $\delta \geqslant 1/3$.

B. RATE OF MASS LOSS

The model computations discussed in Section 3 were based on the assumption of hydrostatic equilibrium. However, beyond age 24 600 years the 1.25 $M_\odot$ star has $\delta \geqslant 1/3$ near the surface. Therefore, all model computations from then on are based on non-physical surface boundary conditions. In effect, B is forced to be negative by the application of Equation (1). In the future we will include hydrodynamic boundary conditions at the surface based on the above development and repeat the model computations. It should be noted that the condition for mass flow arises before the onset of surface convection, and it will be interesting to see how convection will be modified when the hydrodynamic boundary conditions are included.

Here we shall attempt only a crude estimate of the rate of mass loss, $A = 4\pi r^2 \rho v$, based on the static models of the 1.25 $M_\odot$ star. For v we use the expansion rate of the innermost mass shell, q, at which both v is positive and $\delta \geqslant 1/3$. In Table IV we list q and A as functions of age for selected models. The expansion rate, v, is of the order of 0.1 cm/sec for these models. The total mass loss up to age 27 270 years is 2×10^{-3} $M_\odot$.

TABLE IV

Rate of mass flow and fitting points of selected models, 1.25 $M_\odot$ star

Age (years)	q	A ($M_\odot$/year)
25 000	1–3.(-7)[a]	3.2 (-9)
25 710	1–2.(-5)	8.5 (-8)
26 220	0.999	3.3 (-7)
26 650	0.969	2.5 (-6)
26 920	0.957	1.5 (-6)
27 270	0.957	1.2 (-6)

[a] (-7) means 10^{-7}.

This mass loss is quite small, but represents a lower limit for at least three reasons. First, we expect that model computations based on hydrodynamic boundary conditions would lead to larger values of v. Second, the existence of additional opacity sources besides electron scattering would increase the rate of mass loss since $\nabla \cdot \mathfrak{F}$ in Equation (5) would then provide an additional energy source. Third, the presence of a H/He envelope would further enhance mass loss for reasons discussed below.

5. Discussion

The results discussed above indicate that in initially pure C^{12} stars, lying in the mass range from about 1.2 to 1.6 $M_\odot$, the nuclear shell burning advances to within about 0.2 $M_\odot$ of the surface. There the continuously steepening temperature gradient drives the radiative luminosity to large enough values so that the interaction between the radiation field and surface matter produces hydrodynamic conditions which we interpret as mass loss.

We believe that similar conditions in inhomogeneous stars, having a core of carbon and heavier elements and a H/He envelope, may be responsible for the production of planetary nebulae. We shall now discuss our reasons for this belief, examine the availability of additional energy sources in inhomogeneous stars, and compare these theoretical results with some of the observational data for planetary nebulae and nuclei.

Our results are based on highly idealized models. Homogeneous models, consisting of C^{12} with an admixture of O^{16} and heavier elements, would evolve similarly and with only a negligible shift in the mass range as long as the initial C^{12} content is non-negligible. This follows from the high temperature dependence of ε_{CC}. Inhomogeneous models having a C^{12}/O^{16} core of about 1.2–1.6 $M_\odot$ surrounded by a H/He envelope would also be expected to evolve similarly. Even H and/or He shell burning would not affect significantly the evolution of the core. These expectations are based on results by HAYASHI et al. (1962, p. 149) and FAULKNER (1966, p. 983).

The addition of a H/He envelope to a C^{12} core would actually enhance mass loss. First, the increased stellar radius would reduce the gravitational potential per unit mass of the surface layers. Second, the surface convection found in the 1.25 $M_\odot$ C^{12} star would almost certainly carry envelope material into the high-temperature core and thus provide additional energy sources. The burning of .0001 $M_\odot$ of H or .001 $M_\odot$ of He would provide the energy required to carry 0.2 $M_\odot$ from the surface of the 1.25 $M_\odot$ star to infinity, assuming a stellar radius of 1 $R_\odot$.

In Section 4 we found that mass loss occurs when L_{ph} exceeds $1/3 L_0$. In the case of the 1.25 $M_\odot$ star and an envelope composition characterized by $X_1 = 0.625$, we obtain $1/3 L_0 = 10^{4.24} L_\odot$. With L_{ph} at this value it would take 3000 years to drive 0.2 $M_\odot$ into space, assuming 10% efficiency and a stellar radius of 1 $R_\odot$. It is, therefore, not unreasonable to expect that a C^{12} shell source together with the burning of H and/or He could sustain the radiative luminosity at or beyond the critical value, $1/3 L_0$, sufficiently long for the expulsion of 0.2 $M_\odot$.

Such mass loss may be greatly affected by coronal heating due to gravity (also known as internal waves) and acoustic waves created by the convective zone. Using the equations given by STEIN (1967) and the data of Table III (1.25 $M_\odot$, Model 15), we estimate the gravity wave and acoustic luminosity to be of the order of the radiative luminosity. Despite the uncertainties in the numerical constants of Stein's formulas and in the entries of Table III, mass loss analogous to the solar wind must be considered a strong possibility.

The drop in temperature occurring in the escaping matter will provide an additional energy source due to electron recombinations. We estimate that this energy will be of the order of 10^{46} ergs per 1 $M_\odot$ of escaping matter and, hence, will affect the details of mass loss, but will not dominate or cause it.

These considerations suggest that we are dealing here with a mechanism which may indeed explain the large mass loss occurring in the production of planetary nebulae. This suggestion is strengthened by comparisons between our results and the observational data available for planetary nebulae and nuclei.

The stars which we propose to be the parent stars of planetary nebulae have cores in the range from 1.2 to 1.6 $M_\odot$. Therefore, their total masses are probably above 1.3 $M_\odot$, depending on the envelope mass. It is difficult to give an upper limit to the masses on the basis of our theoretical results. Certainly, stars as massive as 2.0 $M_\odot$ cannot be ruled out. The agreement with the observational mass range of 1.0–1.5 $M_\odot$ is not perfect, but appears acceptable in view of the uncertainties present in both theory and observations. The theoretical mass range would be extended to lower masses if stars having initially a He-core would reach critical luminosities for mass loss as the He shell burning comes close to the surface. This possibility is supported by the results SCHWARZSCHILD and HÄRM (1967) obtained for a 1 $M_\odot$ star in which the luminosity increased steadily to $10^{3.5} L_\odot$ as the He shell source advanced to $q = 0.54$.

The present rate at which stars in the mass range from 1.3 to, e.g., 1.8 $M_\odot$ are formed and, hence, become available for the eventual passage through late evolutionary phases is approximately $0.5 \times 10^{-12}/\text{pc}^3$-year (SCHMIDT, 1963). This rate falls within the observational estimate of the rate of formation of planetary nebulae.

Since the planetary nuclei are H-deficient and appear to belong to a carbon sequence (GREENSTEIN and MINKOWSKY, 1964), their evolution should be well approximated by the post-flash evolution of the 1.10 $M_\odot$ C^{12} star. There exists indeed a resemblance between the theoretical and observational tracks in the $\log(L_{ph}/L_\odot)$ vs. $\log(T_{eff})$ diagram except for a shift to higher temperatures of the theoretical tracks (Figure 1). We estimate (SAVEDOFF et al., 1968a) that the addition of $10^{-6} M_\odot$ of H or of .02 $M_\odot$ of He to the 1.1 $M_\odot$ C^{12} star at peak luminosity would change its radius to about 1.5 $R_\odot$ and $\log(T_{eff})$ to about 4.7, providing good agreement with the nuclei of young planetary nebulae.

O'DELL's (1963) estimate for the time scale of evolution of planetary nuclei to the point where they have approximately 145 $L_\odot$ is of the order of 25 000 years, while SEATON's (1966) estimate of their evolution time to the point of 100 $L_\odot$ is of the order of 50 000 years. In comparison, the 1.10 $M_\odot$ C^{12} star contracts from its peak luminosity to 145 $L_\odot$ in 33 000 years and to 100 $L_\odot$ in 40 000 years. Further, there exists close agreement between the radii given by O'Dell and Seaton for evolved planetary nuclei and our results.

Our theoretical results are based on evolutionary sequences computed for the simplest stellar models approximating the cores of what are believed to be the precursors of planetary nebulae. Despite these simplified starting conditions and the uncertainties present in both theory (nuclear and neutrino rates, extrapolation from hydrostatic and homogeneous models) and observations (distance scale, effective temperature of nuclei), the agreement between the theoretical results and observational data appears to be surprisingly good. Therefore, we suggest that the dynamic conditions found in the 1.25 $M_\odot$ C^{12} star do indeed represent a mechanism by which planetary nebulae are formed.

References

BISNOVATYJ-KOGAN, G. S.: 1967, *Prikladnaya Matematika I Mekhanika* 4, 762.
CHIU, H. Y. and MORRISON, P.: 1960, *Phys. Rev. Letters* 5, 573.
CHIU, H. Y. and STABLER, R. C.: 1961, *Phys. Rev.* 122, 1317.
COX, A. N. and STEWART, J. N.: 1965, *Suppl. Astrophys. J.* 11, 22.
FAULKNER, J.: 1966, *Astrophys. J.* 144, 978.
GRASBERGER, W. H.: 1961, *A Partially Degenerate, Relativistic, Ideal Electron Gas*, U.C.R.L. 6196.
GREENSTEIN, J. L. and MINKOWSKI, R.: 1964, *Astrophys. J.* 140, 1601.
HAYASHI, C., HOSHI, R., and SUGIMOTO, D.: 1962, *Suppl. Progr. Theor. Physics* 22.
HENYEY, L. G., FORBES, J. E., and GOULD, N. L.: 1964, *Astrophys. J.* 139, 306.

INMAN, C. L. and RUDERMAN, M. A.: 1964, *Astrophys. J.* **140**, 1025.

KUTTER, G. S.: 1968, *Evolution of Initially Pure C¹² Stars and the Production of Planetary Nebulae*, Thesis, The University of Rochester.

KUTTER, G. S. and SAVEDOFF, M. P.: 1967, *Astron. J.* **72**, 810.

MINORSKY, N.: 1947, *Non-Linear Mechanics* J. W. Edwards, Ann Arbor, Mich.

O'DELL, C. R.: 1963, *Astrophys. J.* **138**, 67.

OSTERBROCK, D. E.: 1966, in *Stellar Evolution* (ed. by R. F. Stein and A. G. W. Cameron), Plenum Press, London, pp. 381–387.

REEVES, H.: 1966, in *Stellar Evolution* (ed. by R. F. Stein and A. G. W. Cameron), Plenum Press, London, pp. 83–122.

SAMPSON, D. H.: 1965, *Radiative Contributions to Energy and Momentum Transport in a Gas*, No. 26 in the series: Interscience Tracts on Physics and Astronomy, John Wiley & Sons, Inc., New York.

SAVEDOFF, M. P., KUTTER, G. S., and VAN HORN, H.: 1968a, in *I.A.U. Symposium No. 34* (ed. by D. E. Osterbrock and C. R. O'Dell), D. Reidel, Dordrecht, pp. 400–406.

SAVEDOFF, M. P., VAN HORN, H., and VILA, S. C.: 1968b, *Astrophys. J.* (in press).

SCHMIDT, M.: 1963, *Astrophys. J.* **137**, 758.

SCHWARZSCHILD, M.: 1958, *Structure and Evolution of the Stars*, Princeton University Press, Princeton, N.J.

SCHWARZSCHILD, M. and HÄRM, R.: 1967, *Astrophys. J.* **150**, 961.

SEATON, M. J.: 1966, *Monthly Notices Roy. Astron. Soc.* **132**, 113.

STEIN, R. F.: 1967, *Solar Phys.* **2**, 385.

TSUDA, H.: 1963, *Progr. Theor. Phys.* **29**, 29.

WEIDEMANN, V.: 1968, in *I.A.U. Symposium No. 34* (ed. by D. E. Osterbrock and C. R. O'Dell), D. Reidel, Dordrecht, pp. 423–424.

VILA, S. C.: 1965, *Study of the Last Stages of Evolution of Solar Mass Stars*, Thesis, The University of Rochester.

VILA, S. C.: 1966, *Astrophys. J.* **146**, 437.

ZAIDI, M. H.: 1965, *Nuovo Cimento* **A40**, 502.

Discussion

Underhill: From your experience with low-mass stars could you comment whether H or He shell burning is near enough to the surface to provide the expanding atmospheres of massive OB and Wolf-Rayet stars could occur?

Kutter: I am quite certain that the nuclear-energy sources of massive OB stars are at or close to the center. With regard to the Wolf-Rayet stars I would rather not venture a guess. However, in the case of planetary nuclei, some of which resemble Wolf-Rayet stars spectroscopically, I am certain that He and/or H shell burning is required to account for the effective temperatures, which are less by a factor of 3 or 4 than predicted by the evolution of pure C^{12} stars. And the addition of a non-burning envelope will not remove the disagreement.

Hall: L. Bautz has similar models for nuclei of planetary nebulae and she feels she cannot make them cool enough to agree with the observed nuclei of planetary nebulae.

Kutter: According to my experience the addition of an envelope of .02 $M_\odot$ He or $10^{-6} M_\odot$ H, will bring our theoretical H-R tracks at peak luminosity into agreement with the observed T_{eff} of planetary nuclei. During the subsequent contraction this envelope will ignite, and I am presently computing models with small envelopes to see if agreement can be reached with observations down to luminosities of about 100 $L_\odot$.

Grzędzielski (1): I did not understand well whether you went through the hydro-dynamics of the process or whether you just based on the radiation pressure argument? In the latter case one might imagine the envelope expanding, becoming transparent and hence no more subjected to radiation pressure, without actually reaching the escape velocity.

Grzędzielski (2): I am not convinced you can properly estimate the rate of mass loss basing on models in hydrostatic equilibrium. There is no time-dependent term in the momentum equation in this case and the real flow may differ drastically from the flow obtained with the hydrostatic models.

Kutter (1): The stellar models that we have computed so far are based on the assumption of hydrostatic equilibrium. However, Don Schuerman is now adding hydrodynamic boundary conditions and, as I have reported, the indications are that mass loss to infinity occurs. These computations are based on electron scattering and mass loss to infinity seems to be an inevitable consequence in our $1.25\,M_{\odot}$ star. I don't think transparency will be a problem because I'd expect that a realistic opacity would go up during expansion and the accompanying cooling, at least during the initial phase of expansion.

Kutter (2): You are quite right. And we are only giving a lower limit to the rate of mass loss. However, Schuerman's work should lead to a more realistic estimate.

Sargent: I have two questions: (a) Where does the 'observed' mass range of 1.0–$1.5\,M_{\odot}$ come from? (b) Does your theory lead to a natural explanation of why the expansion velocity of planetaries is small as compared with the velocity of escape from the central star? Some authors, for example Goldreich and Abell, have inferred that the precursor of a planetary must be a red giant.

Kutter: (a) This mass range I've gotten from Osterbrock's discussion in *Stellar Evolution* (ed. by Stein and Cameron). This mass range is derived from the average distance of planetary nebulae from the galactic plane. (b) We are in complete agree-ment with Goldreich and Abell. Our C^{12} stars represent only the cores of evolved stars. The addition of an envelope in the range of $0.2\,M_{\odot}$ of H/He would place them into the red-giant region. Our mechanism for mass loss is not explosive, but slow and driven by the pressure gradient. Thus the expansion velocity should be approximated quite well by the sound velocity in the envelope, which at a few 100000 K is of the order of 30 km/sec.

Hall: If the envelope has already been ejected, how can you still add a helium envelope to make your models cooler and thereby agree with observed nuclei of planetary nebulas?

Kutter: Your question can be answered in detail only after a considerable amount of additional computing. However, off hand, I would not expect that the ejection of the envelope would mean that every gram of He or H has gone. There may well be left enough material, say a few percent of the original envelope, to lower T_{eff} sufficiently.

Dallaporta: Have I rightly understood your scheme if I say that neutrino losses are critical for it, in the sense that without neutrino losses you should not get a burning shell?

Kutter: Without neutrinos ignition would occur centrally. Of course, in this case shell burning will result eventually too when the fuel at the center is exhausted. But judging from the non-burning models computed by Vila and from our $1.04\,M_{\odot}$ carbon star, I think neutrino emission is required to obtain the high luminosities and short time-scales observed for the planetary nuclei.

PHOTOELECTRIC OBSERVATIONS OF CH CYGNI
DURING THE EXPLOSIVE PHASES OF 1967–68

BRUNO CESTER

Osservatorio Astronomico, Trieste, Italy

Abstract. Observations made at Trieste in 1967 and 1968 of CH Cygni reveal a strong variability, particularly in U colour. From the observed colours, a suggestion is made about the existence of a hot companion which could be a subdwarf.

The observations of the SRa variable CH Cygni ($\alpha = 19^h\ 22^m\ \delta = 50°3' : 1900.0$) were begun at Trieste Observatory on June 30, 1967 after the announcement made by Deutsch that a hot blue continuum was superimposed on the ordinary M6-spectrum of the star. They were continued during the same year, and a short report of the first results has already been made (CESTER, 1967b). They were then taken afresh

TABLE I

N	J. D.	V	B-V	U-B	N	J. D.	V	B-V	U-B
1	2439 672.44	7.42	+1.22	−0.17	30	2439 784.28	7.78	+1.14	−0.26
2	673.42	7.41	1.20	.22	31	785.28	7.80	1.11	.30:
3	675.48	7.30:	1.07	.31	32	787.31	7.80	1.09	.25
4	678.47	7.39	1.06	.39	33	791.25	7.96	1.25	.05
5	682.48	7.53	1.18	.25	34	794.30	7.98	1.16	.32
6	683.45	7.54:	1.14	.28	35	801.28	8.02	1.11	.31
7	687.47	7.61	1.12	.25	36	808.24	8.04	1.12	.40
8	690.49	7.58	1.07	.30	37	815.24	8.10	1.18	.18
9	694.42	7.62	1.10	.18	38	816.25	8.02	1.09	.20
10	700.53	7.60	1.14	.20	39	817.24	8.05	0.96	.15:
11	705.42	7.57	0.99	.39	40	818.23	7.96	1.02	.40
12	706.40	7.61	1.04:	.28:	41	819.23	8.10	1.10	.10
13	710.43	7.64	1.08	.24	42	834.22	7.92	1.28	.06
14	712.43	7.49	1.00	.37	43	968.50	7.14	1.25	.11
15	714.45	7.45	0.95	.35	44	970.50	7.07	1.19	.21
16	718.39	7.48	1.04	.33	45	971.46	7.07	1.23	.15
17	721.42	7.34	0.97	.37	46	2440 035.44	6.65	1.08	.35:
18	724.41	7.38	1.04	.34	47	048.43	6.59	1.07	.35:
19	725.39	7.35	0.96	.26	48	051.44	6.60	1.07	.41
20	746.41	7.59	1.21	.14:	49	065.44	6.67	1.07	.48
21	757.35	7.65	1.23	.12	50	066.39	6.63	1.07	.58
22	758.35	7.61	1.19	.21	51	068.39	6.70	1.10	.46
23	763.31	7.58	1.29	.08	52	069.45	6.70	1.02	.51
24	765.37	7.60	1.24	.05	53	085.39	6.49	1.01	.63
25	769.35	7.56	1.13	.13	54	088.34	6.54	1.06	.56
26	770.35	7.62	1.20	.14	55	089.37	6.51	0.99	.57
27	771.33	7.62	1.18	.13	56	090.34	6.53	0.93	.63
28	772.33	7.68	1.21	.20:	57	093.44	6.63	1.02	.55
29	2439 783.30	7.78	1.11	−0.30:	58	2440 094.38	6.53	1.04	−0.57

M. Hack (ed.), Mass Loss from Stars. All rights reserved

in 1968, but unfortunately not with the frequency which would be desirable in order to have a complete picture of the phenomenon.

The first series was carried out with a conventional d.c. photomultiplier equipped with an EMI 6256 photomultiplier attached to the 30-cm Cassegrain reflector. The colours, matching the UBV system, are inserted in the optical beam in cyclic succession. A difficulty arose in the definition of the colour indices, as the light of CH Cygni is continuously and quite irregularly variable. They were subjected to an extrapolation of data in the sense that each colour was to be referred to the previous one, which in the meantime could have varied. Fortunately the v-light varied very little during each set of observations, whereas b and u showed higher oscillations.

For these reasons during the winter gap a new photometric head was constructed, in which a beam splitter was used, followed by two similar photomultipliers. In this manner two colours at least could be observed simultaneously. It was attached to the new 50-cm reflector. Because of the greater variability of b and u light, most of the observations were performed in these colours, alternating b with v usually for a short time.

Table I gives the list of the observations in the usual manner; the comparison stars, with their adopted magnitudes and colours, are the following:

$$a = BD + 50°2781: \quad V = 7.23 \quad B\text{-}V = +1.65 \quad U\text{-}B = +2.04;$$
$$b = BD + 49°3009: \qquad 7.96 \qquad\quad +0.93 \qquad\quad +0.62;$$
$$c = BD + 50°2794: \qquad 8.11 \qquad\quad +0.58 \qquad\quad +0.09.$$

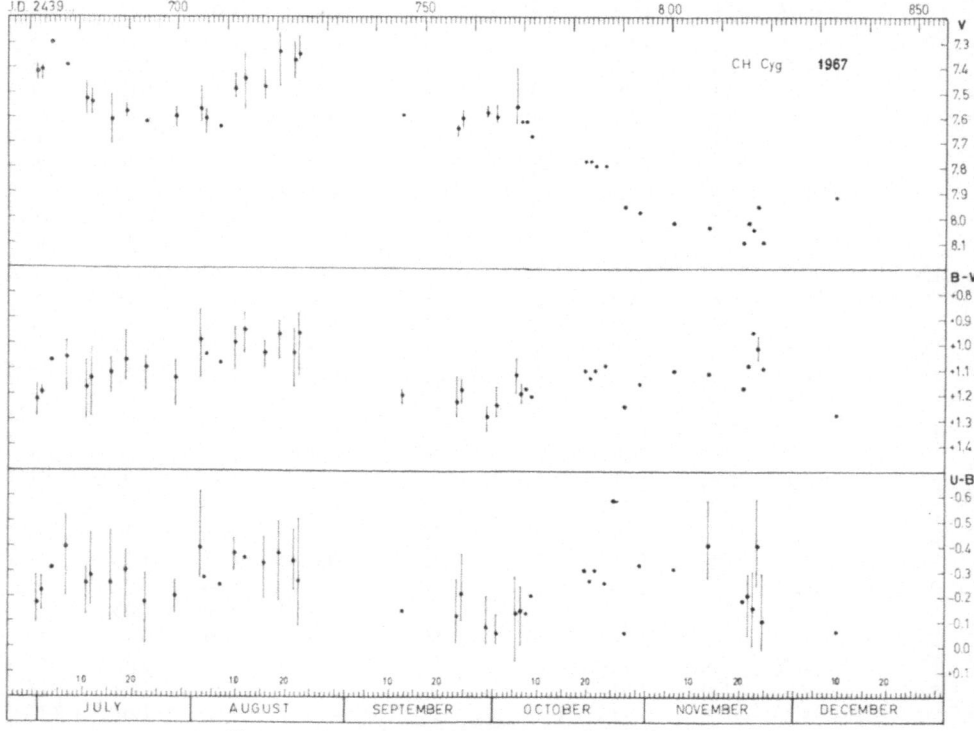

Fig. 1.   Magnitude and colours of CH Cygni in 1967.

From our observations the following features can be put in evidence:

(a) A variation of light in a time-scale longer than the 100-days period given by other observers (YAMAMOTO, 1934; GAPOSCHKIN, 1952) is present; it can be connected with two outbursts which happened approximately in June 1967 and July 1968 but an intrinsic long-period variation of the star, to which the outbursts are superimposed, is not completely to be rejected. A communication of the second new increase in luminosity, more conspicuous than the first one, has been given elsewhere (CESTER, 1968).

(b) A variety of wave of the order of about 46 days (i.e. half of the period quoted above) appears after the outburst of June 1967, but many more observations covering the unavoidable gaps are needed in order to confirm definitely this interpretation.

Both features can be seen in Figures 1 and 2 representing the observations of 1967 and 1968; the dots are the simple mean of all observations made in one night; the little vertical dashes crossing each dot give the range of the variation measured in each night. The reductions to the UBV system allowing for the atmospheric extinction have been made in the usual manner (HARDIE, 1962).

(c) The colour indices are obviously quite different from those pertaining to an M6 star: B-V oscillated in 1967 around +1.1 and U-B around −0.2 (with peaks up to −0.4 in August). In 1968 both colours changed toward even less values. Another

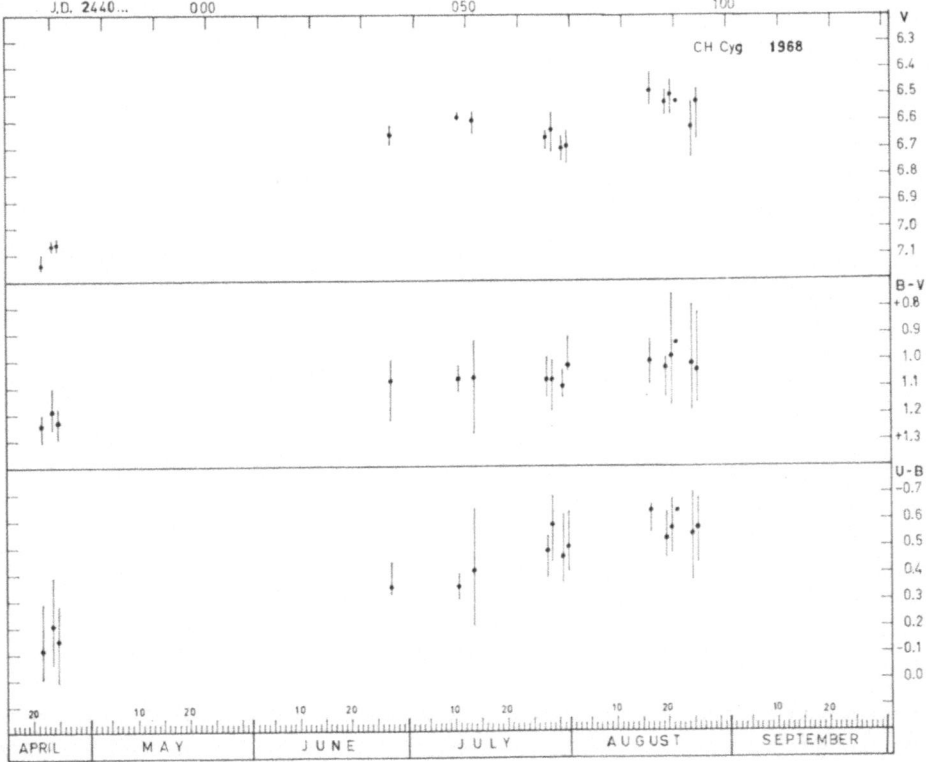

Fig. 2. Magnitude and colours of CH Cygni in 1968.

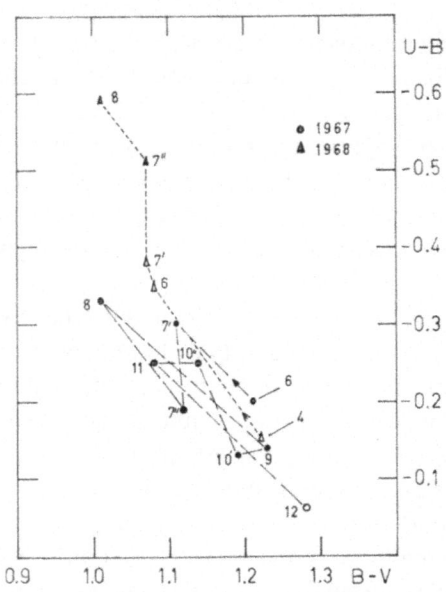

Fig. 3.   Colour-colour diagram for CH Cygni in 1967 and 1968.

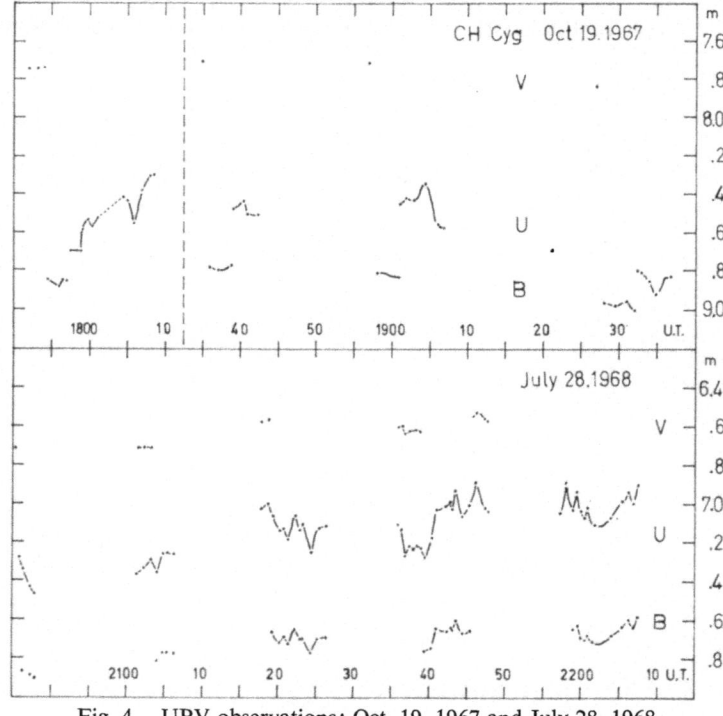

Fig. 4.   UBV observations: Oct. 19, 1967 and July 28, 1968.

distinction of the behaviour of the star during the 2 years is revealed in the colour-colour diagram (Figure 3). While the variations of both colours in 1967 are in the ratio of about one to one, U-B is varying in 1968 much more than B-V.

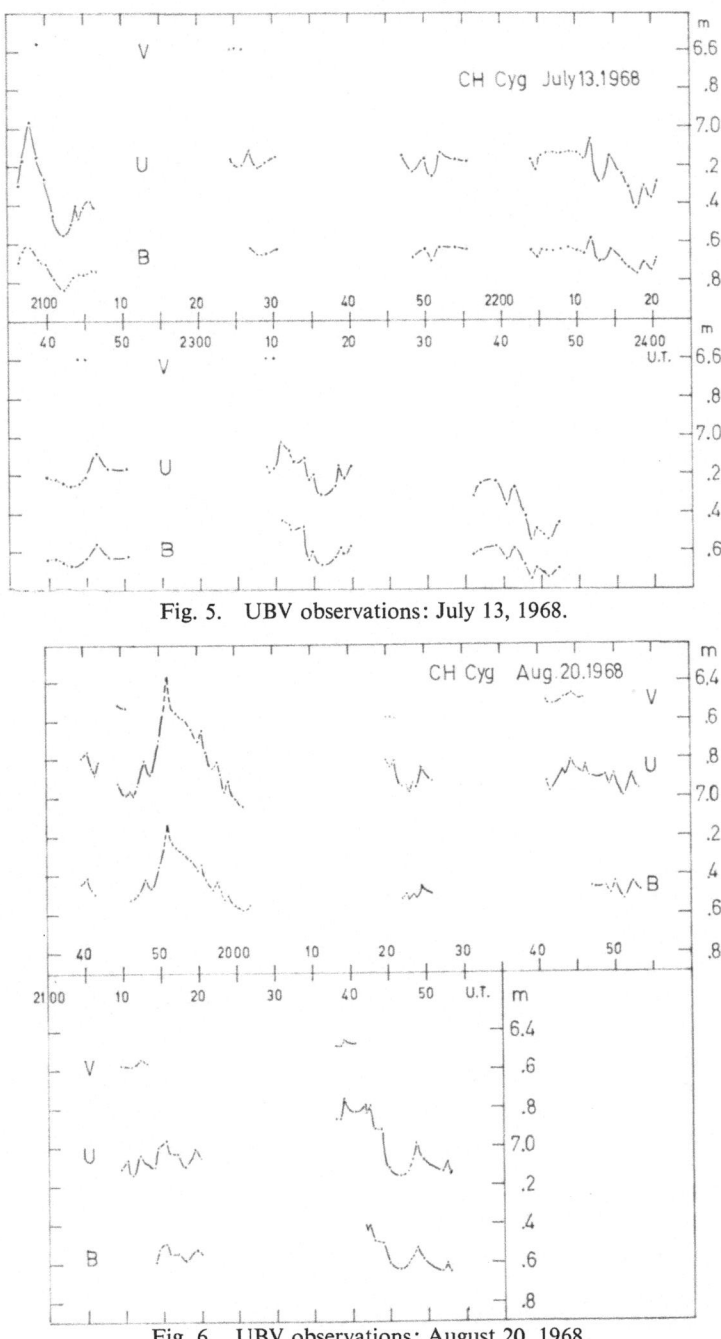

Fig. 5.   UBV observations: July 13, 1968.

Fig. 6.   UBV observations: August 20, 1968.

(d) The most peculiar feature is certainly that the star underwent rapid and irregular fluctuations, strongly correlated in the different colours (although a very accurate check of a possible lag has not been made). The variations increase from V

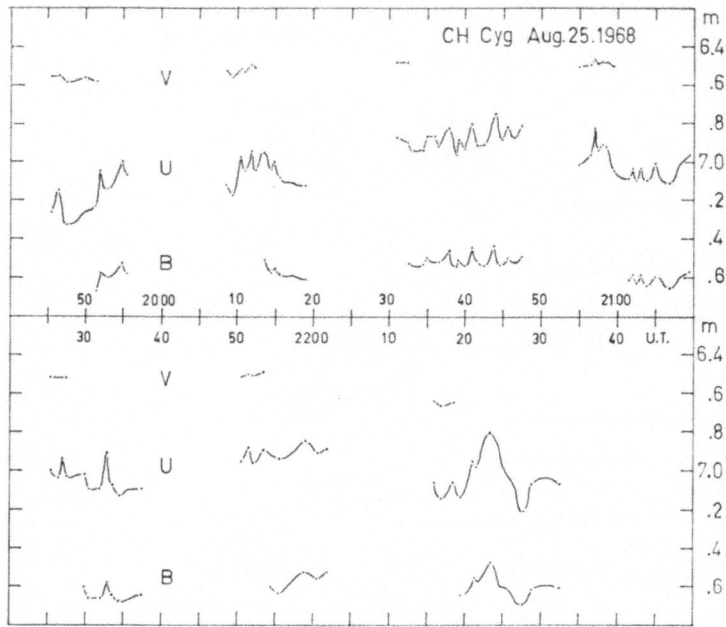

Fig. 7.   UBV observations: August 24, 1968.

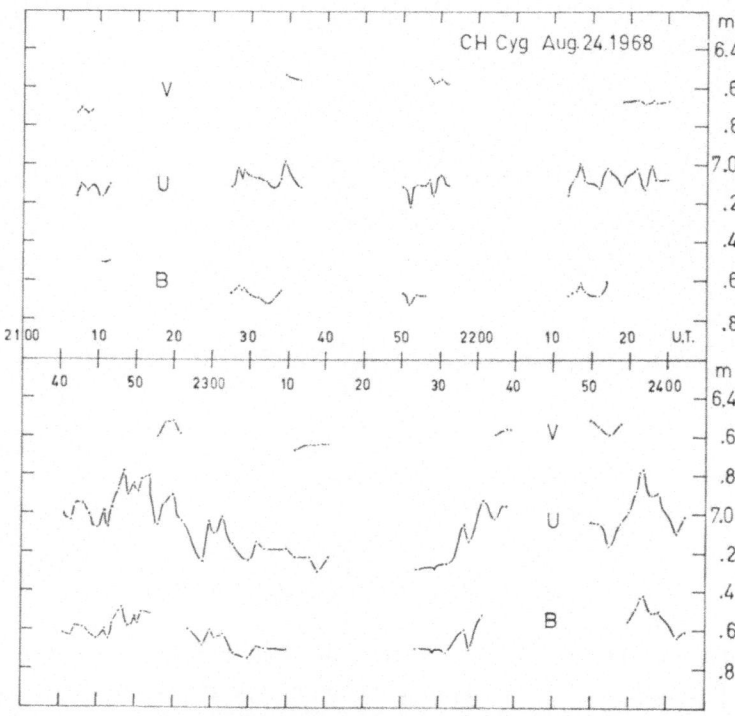

Fig. 8.   UBV observations: August 25, 1968.

through B to the U magnitude, in which they are very conspicuous. Sometimes during one night the variations remained small but on other nights they reached some tenths of magnitude in a few minutes (CESTER, 1967a). Confirmation of these variations has been given by other observers (MARTEL-CHOSSAT, 1967; WALLERSTEIN, 1968). Some selected observations are shown in Figures 4–8, in which the magnitudes are plotted against the time.

If we try to guess about the nature of CH Cygni, a possible suggestion is that the star is double; assuming then reasonable values for the M6-star (i.e. $B-V = +1.8$ and $U-B = +1.9$) and either the mean observed colours $B-V = +1.1$, $U-B = -0.3$ or the lowest ones ($+1.4$; 0.0) one can derive for the hypothetical companion $B-V = -0.3$ and $U-B = -1.0$, corresponding to a B1V star; but the difference in V in these cases is $+2.0$ and $+2.6$ magnitudes. Accepting for the M star the absolute magnitude $M_v = -1.0$, the companion should be a hot subdwarf (like e.g. the hydrogen-poor star BD$+13°3224$; HACK, 1967).

References

CESTER, B.: 1967a, *Comm. 27 I.A.U. Inf. Bull. Var. Stars* **219**.
CESTER, B.: 1967b, in *Atti XI Convegno Soc. Astron. Ital.* p. 245.
CESTER, B.: 1968, *Comm. 27 I.A.U. Inf. Bull. Var. Stars* **291**.
GAPOSCHKIN, S.: 1952, *Harv. Annals* **118**, 155.
HACK, M.: 1967, 'Hydrogen-Poor Stars', in *Modern Astrophysics*, Gauthier-Villars, Paris, p. 163.
HARDIE, R. H.: 1962, 'Photoelectric Reductions', in *Astronomical Techniques* (ed. by W. A. Hiltner), Univ. of Chicago Press, Chicago, p. 178.
MARTEL-CHOSSAT, M. T.: 1967, private communication.
WALLERSTEIN, G. N.: 1968, *The Observatory* **88**, 111.
YAMAMOTO, I.: 1934, *Kyoto Bull.* **3**, 285.

SPECTROSCOPIC EVIDENCE FOR MASS LOSS
FROM CH CYGNI

R. FARAGGIANA and M. HACK

Astronomical Observatory of Trieste, Italy

Abstract. Some high-dispersion spectrograms of CH Cygni taken in epochs at which only the M6 spectrum is visible and after the explosions of June 1967 and July 1968 have been studied. The strong negative radial velocities of the Ca II chromospheric absorptions and the turbulent motions broadening the emission lines of H I, He I, Fe II, [Fe II] prove that mass is lost from the star at a rate of the order of 10^{-8} solar masses per year. Arguments are given in favour of the hypothesis that CH Cygni is a close binary composed of an M giant and a blue unstable subdwarf.

CH Cygni is classified as an M6 semi-regular variable star, according to the *Kukarkin Catalogue*. In September 1963 DEUTSCH (1964) observed for the first time that the spectrum of CH Cygni showed a composite spectrum: A hot blue continuum super-posed over the late-type spectrum, emission lines of H I, He I, Ca II, Fe II, [Fe II], [S II]. Deutsch reports that in March 1961 and in September 1966 the hot spectrum was invisible. We have studied one spectrogram (12.4 Å/mm) taken at the Haute-Provence Observatory in August 1965 by Bouigue and kindly lent to us: it is that of a normal M6 giant star. The only proof of the existence of an extended envelope is given by the P Cygni contour of Hα and Hβ; trace of emission is visible also at Hγ. In June 1967 DEUTSCH (1967) observed again a blue continuum and the same emission fea-tures. Another spectrogram (9.4 Å/mm) was obtained by Faraggiana on July 14, 1967 (FARAGGIANA, 1968) characterized by P Cygni contours for the Ca II lines, the strongest Ti II lines, the Balmer lines until H 18 (absorption cores are visible until H 24), broad He I emissions, sharp emissions of Fe II, [Fe II], [S II].

It was impossible to obtain other high-dispersion spectrograms until July and August 1968. At this epoch the star had increased in luminosity and the colour has become bluer (Table I). The more apparent spectral variations from July 1967 to 1968 have the appearance of a second violet-shifted absorption at H and K of Ca II, and of several strong absorption lines originated by low excitation levels of ionized metals in the ultraviolet ($\lambda\lambda3760$–3400). These features resemble closely the strongest chromo-spheric lines due to the extended atmospheres of the atmospheric eclipsing binaries like ζ Aurigae (Figure 1). These chromospheric lines become more intense from July to August 1968 (Figure 2).

All the M6 absorption lines are veiled by the blue continuum, the veiling increasing appreciably from $\lambda5000$ to $\lambda3700$. Figure 3 shows clearly that $\lambda3758$ Fe I 21 is filled by the blue emission continuum, while $\lambda\lambda3759$ and 3761 Ti II 13 absorb from the blue continuum.

The absorption lines of the M6 star present small variations of radial velocity (Figure 4). The strong variations in radial velocity of the broad emissions of He I

TABLE I

Available high-dispersion spectrograms of CH Cygni

Date	Number of H.P.O. spectrograms	Spectral features	U-B	V	Reference or observer
March 1961		M6			DEUTSCH (1964)
Sept. 1963		M6 + blue cont. + emission features			DEUTSCH (1964)
August 1963	W 2723	M6 (Hα and Hβ with P Cyg contour)			Bouigue
Sept. 1966		M6			DEUTSCH (1964)
June 1967		M6 + blue cont. + emission features			DEUTSCH (1967)
July 15, 1967	W 3599	M6 + blue cont. + emission features	~ -0.2	$\sim 7^{m}$	Cester, Faraggiana
July 8, 1968	W 4363	M6 + blue cont. + emission features	~ -0.6	~ 6.5 quiet	Cester, Faraggiana, Martel
	W 4365				
July 11, 1968	W 4372	M6 + blue cont. + emission features	~ -0.6	~ 6.5 flares	Cester, Faraggiana, Martel
August 14, 1968	W 4468	M6 + blue cont. + emission features	~ -0.6	~ 6.5	Cester, Hack

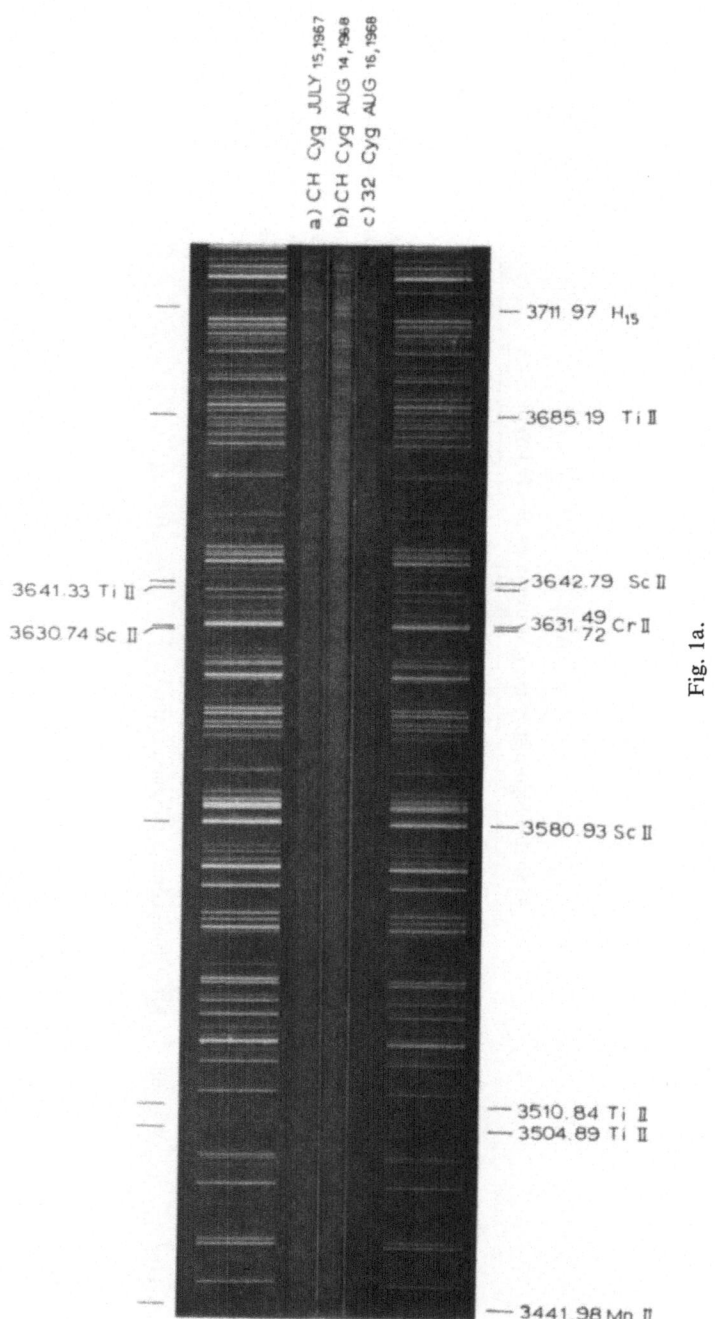

Fig. 1a.

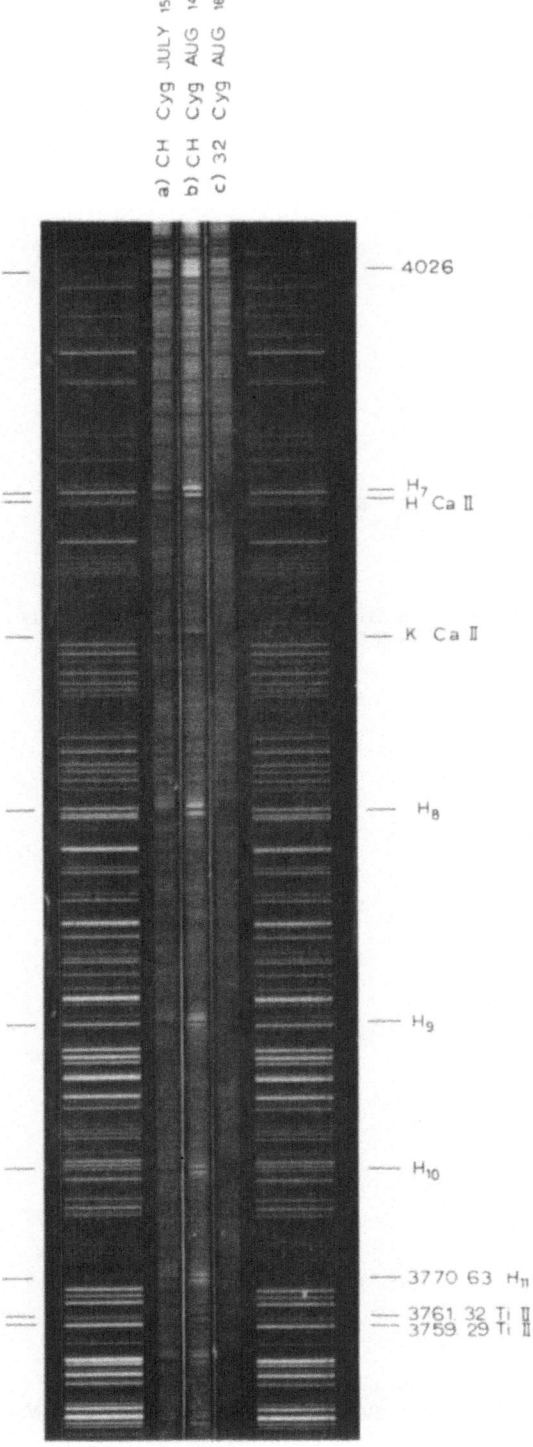

Fig. 1b.   Spectrograms of CH Cygni in July 1967 (a) and August 1968 (b) and of 32 Cygni during the atmospheric phase of eclipse (c) on August 16, 1968, about 27 days before totality. The absorption lines in the UV are missing in the 1967 spectrogram, but are clearly visible in the 1968 spectrogram; the doubling of the H and K absorptions which occurred in 1968 is evident.

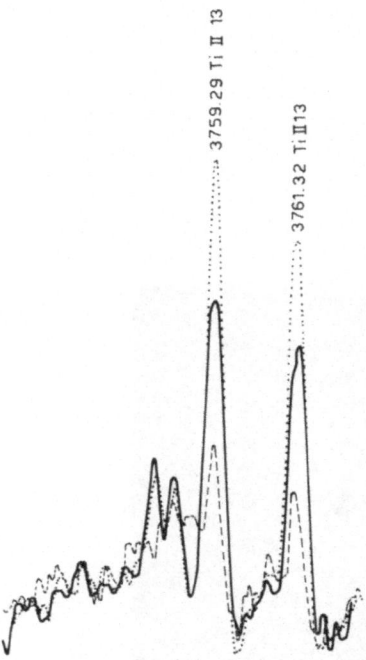

Fig. 2.   The Ti II doublet at λ3761 and λ3759 in July 1967 (broken line), July 1968 (full line), and August 1968 (dotted line).

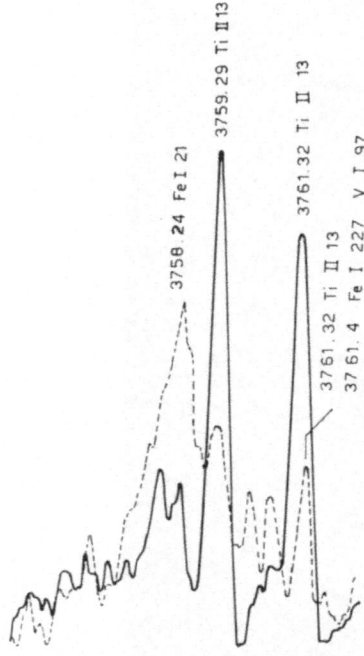

Fig. 3.   An example of chromospheric (λ3759 and λ3761 Ti II) and stellar (λ3761.4 Fe I, λ3758.2 Fe I) lines. Full line: CH Cygni in August 1968; broken line: the normal M6 III star 30 Her.

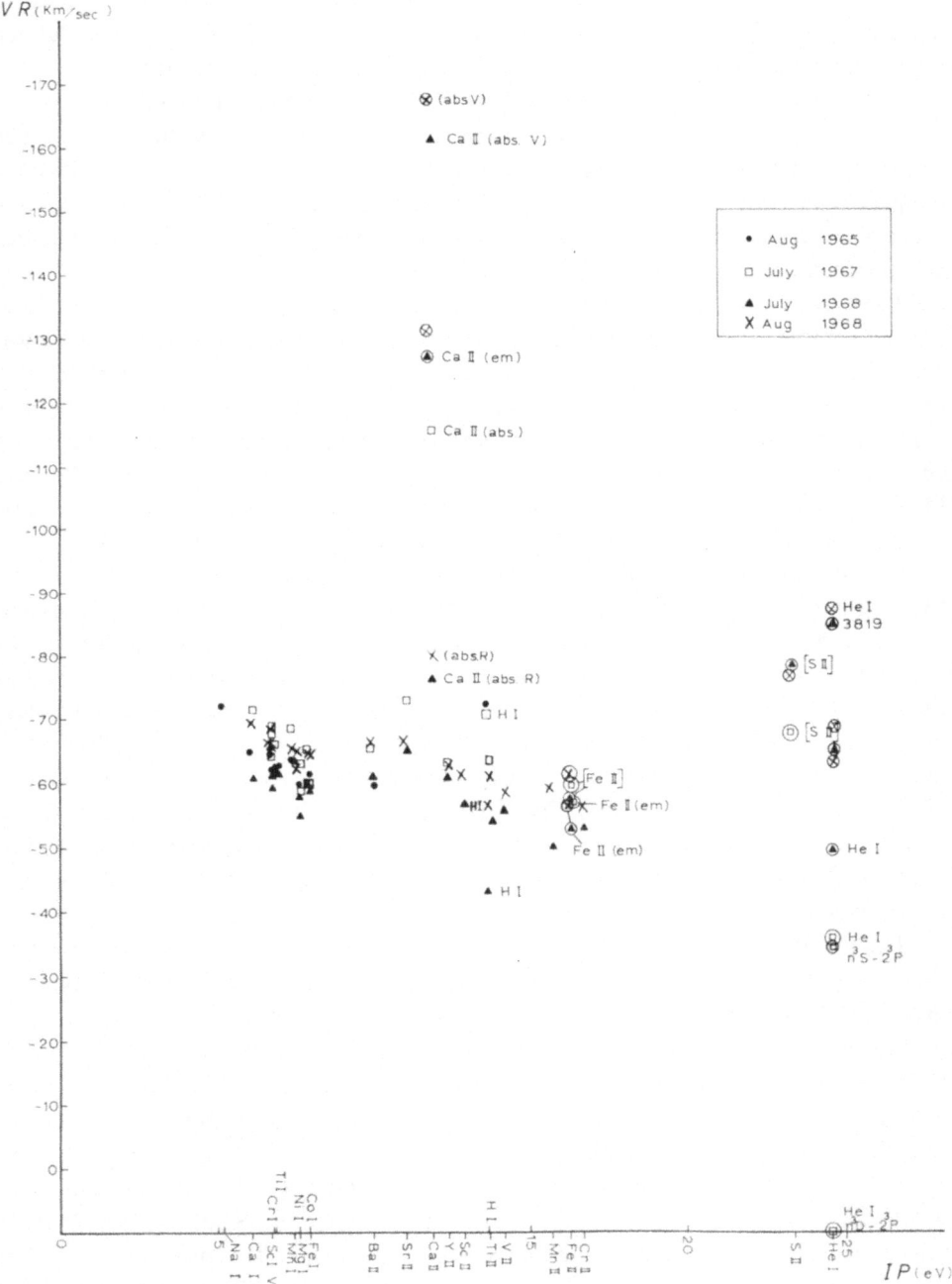

Fig. 4.   The radial velocities measured on the CH Cygni spectrograms at different epochs, vs. the ionization potential. The symbols enclosed in a circle are relative to emission lines.

suggest that they could be associated with an envelope surrounding a hot companion. It seems improbable that the Balmer cores are formed in the same envelope because of the strong difference in radial velocity between the lines of H I and He I. The Balmer lines seem rather to be associated with the chromosphere of the M star.

Of the three spectrograms taken in July 1968, two were taken on July 8 when the photoelectric observations showed that the star was at constant brightness; another was taken during the night of July 11, when the photometric observations made contemporaneously by Mme Martel at Haute-Provence Observatory showed that the star was enduring a continuous series of flares. During this last night the radial velocities were systematically more negative of 2 or 3 km/sec.

These facts suggest the existence of a hot, unstable source originating the short duration flares, and an envelope associated with it where the broad He I emission originates. The Ca II emissions and the Balmer emissions could be originated in a relatively hot chromosphere of the M star. The H and K absorption cores, the Balmer cores and the strong unveiled metallic absorptions could be originated above the chromosphere in an outer envelope surrounding the whole system, since all these lines absorb from the chromospheric emissions and from the blue continuum; two or three components are visible, the separations being from 7 to 15 km/sec.

It seems difficult to explain the observations with a model of a single star formed by a hot core shining through an extended cool envelope, as was suggested, for example, by SOBOLEV (1960) in order to explain symbiotic stars. In this case we should expect that all the M6 lines absorb from the blue continuum of the hot core and therefore that they are unveiled. We observe that when the blue continuum is less blue – (U-B) = -0.2 (CESTER, 1967) – as at the epoch of our observations in July 1967, the envelope is less dense, since no strong ultraviolet absorptions are visible, and in 1968 when the blue continuum is bluer – (U-B) = -0.6 (CESTER, 1969) – the envelope is more dense, as suggested by the appearance of the strong ultraviolet absorption lines, suggesting an association between flares of the blue companion and density of the envelope surrounding the whole system.

The variation of the radial velocity of the broad He I emissions, which are sometimes very different from those of the Balmer lines, is probably the most important proof in favour of the existence of the blue star.

We have attempted to evaluate the rate of mass loss from the envelope of CH Cygni during the explosive phases. The mass loss is given by the relation $dM/dt = $ $= 4\pi R_s^2 V\rho$, where R_s is the radius of the envelope and ρ is the density. In order to find the density of the envelope where the H and K violet-shifted absorptions are formed, we have computed the number of Ca II atoms by plotting the equivalent widths of these lines on the curve of growth computed by STRÖMGREN (1948). A microturbulence of about 15 km/sec is found, a value which is comparable with that found in the high chromosphere of the ζ Aurigae-type systems. We have assumed that all the calcium is once ionized – a reasonable assumption since no trace of sharp absorption is visible at $\lambda 4227$ Ca I; moreover, judging from the other lines present in the hot spectrum, it seems improbable that a strong percentage of Ca III (I.P. = 51 eV) exists in the envelope. Assuming that the chemical composition is normal (log H/Ca = 6), as the comparison with the spectrum of 30 Her suggests, we find for the number of hydrogen atoms contained in a column of height h and base 1 cm^2:

$$\log N = 20.7 \text{ (July 1967)}$$
$$\log N = 19.4 \text{ (from the new violet-shifted absorption which appeared in the July–August 1968 spectra).}$$

To derive the envelope density, we need to know the height h at which these lines are formed. This is the most uncertain assumption we must make. Since the H and K lines are so strictly similar to the chromospheric lines in the ζ Aurigae binaries, we have assumed that they are formed at the same height over the photosphere found for 31 Cygni for a value of the intensity of the chromospheric H and K equal to those in CH Cygni, i.e. about 3×10^{12} cm (WRIGHT, 1959). It follows $\rho \simeq 10^{-16}$ g cm^{-3}, and assuming for the stellar radius $R = 200\ R_\odot = 1.4 \times 10^{13}$ cm, we find

$$dM/dt = 5\ \ \times 10^{-8}\ M_\odot/\text{year (July 1967)}$$
$$dM/dt = 1.5 \times 10^{-8}\ M_\odot/\text{year (from the two new absorptions at H and K present in July–August 1968).}$$

Another argument in favour of mass loss from this star is the width of the shell absorptions indicating a macroturbulence of about 40 km/sec. If the velocity distribution is gaussian, we can expect a mass loss for evaporation (since the escape velocity can be estimated of the order of 70 or 100 km/sec). The turbulent velocity of the M6 chromosphere and of the envelope of the hot object, indicated by the width of the emission lines is much greater than the escape velocity – 250–300 km/sec for the Balmer emissions, about 200 for the He I emissions; in the higher layers where the forbidden lines are formed the broadening indicates a turbulence of about 40 km/sec.

Acknowledgements

The authors wish to thank Prof. Ch. Fehrenbach for having given the possibility to use the 193 coudé telescope of Haute-Provence Observatory, Prof. R. Bouigue for lending one spectrogram of CH Cygni taken during a non-explosive phase, and Mr. U. Flora for having made the microphotometric tracings and collaborated in their reduction.

This research has been made with a contribution of the Consiglio Nazionale delle Ricerche.

References

CESTER, B.: 1967, in *Atti dell'XI Convegno della Società Astronomica Italiana*, p. 245.
CESTER, B.: 1969, in *Colloquium on Mass Loss from Stars, Trieste, 1968*, p. 329.

DEUTSCH, A. J.: 1964, *Annual Report of the Mount Wilson and Palomar Obs.* **11.**
DEUTSCH, A. J.: 1967, *IAU Circular No. 2020.*
FARAGGIANA, R.: 1968, *Mem. Soc. astron. ital.* **39,** 291.
SOBOLEV, V. V.: 1960, *Moving Envelopes of Stars,* Harvard University Press, Cambridge, Mass., p. 82.
STRÖMGREN, B.: 1948, *Astroph. J.* **108,** 242.
WRIGHT, K. O.: 1959, *Publ. Dominion Astrophys. Obs., Victoria, B.C.*

Discussion

Hutchings: A paper now in press by Walker and Morris reports results of photoelectric scans of CH Cygni made in 1967. By subtracting the spectrum of a comparison M star they found a strong UV continuum and a weaker blue continuum which varied in strength, in phase, with times of a few minutes. This agrees well with the photoelectric results just given.

Lortet-Zuckermann: I wish to make two remarks. First, I could see a portion of the continuous runs obtained in July 1967 by Mrs. Martel and coworkers. There is a striking similarity between the rapid brightness fluctuations observed in CH Cygni and those observed in U Gem stars, old novae and Sco XR-1 objects. I don't mean that the physical phenomena are exactly the same in all these cases, but I think it is of utmost interest to find these fluctuations in a star as bright as CH Cygni, as it might allow quite new methods of observations. My second remark is somewhat related to the first. I don't agree with the conclusion of Mrs. Hack that the blue and ultraviolet continuum arises from a blue star. Free-free emission of hydrogen may also produce a strong increase of intensity towards shorter wavelengths. If there is a strong free-free and bound-free emission, it might be possible to detect radioemission, if the extension of the gas is great enough (larger than 10^{13} or 10^{14} cm) in order that the radioemission is not self-absorbed; in this case a radiation of the order of 1 FU $(10^{-26} \text{ W m}^{-2} \text{ s}^{-1} \text{ Hz}^{-1})$ could be expected. Such observations have been made at Nançay by Kazès at the wavelength of 11 cm. He has found that the intensity of the eventual radioemission is less than 0.5 FU. New observations are in course in order to see if there is emission more intense than 0.2 FU.

Hack: The reason I think there is a blue companion is mainly the following: the broad emissions of He I have radial velocity changing from about -30 km/sec to -70 km/sec at the different epochs of our observations. The He I emissions are the only lines which present RV differing by those of the metallic line absorptions by more than 30 km/sec in July 1967, and the only ones which present a variation in RV of 40 km/sec from July 1967 to August 1968 (with the exception, of course, of the H and K circumstellar cores). If the blue continuum is originated by the photosphere of the M star, we could expect that all the lines absorb from this continuum. If, on the contrary, the blue continuum is originated by an outer very extended envelope also the lines like Ti II $\lambda\lambda$ 3685, 3759 and 3761, and the other low-excitation lines of ionized metals observable in the UV spectrum of 1968, should be filled by this continuum, since these lines are usually formed at a moderate height over the photosphere (of about 0.2 or 0.3 stellar radii, i.e. about 3×10^{12} cm over the photosphere, by analogy with 32 Cygni, since these UV absorptions are very similar to the chromospheric spectrum of this star). Instead there are two kinds of absorption: the typical M spectrum lines which are filled by the blue continuum, and the chromospheric and circumstellar lines, which absorb from it.